The KNOWLEDGE OF NATURE

and the NATURE OF KNOWLEDGE

in EARLY MODERN JAPAN

STUDIES OF THE

WEATHERHEAD EAST ASIAN INSTITUTE,

COLUMBIA UNIVERSITY

The Studies of the Weatherhead East Asian Institute of Columbia University were inaugurated in 1962 to bring to a wider public the results of significant new research on modern and contemporary East Asia.

The KNOWLEDGE OF NATURE

and the NATURE OF KNOWLEDGE

in EARLY MODERN JAPAN

Federico Marcon

The University of Chicago Press *Chicago and London*

The University of Chicago Press, Chicago 60637
The University of Chicago Press, Ltd., London
© 2015 by the University of Chicago
All rights reserved. Published 2015.
Paperback edition 2017
Printed in the United States of America

24 23 22 21 20 19 18 17   2 3 4 5 6

ISBN-13: 978-0-226-25190-5 (cloth)
ISBN-13: 978-0-226-47903-3 (paper)
ISBN-13: 978-0-226-25206-3 (e-book)
DOI: 10.7208/chicago/9780226252063.001.0001

Library of Congress Cataloging-in-Publication Data

Marcon, Federico, 1972– author.
    The knowledge of nature and the nature of knowledge in early modern Japan /
Federico Marcon.
        pages cm
    Includes bibliographical references and index.
        ISBN 978-0-226-25190-5 (cloth : alkaline paper) — ISBN 978-0-226-25206-3
(ebook) 1. Nature study—Japan—History. 2. Science—Japan—History. 3. Japan—
History—Tokugawa period, 1600–1868. I. Title.
    QH21.J3M7 2015
    508.52—dc23

                                                                        2014047538

♾ This paper meets the requirements of ANSI/NISO Z39.48-1992
(Permanence of Paper).

A philosophical interpretation of world history
would have to show how the rational domination
of nature comes increasingly to win the day, in spite
of all deviations and resistance, and integrates
all human characteristics.
— Max Horkheimer and Theodor W. Adorno

Contents

Prologue

. . . .

The Sanjō Guest House, where I spent the summer of 2013, is located in the middle of the Hongō campus of Tokyo University. It stands near a garden, the Ikutokuen, which was built in the 1630s in what was then the Edo residence of the powerful Maeda Toshitsune, domainal lord (*daimyō*) of one of the wealthiest regions of Tokugawa Japan, Kaga.[1] The garden expands around a pond shaped like the character 心, or *kokoro* ("heart-mind"), today known by most as "Sanshirō's pond" (Sanshirōike) from the name of the protagonist of Natsume Sōseki's novel.[2] The vegetation around the pond is so exuberant that one cannot help but perceive a sense of disordered and disquieting wilderness. At least, that is what I usually felt when I walked the narrow and uneven paths around the lake. It is populated by a variety of birds: crows, cuckoos, thrushes, woodpeckers, ibises, kingfishers, bushwarblers, rufous turtledoves, and a bevy of green parrots. One night I even met a Japanese raccoon dog (*Nyctereutes procyonoides viverrinus*), or *tanuki*, as it is called here, protagonists of a mass of folktales that describe them as creatures endowed with supernatural powers—mischievous tricksters, masters of disguise, often portrayed with portentously huge testicles.

At a closer look, however, the disordered luxuriance of the garden is far from being a sign of its wilderness. In the trunk of many trees and among short herbs, in fact, one can spot plastic labels with the names of many of the plants growing there. These tags catalog the vegetation of the garden in a precise inventory of its natural riches. They represent an odd contrast with the first impression of wilderness it gives. They suggest design, planning, artifice, and, most important, dominion over nature. If you visit the parks of Tokyo in search of an improbable relief from the sultriness of Japanese summer, you will have the same odd experience: a sense of disordered wilderness that vanquishes as soon as you notice labels bearing the name of trees and herbs, sometimes with even the Latin scientific name attached.

Knowledge of the natural world is as old as human beings. Information on the nutritional, curative, and venomous properties of plants constituted a matter of life or death for early *Homo sapiens*. Even today, biologists routinely use the "botanical" knowledge of tribes of hunter-gatherers in Southeast Asia, Africa, and South America to explore the remotest recesses of the last surviving rainforests. But the kind of knowledge natural sciences like botany and zoology produce is distinct. It parcels an ecosystem in discrete elements, which are isolated, decontextualized, analyzed, sectioned, objectified as pictorial, dried, or embalmed samples, experimented upon, manipulated, transformed, copyrighted, and often reproduced and commercialized in mass quantities.

Although in the last decades a variety of "green" thinkers and movements have underlined the inseparability and imbrication of human societies and the environment, we are still largely confident of the modern paradigm that sees human beings as distinctly separated from the natural world. In the age of the Anthropocene, the disavowal of our embeddedness with nature prevails. We see ourselves as destined to exercise our dominion over nature. And in spite of concrete evidences of the catastrophic impact we have on the environment, today "the fully enlightened earth radiates disaster triumphant."[3]

Historians of science locates the origin of this modern paradigm in the Renaissance period, part of that long and complex ensemble of social and intellectual processes clumped together in the rubric of the Scientific Revolution. Natural philosophers of early modern Europe increasingly isolated species from their ecosystems, objectified them in atlases and breeding experiments, and commodified them as resources for culinary consumption, pharmacology, agriculture, industry, and entertainment. According to this canonical view, with the expansion of European power during the age of empires this paradigm globalized as traditional (meaning "backward") cultures like, for example, Japan and China embraced the Western sciences as integral part of their modernizing efforts in the late nineteenth century. As a result, whether to glorify or denounce the revolutionary effects of scientific modernity in the last two centuries, the "enlightening" of the world is always and indisputably a *Western* and, in particular, European undertaking.

This book aims to correct this assumption. It demonstrates that well before the modern age, during the Tokugawa period (1600–1868), Japan began a process of desacralization of the natural environment in the form of a systematic study of natural objects that was surprisingly similar to European natural history without being directly influenced by it. This process was carried out by scholars invading pristine regions to survey the vegetal and animal species living in Japan and classify them as discrete entries of dictionaries and encyclopedias or as objects to collect, analyze, exchange, exhibit, or consume as cognitive, aesthetic, or entertaining commodities. Originally framed as *honzōgaku*—a field of study of Chinese origins ancillary to medicine, devoted to the pharmacological properties of minerals, plants, and animals—this discipline evolved into a very eclectic field encompassing vast arrays of practices, theories, conceptualizations, and goals. Its evolution, I here argue, derived from its internal development as much as from the deep transformations of Tokugawa society and of the socioprofessional trajectories of scholars in that society. Many of the practices, institutions, and knowledges of *honzōgaku* were not lost or abandoned when the Western sciences were introduced in the Meiji period (1868–1912) to sustain the modernization of Japan but would be rather translated, adapted, and incorporated in the language and forms of the new disciplines of botany, zoology, and biology.

When the Maeda compound in Hongō was turned into public land and given to the Ministry of Education to edify the new facilities of the Tokyo Igakkō and the Tokyo Kaisei Gakkō—soon to be fused in 1877 as the University of Tokyo—the Ikutokuen was a wasteland. It would be progressively reduced to its actual size and the maintenance of its vegetation put under the guidance of the center for botanical research of the university along with the Koishikawa garden in Hakusan. In all probability, the tags domesticating the wilderness of its trees and plants were placed then. However, the Maeda were domainal lords who in the Tokugawa period also practiced as amateur scholars of *honzōgaku*. Who knows if they tagged the vegetation around the heart-shaped pond too?

PART I *Introduction*

Philosophy persistently and with the claim of truth, must proceed interpretively without ever possessing a sure key to interpretation. . . . The text which philosophy has to read is incomplete, contradictory and fragmentary.
— *Theodor W. Adorno, "The Actuality of Philosophy"*

1   *Nature without Nature*
Prolegomena to a History of Nature Studies in Early Modern Japan

. . . .

DISENCHANTING NATURE

The *Hitachi no kuni fudoki* (Gazetteer of the Hitachi Province, 721 CE) narrates that during the reign of the emperor Keitai, who ruled at the beginning of the sixth century,

> there was a man called Yahazu of the Matachi clan who reclaimed a marshy land in a valley west of the District Office to open up new rice-fields. Soon, a troop of Yatsu no kami gathered together to obstruct his work and hinder the opening of new fields to cultivation. According to the people of the area, Yatsu no kami are snake-like spirits. They have bodies of snakes and horns on their heads. If anyone turns to look at them while escaping, his household would be ruined and he would die without heirs. Fields neighboring the District Office harbor many of these creatures.
>
> Matachi was furious: he put on his armor, took his halberd, and killed many of them, driving the rest away. Later, he went to the mountain entrance and erected some sticks to mark a boundary. He then said to the Yatsu no kami: "The Deities shall live in the lands above this point towards the mountain; humans shall instead cultivate their fields below it. From now on I shall become a priest to worship you and so will my descendants do forever and ever. I beseech you not to curse me or bear a grudge against me."
>
> He built a shrine and conducted the first ceremony. Then, he opened up more than ten *cho* of rice-fields. His descendants continued generation after generation to worship these deities, and so are still doing today.[1]

Matachi's tale discloses a pattern that characterized the religious beliefs and mythological stories of various communities of early Japan, consisting of a fundamental division of human and sacred space. On one side, human dwellings and cultivated lands formed the symbiotic domain of a nature domesticated to human needs; on the other, dark forests and impenetrable mountains were the realm of wild animals and untamed spirits, a nature inhospitable to humans. In between, shrines and wooden *torii* gates marked the borders that kept the two worlds separated: they signaled to human wanderers that if they proceeded further, they entered at their own risk in the hostile reign of wild animals and *kami* (spirits), and they reminded deities of a pact of coexistence with human beings—a pact that usually originated, as in the story from the *Hitachi no kuni fudoki*, from a violent clash of human communities and natural spirits.

The sacred space of untamed nature is inhospitable and alien to human beings. It is populated by a variety of trees, herbs, and animals but also by *kami* and other monstrous creatures like *kappa* (water goblins), *tengu* (long-nosed demons), *kitsune* (supernatural foxes), and *tanuki* (raccoon dogs), which tricked, kidnapped, challenged in wrestling matches, and sometimes killed all those humans who dared to enter their realm.[2] Folk tales abound of such creatures, and many Shinto shrines worship one or the other of these supernatural creatures "pacified" to allow human beings to enjoy the natural riches of an area—like in the case of Matachi's story.[3] Shrines protect the impermeability of the borders between human and sacred space, and they control through ritual protocols of worship the divine forces that humans had subdued. Shinto rituals include *tama shizume* (pacification of spirits) and *jijinsai* (pacification of place), ceremonies performed even today before constructing new buildings or opening up new lands for cultivation. Shinto shrines thus function as signposts of a separation of the human and the natural worlds.

Anthropologists and ethnographers of Japan have often argued that the mountainous landscape of the major Japanese islands reproduces a similar division of space: villages and cities surrounded by cultivated lands and encircled by village mountains (*satoyama*)—reservoirs of timber, firewood, and hunting and gathering resources—constitute the human space; outside it, forest-covered mountains with sparse or no human presence. Shinto shrines punctuate this concentric organization of space with a system of village shrines (*sato no miya*), field shrines (*ta no miya*), and mountain shrines (*yama no miya*), in which people worshiped those spirits enraged by the opening up of new lands for human exploitation; small shrines (*oku no miya*), built to mark the borderline between the human space and the sacred space of untamed nature, lie on the outer ring.[4]

In the last century, scholars of various disciplines—from Yanagita Kunio's ethnographical research to Nakao Sasuke's biogeography and Sonoda Minoru's Shinto environmentalism—have mobilized this pattern of spatial division to create and sustain an ideological constellation of beliefs according to which a unified and unique Japanese identity in empathic relation with the natural environment have always existed substantially unaffected by historical change.[5] Even though evidences suggest that a ritual separation of natural (or sacred) and human spheres can be retrieved throughout the history of Japan,[6] recent research on the environmental history of the archipelago have shown how, in actual fact, the destruction of the environment has been a constant feature of Japanese history since its earliest times precisely *because*, rather than *despite*, the existence of a religious rituality that sublimated human intervention on nature.[7] For example, economic and purely damage control–related concerns, rather than religious "love for nature," were at the core of the reforestation policies of the Tokugawa shogunate in the eighteenth and nineteenth centuries.[8]

This book reconstructs various processes that slowly but relentlessly demolished this traditional division of space and the worldviews that sustained it.[9] It shows how a series of disparate intellectual and manual practices, spanning from the late sixteenth to the mid-nineteenth century, on the one hand favored the expansion of the human domain to include pristine mountains and forests as reservoirs of material and cognitive resources and on the other hand developed new ways of conceptualizing natural species and their environment. The vanishing of the invisible texture of metaphysical relations that held together sacred and human spaces was the effect of a long, unplanned, and contingent series of intellectual, economic, political, and cultural processes. A variety of disciplines of nature studies more or less directly associated with *honzōgaku*—a scholarly field that encompassed subjects ranging from *materia medica* and agronomy to natural history—knitted together these processes. The ensuing secularization of nature sprang from a parceling of nature in myriads of discrete objects to be described, analyzed, consumed, or accumulated in the form of standardized and quantifiable units as products, natural species, or collectibles.[10]

Because I am mainly concerned with the changing attitudes toward the material environment in early modern Japan, I do not develop here a comprehensive history of the discipline of *honzōgaku*.[11] Rather, I follow some genealogies of its complex history to reveal its impact on the professionalization of specialized scholars; agricultural advancements; the development of new economic policies in shogunal and domainal administrations; and the formation of various forms of popular entertainments, refined pastimes, artistic creations, and intellectual discourses related to plants and animals. I claim that these were all interrelated processes that played a critical role in the secularization of nature and the objectification of the natural species that populated Japan. On the one hand, the increasing commercialization of agricultural production—which included the farming of rice and other grains as well as protoindustrial activities like fishery, textiles, sake brewing, dyeing, mining, and so on—led to the commodification of plants and animals and their transformation in resources to inventory and accumulate for the needs of agricultural growth and the demands of the expanding market of medicinal substances. On the other hand, *honzōgaku* scholars and amateurs tended to examine plants and animals as intellectual commodities in isolation from their ecosystems, to be cataloged as concrete samples of abstract species in encyclopedias, atlases, monographs, and collections. This tendency derived in part from the adherence of early *honzōgaku* scholars to the explanatory style of canonical texts like *Honzō kōmoku*, which tended to treat mineral, vegetal, and animal species as *meibutsu* (names) in the form of discrete encyclopedic entries revealing their pharmacological properties. The description and manipulation of individual specimens disconnected from their ecosystem developed in concomitance with the recruitment of naturalists in state-funded missions to inventory plants and animals as *sanbutsu* (products, resources) and to experiment with medicinal herbs (*yakuhin*) and pest-resistant crops in botanical gardens. A burgeoning popular interest in plants and animals contributed to their transformation into objects of curiosity (*kōbutsu*) to be collected, admired, exchanged, and exhibited as spectacles (*misemono*).

Whether to expand the encyclopedic reach of human knowledge or to be engaged in a morally uplifting practice, whether to improve agricultural production or to generate new aesthetically appealing images, or whether for social utility or simply for fun, professional and amateur scholars collected, observed, bred, exchanged, analyzed, compared, depicted, described, fantasized on, and classified the most varied assortments of insects and fish, herbs and mushrooms, and trees and flowers, following either theoretical or practical protocols in both solitary and collegial enterprises. A majority of *honzōgaku* scholars favored a lexicographical approach, accumulating knowledge and gathering information from books and encyclopedias. But from the beginning of the eighteenth century, a growing number of naturalists started to invade pristine and uncharted forests, mountains, and ravines to make complete inventories of the vegetal and animal species they contained. To these scholars, nature was no longer conceivable as an organic, meaningful, and homopoietic space of supernatural and mystifying relations but as a multitude of objects—myriads of things (*banbutsu*). As one of these naturalists, Kaibara Ekiken, put it, "I climbed tall mountains. I penetrated into deep valleys. I followed steep paths and walked through dangerous grounds. I have been drenched by rains and lost my way in fog. I endured the coldest winds and the hottest sun. But I was able to observe the natural environment of more than eight hundred villages."[12]

At once material and intellectual commodities, plants and animals constituted, as *specimens*, the myriads of things that populated the world (*banbutsu*). But as *species* to be studied, produced, or exchanged, plants and animals became concrete bearers of abstract characteristics. Different social practices converted them into names (*meibutsu*) of natural kinds in encyclopedias and atlases, into products (*sanbutsu*) in agronomical manuals and agricultural enterprises, into medicinal substances (*yakuhin*) in pharmacopoeias, or into curiosities (*kōbutsu, misemono*) in various forms of popular entertainments or amateurish research. These social practices—intellectual, artistic, political, and economic, but more often a mixture of them all—secularized nature by transforming what was once the enchanted realm of unfathomable divine forces and metaphysical principles into a multiplicity of "objects" that could be grasped and manipulated through protocols of observational, descriptive, representational, and reproductive techniques.

In particular, the effort to produce faithful pictorial representations of plants and animals that characterized *honzōgaku* practices in the second half of the Tokugawa period had the precise aim of abstracting from the material appearances of individual plants and animals those morphological traits that were supposed to be the defining marks of species (*shu*). This was a remarkable development in East Asian natural history. In Chinese pharmacopoeias, the usage of concepts like *zhong* (*shu* in Japanese) and *lei* (*rui*)—the standard renderings respectively of species and genera in modern biological terminology—was rather loose. Both terms designated a discrete kind or group of minerals, plants, or animals in a system of signification distinctive of a text or a series of texts but without an overarching systematic consistency.[13] Li Shizhen's *Bencao gangmu* (*Honzō kōmoku* in Japanese)—throughout the period the most influential canonical source—attempted to develop a coherent two-layered hierarchical order of species and genera, the *gangmu* (*kōmoku*) system.[14] The identification of each species depended on the names of the various plants and animals in different regions, but their grouping in a more refined taxonomy was largely based on morphological resemblances. Li inscribed his work in a Neo-Confucian framework, and therefore he theoretically assumed that the classificatory order of all species depended on their metaphysical constituents.

When at the beginning of the seventeenth century *Honzō kōmoku* was introduced in Japan, it became the foundational text of the field of *honzōgaku* until the formal institutional adoption of Western medicine by the Meiji state after 1868. Its classificatory system persisted unchallenged in its essential features and was only subject to updates and corrections until its demise in the second half of the nineteenth century. True-to-nature illustrations, which attempted to abstract essential morphological features distinctive of each species, were also inserted in a framework that largely accepted *Honzō kōmoku*'s taxonomy or, at the very least, its notion of species. What was lost was any appeal to metaphysical foundations.

True-to-nature illustrations in atlases and encyclopedias had three basic functions. First, they distilled the results of shared protocols of observation to pictorially represent the concrete specimens under investigation in specific circumstances. Second, they aimed at revealing species-specific morphological characteristics of each species, not unlike the illustrations in many of today's bird-watcher manuals. Third, as epistemic paradigms, they trained the naturalist's eye to recognize species in nature by distinguishing morphological and quantitative features of plants and animals, which is also what bird-watcher manuals do. The practical function of these illustrations was to precisely identify plants and animals, a difficult task for scholars who had to negotiate between multilingual sources (in Chinese, Japanese, and, later, Dutch) and regional and dialectical variations in nomenclature. As such, they had the same impact of Linnaeus's *systema naturae* in teaching naturalists to "see systematically," as Michel Foucault put it, by molding the expert's observing gaze to the necessities of the system.[15] Their institution as the predominant form of description and explanation of natural species in the second half of the Tokugawa period represented a great change in scholars' cognitive practices, but it did not openly challenge *Honzō kōmoku*'s classificatory system.

On the one hand, we may say that there was no revolutionary paradigm shift in Tokugawa period's nature studies. These representational practices did not dismantle the taxonomical apparatus of *Honzō kōmoku* nor did they *openly* question the Neo-Confucian metaphysics that sustained its classificatory divisions. On the contrary, *honzōgaku* scholars often struggled to make their descriptions fit the established definitions of plants and animals in canonical encyclopedias, preferring to follow the conventional naming of species even when their pictorial descriptions were clearly at odds with the received knowledge.[16] But, on the other hand, the emphasis on morphological descriptions and juxtapositions that characterized *honzōgaku* practices in the second half of the Tokugawa period affected the way in which natural objects, qua object of knowledge, were conceived and constructed. Now manuals, encyclopedias, field notes, and monographs no longer deductively derived the properties of minerals, plants, and animals from an overarching metaphysical system, but these were inductively inferred from a functionalistic description of their own objective materiality, without any open and direct questioning of the previous metaphysical presuppositions. In other words, Neo-Confucian metaphysics became simply irrelevant. Only visible objects remained: rocks, plants, and animals were material—they could be touched, gathered, collected, reproduced, exchanged—but abstract, insofar as they could be distilled through social practices into universal bearers of abstract qualities of species.

Through their collecting, observing, and representing, *honzōgaku* scholars reduced the material environment to a collection of material objects that were manipulated to manifest increasingly abstract qualities, species-essential morphological properties that were abstracted from the variety of individual specimens. Even mysterious and awe-inspiring supernatural creatures like *kappa*, *ningyo* (sirens), and *tengu*, once protecting the inviolability of the sacred space of untamed forests, were naturalized and treated like any other animal, with maps charting regional morphological differences and distribution.

WHAT IS A SPECIES?

One assumption behind true-to-nature illustrations was that the bodies of plants and animals revealed their membership in a precise species. Hence their accurate pictorial description in atlases and manuals and their preservation through increasingly refined drying and embalming techniques became standardized instruments to represent species. This accompanied and, in certain instances, replaced the traditional lexicographical approach of early *honzōgaku* scholars.

The transformation of individual, concrete specimens into material bearers of a set of universal, species-specific abstract features reified epistemological norms. The assiduous labor of *honzōgaku* scholars and their collaborators involved inventorying, collecting, growing, breeding, exchanging, drying, storing, cataloging, painting, and describing. Specimens-as-species representatives embodied intellectual and material practices contrived to convert particular examples into universal exemplars, which as "social hieroglyphics" concealed social labor.[17] Specimens-as-species representatives were "sensuous things which [were] at the same time supra-sensible or social."[18] Their homology with the commodity-form is revealing. Like commodities, species were a product of human labor, an abstraction performed through an array of intellectual and manual practices. Species—unless perhaps considered from their cladogenetic history, which renders the very notion of species meaningless—are by no means natural kinds but social constructs.[19] In that sense, the operations performed by *honzōgaku* scholars were not qualitatively different from those of contemporary Linnaean naturalists.[20]

Furthermore, the search for accurate, true-to-nature pictorial representations of plants and animals, in Europe as well as in Japan, concealed in the lifelike appearance of specimens-as-species representatives in atlas illustrations the fact that those representations were the result of human labor performed under historically determined social conditions. In other words, the "realism" of these illustrations masked the fact that species were the product of an operation performed by scholars under determined social circumstances, following specific protocols and shared standards of credibility, under the legitimating umbrella of determined political authorities, and for certain purposes. The same mechanism is at work in dried or embalmed samples in private collections or public expositions.

This book defends the view that this process of reification of nature—the tendency of conceptual knowledge to objectify what it seeks to describe—was coeval with and connected to deep structural transformations in the mode of production that occurred during the early modern period of Japanese history: the commodification of agriculture, the monetization of society, and the development of market-oriented mechanisms of commodity exchange. The role of scholars in this process of reification and disenchantment of the natural sphere was central.

A SOCIAL HISTORY OF SCHOLARS

In fact, the various forms of natural research, agronomical experimentations, and cultural divertissements that involved collecting and describing plants and animals were all activities mediated by professional scholars. These scholars did not act in a vacuum but were subjected to various forms of social dominations. In narrating the development of nature studies in Tokugawa Japan, I therefore pay great attention to the social conditions of intellectual production and give a brief biographical introduction to the main *honzōgaku* specialists. These sketches, however, are not to be taken as indulgences in informational punctiliousness but have the purpose of shedding light to their sociohistorical trajectories. Indeed, like any other discipline, *honzōgaku* was a system of meanings structured by its social situatedness at the same time that it contributed to structure the society that engendered it. Its historical development mirrored changes in the structure of Tokugawa society as both symptom and cause. Moreover, *honzōgaku* scholars functioned as mediators between nature and society: the knowledge they produced gave order and meaning to the various experiences of the natural world one could have through atlases, cultural circles, public exhibition, herbal expeditions, and the like. In Hilary Putnam's terms, in the division of linguistic labor distinctive of Tokugawa society, scholars organized and legitimated the "clear" and "distinct" standards of understanding.[21]

Intellectual production, like any other human activity, is subject to social constraints. Forms of scholars' organization, legitimating institutions, patronage venues, networks of information exchange, instruments of intellectual research, and markets for cultural products are all historically contingent social dynamics that help not only shape scholars' assumptions, worldviews, and dispositions—that is, the standards of social acceptability for how, what, and in which circumstances they can think, write, and inquire about—but also mold the self-fashioning strategies that scholars consciously or unconsciously adopt to establish their socioprofessional identity. "All men are intellectuals, but not all men have in society the function of intellectuals."[22] If intellectual and scholarly production coincides with the history of literate human societies, professional scholars emerged as socially recognizable personae only because of a heterogeneous ensemble of contingent historical events and processes in the early modern world. In Europe, the history of professional scholars active in universities, academies, or princely courts coincides with the history of modernity from the Renaissance to the present.[23] In Japan, there were no professional scholars socially recognizable as such before the seventeenth century. Before the Tokugawa period, intellectual production was an exclusive province of Buddhist monks, court aristocrats, and samurai elites. The history of *honzōgaku* is therefore simultaneous with the history of the emergence of professional scholars, and as such it reflects the negotiations and compromises of its specialists in establishing themselves and their credibility in the larger Tokugawa society. Hayashi Razan was the first promoter of *Honzō kōmoku* in Japan at the beginning of the seventeenth century, at the same time that he was struggling to invent a socioprofessional identity of scholar for himself (chapter 3). Kaibara Ekiken, a retainer of the Kuroda household of domainal lords, conceived and envisioned his activity as scholar and the entirety of his scholarly production as a form of service he owed, as a loyal samurai, to his lord and the people of his domain (chapter 5). Inō Jakusui's encyclopedic work aimed at boosting the glory of the Tokugawa's regime (also in chapter 5). Niwa Shōhaku's and Tamura Ransui's research was an integral part of shogunal economic policies, and as scholars they were organic members of the Tokugawa's bureaucratic apparatus (chapters 7 and 10). Ono Ranzan and Kurimoto Tanshū developed their scholarly careers negotiating among cultural circles, the market for botanical manuals, and state support (chapters 8 and 9). Hiraga Gennai saw in scholarly excellence the only venue for upward social mobility (chapter 10). In other words, a history of *honzōgaku* is tantamount to a history of the emergence and establishment of professional *honzōgaku* scholars. This is the reason a large portion of this book on the knowledge of nature is devoted to reconstructing the nature of knowledge in early modern Japan through the self-fashioning tactics, negotiations, and struggles for social recognition of *honzōgaku* scholars.

MANUFACTURING KNOWLEDGE, CHANGING THE WORLD: EPISTEMOLOGICAL FOUNDATIONS

Nonetheless, it would be a mistake to reduce the knowledge that these scholars produced to the social conditions of that production. Knowledge is a social creation that depends as much on the objective structure of what is to be known as on the system of objective social relations that organizes, supports, and legitimates the production, circulation, and accumulation of knowledge itself and on the constellation of subjective discursive and conceptual choices of scholars. In different degrees and forms, this applies to every discipline that emerged in human history, *honzōgaku* included. The social production of nature knowledge has a history that can be regarded as a process that is both cumulative and punctuated by radical paradigm shifts because the social conditions of scholarly production change over time and because the accumulation of knowledge involves manipulations and interventions on the world that have enduring effects on the world itself. In a sense, knowledge, once produced, has the capacity to live its own life autonomously from the original intentions of those who produced it in the first place.[24]

For a history of any field of knowledge to rightfully advance claims of completeness, it should therefore have both an internalist and externalist approach, retrieving the epistemological practices and discourses as well as the strategies of self-description and justification, the social conditions of its production, its effects on the objects of cognition, and its effects on the world. While I do not nourish any ambition of completeness, in my account of the development of natural history in Tokugawa Japan, I analyze the changing practices and discourses of its practitioners between the seventeenth and nineteenth centuries, the various legitimating and self-fashioning strategies they adopted, their different social trajectories and networks of aggregation and information exchange, their conception of species, and the descriptive and manipulative techniques they adopted. I then reconstruct the intricate ways in which all these tendencies and forces interacted with each other to transform natural things into objects of study and to intervene—sometimes with dramatic effects—in the natural environment. These transformations assumed different forms in different historical moments, often in accordance with the different purposes that sustained the production of knowledge.

Let me clarify this with an example. A storyline in this book is the passage from a focus on systematizing the names of natural species (*meibutsu*), in the early phase of the development of *honzōgaku*, to a focus on surveys and experiments on natural species as material products (*sanbutsu*) by naturalists often employed by the shogunal or domainal administrations. While the two forms of knowledge of natural objects were not distinct but continued to influence and assist each other, the different aims that motivated naturalists' research, their different institutional affiliation and legitimating forces, the different formats chosen to convey the results of their research, the different techniques of individuation and description of natural species (verbal and pictorial), and the like created two distinct "objects" of knowledge out of the same material reality. In other words, concrete vegetal and animal specimens became generic members of a "species," as either names (*meibutsu*) or products (*sanbutsu*), as a result of distinct conceptual operations. As material objects, they were bearers of a variety of properties that, on the one hand, responded directly to the epistemological practices that curtailed them as names or products but, on the other, they could be appropriated to respond to different social demands, classificatory or purely epistemological in the case of names, economic in the case of products (medicinal, agricultural, culinary, etc.).

This dynamic mediation of subject and object of cognition has been a central concern of the philosophy of knowledge—and of scientific knowledge, in particular—in the last two centuries, and in developing my argument, I was influenced by a great deal of research. Of particular interest to me was Adorno's attempt, in his spectacularly complex *Negative Dialectic*, to conceptualize as precisely as possible the nonidentity relation of the material and the conceptual (in the larger sense of including both discursive and material practices) and the priority he accorded to the material as being always and necessarily in excess with respect to the conceptual. This position has a number of advantages. First, it avoids the naïve belief in the possibility of unmediated knowledge, which tends to reduce the history of science into a description of the slow path toward an increasingly refined approximation to a fixed and unchanging external reality. This approximation usually coincides, especially in Japanese historiography, with Western sciences and has the double effect of dogmatically dehistoricizing modern science and transforming its non-Western precursors either in immature forms of protoscientific knowledge or in irrationalistic forms of nationally or ethnically exclusive sensitivity toward the natural world. Second, the relation of nonidentity between the material (object) and the conceptual (subject) gives a more nuanced and nonreductionist understanding of history and nature, which allows "to comprehend historical being in its most extreme historical determinacy, where it is most historical, as natural being, or . . . to comprehend nature as a historical being where it seems to rest most deeply in itself as nature."[25] That is, the material environment and society are entwined because the material world is both and at the same time natural and social: if human beings have evolved their social and intellectual instruments as a result of evolutionary pressures, "nature is always bound up with the historically and socially conditioned concepts and practices that we use to grasp and manipulate it."[26] Third, it reveals the historical and social situatedness of natural knowledge, whereby, as Deborah Cook put it, "concepts are entwined with a non-conceptual whole because what survives in them by dint of their meaning is their non-conceptual conveyance or transmission under specific historical conditions."[27]

Far from being just a tedious exercise in theoretical speculation, the awareness of the nonidentity of the material and the conceptual and of the dynamics that this nonidentitarian association puts in motion is the precondition for my examination of the connections of knowledge, society, and the material world. In our post-Kantian and postdogmatic situation, what happens to the knowing subject and to the known object in the process of cognition is no longer self-evident, but it is precisely what needs to be explained. On the one hand, it is the cognitive process itself that creates its objects of cognition by forcing what is to be known into a conceptual framework of discourses and practices that is essentially distinct from it but renders it intelligible. Material things like plants, insects, trees, or viruses, as well as natural phenomena like earthquakes, metamorphoses, or snowflakes become natural objects as a result of historically situated cognitive procedures that reduced them into specific conceptual apparatuses that do not exhaust their material reality but that make them nonetheless intelligible, controllable, and manipulatable. On the other hand, these material objects, insofar as, qua objects, they become intelligible, controllable, and manipulatable, are mobilized to satisfy a variety of historically situated human wants and needs (cognitive, aesthetic, or economic), and, as a result, they intervene in social dynamics that can have enduring effects in human societies, in the objects themselves, and in the environment. This is what I meant when I claimed, echoing Adorno, that nature and history cannot be separated but must be understood in their dialectic interconnectedness. As corollary, this heuristic move implies that the knowledge that the *honzōgaku* scholars produced cannot be measured against what we now know as a result of the scientific discoveries of the last two centuries—as both *honzōgaku* and modern science are sociohistorically situated—but has to be reconstructed in its own terms, as it immanently unfolded in Tokugawa society.

Take, for example, the case of ginseng (*Panax ginseng*)—*ninjin* in Japanese and *renshen* in Chinese. This is a name given to bitter roots that originally grew only in two cool-temperate regions of the world: the northeastern portion of North America and an area comprising southern Manchuria and the Korean peninsula. Conceptualized as a panacea instilled with miraculous medicinal properties through textual authorities, mythological tales, institutions, physicians, apothecaries, herbalists, and so on, any root that was acknowledged, via specific procedures of verification, to be member of that species became an object that could be mobilized for a number of cognitive, medicinal, social, cultural, economic, and political practices (cultivation, marketization, exchange, smuggling, powdering, observation, description, consumption, etc.) that affected human beings but also affected these material objects (today's botanists acknowledge seven species to be members of the genus *Panax*, mostly the result of human selection) and the environment as well (by the clearing up, for example, of terrains for its cultivation in different parts of the world, from Japan to Germany). In short, the knowing process is never neutral, but it necessarily affects at the same time the knowing subject, the known object, and the world that contained them.

In sum, we know the world by changing it. The expansion of natural knowledge in early modern Europe and Japan entailed an array of practices, from collecting, dissecting, planting, interbreeding, and displacing to the drying, embalming, cataloging, and introduction of alien species in different ecosystems. These practices, far from being a simple collection of empirical data, emerge from a constant manipulation of material entities. Moreover, we change the world by knowing it. The age of exploration brought dramatic changes in the natural environments of all continents: horses, cattle, and a variety of bacteria and viruses were introduced in the Americas at the same time that tomatoes, potatoes, and tobacco invaded the Eurasian continent.[28] This effect of human knowledge on the earth has reached dramatic magnitudes in the last century: genetic engineering has introduced man-made species in the world and is now forcing us to drastically reconsider the notion of life itself. Nobel Prize–winning chemist Paul Crutzen has proposed the neologism "Anthropocene" to refer to the last three centuries of human history in consideration of the global impact of human beings in the ecosystem.[29] Global warming, pollution, and the destruction of wilderness are human causes to vegetal and animal extinctions of geological magnitude as great as earthquakes, meteorites, and ice ages.[30]

As Hilary Putnam—before he joined the antirealist camp—put it, "The mind and the world jointly make up the mind and the world."[31] In that sense, I reject both the naïve realism that conceives of knowledge as the unveiling of an unchanging nature and the skeptical antirealism that reduces knowledge to social constructions, as in David Bloor, or to instrumentalist and pragmatic fictions, as in Wolfgang Stegmüller and Richard Rorty.[32] For the naïve realist, the history of knowledge is the progressive development of more refined instrumental and conceptual apparatuses that allow a deeper understanding of the laws of nature, existing independently from and unaffected by human activity. For the instrumentalist or idealist antirealist, the history of knowledge is an endless reconfiguration of representational fictions that aptly respond to transformations in social power relations. Contrary to these two positions, I embrace the critical realist (or critical materialist) stance that conceives of knowledge as an active and mutual making of both the community of inquirers and their objects of study. In the case of nature knowledge, it means that nature and the naturalist continuously "make" each other. A history of natural knowledge, therefore, cannot be confined to the reconstruction of all discursive and manual practices shared by a community of naturalists. It should also retrieve the processes of legitimation of natural knowledge in the larger society; it should uncover the socially accepted scopes and functions of natural knowledge in different periods; and it should measure the effect of those cognitive practices not only in the social, political, and economic spheres but also in the natural environment. This is why I hope that this book, although narrowly focused on the development of natural history in Tokugawa Japan, will address issues common to historians of other areas and specializations.

NATURE WITHOUT NATURE

One of the major difficulties in reconstructing the activities of Tokugawa naturalists does not lie in understanding their conceptualizations of the objects they studied or the environment that contained them but rather in the semantic intricacies of the English concept of nature itself. Raymond Williams defined "nature" as "perhaps the most complex word in the language."[33] Arthur O. Lovejoy equated the development of its meanings to the entire history of Western thought.[34] Its semantic capacity is staggering:[35] I can call "nature" the environment that surrounds me, the incontrollable impulses inside me, the laws that sustain physical reality, all that exists in a metaphysical sense, the inner essence of things, the concept of being, God, or all of the above at the same time. Even more intimidating is to reconstruct a history of the conceptions of nature in a non-Western cultural sphere like Japan, where a single concept with a semantic capacity equivalent to "nature" did not exist until the 1880s, when the Japanese *shizen* (*ziran* in Chinese) was adopted to translate the German *Natur*.[36] In its place a constellation of different terms—such as *tenchi* (heaven and earth), *sansui* (mountains and waters), *shinrabanshō* (all things in the universe), *banbutsu* (ten thousand things), *honzō* (the fundamental herbs), *yakusō* (medicinal herbs), *sanbutsu* (resources), and the like—were utilized to express different aspects of the environment, material reality, natural objects, and the laws that regulated them.[37] Does the term "nature"—and the modern Japanese *shizen* that translates it—convey such a universal idea as to justify the assumption that the sum of the Tokugawa expressions pointed to nothing other than the same human experience of reality, justifying therefore their unqualified translation with "nature"? I do not think so. There is nothing *natural* in our conceptions of "nature." As Graham Harman put it, "Nature is not natural and can never be naturalized. . . . *Nature is unnatural*, if the world 'nature' is supposed to describe the status of extant slabs of inert matter."[38] Besides, if we subsume under the semantic umbrella of the modern English "nature" the historically specific understanding of the relationship between human beings and the environment that the constellation of pre- and early modern Japanese terms expressed, do we not risk imposing to large chunks of the ideas and practices that defined that society meanings that are alien to it?

"What is nature?" is a question that seems impossible to answer. The challenge to fathom, even in the most general and preliminary sense, what exactly the *what* of the question refers to—A thing? A process? A logic? A field? A concept? A metaconcept? A trope? A condition? Being itself?—is daunting enough to bring to mind Augustine of Hippo's answer to the riddle of time: "What is therefore time? If nobody asks me, I know; but if I am asked, I would like to explain it, but I can not."[39] "Nature" is one of the most important concepts in the intellectual history of the Western world. And yet, if we were to look at its semantic palimpsest in one glimpse, we would discover it crammed with contradictions. When we talk about "nature," we conjure up something that is at the same time concrete and abstract, material and conceptual, physical and metaphysical. To the modern person, "nature" can evoke breathtaking landscapes, the thick of a rainforest, or awe-inspiring natural phenomena.[40] And yet it stands for those landscapes—particular, material, and tangible—also as a whole, as a totality abstracted from their concrete appearance. "Nature" encompasses the objects that populate those landscapes as well as the invisible forces that move them. "Nature" designates the essence of things, the immutable quid that makes things what they are, and contains connotations of eternity, changelessness, and ahistoricity. And yet it changes: nature evolves, unremittingly producing and extinguishing populations, species, and ecosystems. It is at the same time alien and familiar, a perfect example of that which Sigmund Freud called "*das Unheimliche*"—"the uncanny."[41] "Nature" loves to hide its secrets—as in Heraclitus's famous aphorism[42]—but it is also a perfectly intelligible "book," "written in mathematical language, and the letters are triangles, circles and other geometrical figures, without which means it is humanly impossible to comprehend a single word."[43] Nature is the mysterious goddess Isis, Spinoza's God, and benign Mother, but it is also "red in tooth and claw."[44] It is a harmonious, autopoietic, and self-healing organism[45] *and* a field of conflicting and destructive forces. It is both within and without us. It is particular: it defines what kind of human beings we are as individuals, with our peculiar attitudes, vices, and virtues, but it is also universal, defining what it means to be a human being, endowed with inalienable rights. Human beings, for some philosophical traditions, are an integral part of nature,[46] while other thinkers, from Aristotle to Heidegger, via, needless to say, René Descartes, have struggled to demonstrate our substantial distinction and separateness from it. The state of nature is for human beings at the same time a nightmarish condition of continuous warfare (Hobbes) and a blissful brotherhood with the surrounding environment (Rousseau). The list of the oxymora of "nature" can be even longer. They are the symptom of the complex history of this term, passing through successive translations—from the Greek φύσις to the Latin *natura*[47] and then to the Indo-European vernacular variations—and successive reconfigurations in different philosophical schools, cultural practices, religious traditions, and socioeconomic processes. Meanings and connotations added up rather than erasing each other, thus contributing to the semantic stratification of "nature" into a palimpsest that is difficult to break apart.

The semantic complexities of the English "nature" affect our understanding of those societies that did not develop an analogous concept. Words do not merely describe but also prescribe the world we live in. This is not simply a matter of semantic punctiliousness. Even today "nature" is constantly mobilized to justify the most varied beliefs and practices. From human rights, competitive instincts, and free-market liberalism to sexual orientations, family organization, national identity, and so on, political leaders, think tanks, and media "intellectuals" legitimate their own views on these fundamental issues by appealing to their *naturalness*—attempting, that is, to exclude them from becoming a matter of debate or criticism.[48]

What I want to emphasize here is not the *lack* of a term equivalent to "nature" in traditional East Asia but rather its semantic and ideological *excesses*. In fact, "nature," while referring to the material, *physical* environment, also stands, often without us acknowledging it, for the *metaphysical* assumptions that have been associated to it in the course of its history and are now an organic part of its semantic palimpsest. When we say that something is "natural," in other words, we conceive of it as existing independently from human will or as standing for what is normal, what cannot be otherwise than what it appears to be; saying that something (an event, an object) is natural is attributing to it a sense of originality and authenticity. In early modern Japan, the terms expressing these connotations of "nature" did not have any semantic affinity with those that referred to the material environment and its laws. That is why to me "nature," rather than an empty signifier—as Jean Baudrillard put it, deprived of any originality and authenticity[49]—is in fact overloaded with meanings that surreptitiously summon each other up: physical, metaphysical, aesthetic, religious, cognitive, economic, ethical, and political. These meanings are not eternal or universal but historically situated and socially conditioned. Very often, appeals to "nature" have ideological overtones. It suffices to think, for example, of the idea of nature as an organic, self-regulating, and homopoietic totality so common in popular culture and political discourse, from "deep ecology" environmentalists to New Age pundits.[50] To appropriate the political dimension of the metabolic relation of humans with the environment—to "democratize" nature, in Bruno Latour's parlance[51]—we must then emancipate ourselves from the mystifying power of "nature," as I believe, with Adorno, that "people are themselves dominated by nature: by that hollow and questionable concept of nature."[52] "Nature" has acquired such an influential ideological force that some philosophers and social theorists have begun to defend an "ecology without nature."[53] In short, it seems that today "nature" must die so that the environment can live.[54] Accordingly, when the term "nature" appears in this book—as it is a term that we cannot easily dispose of—it has to be understood with the awareness of its semantic and ideological complexity.

No single equivalent to the English "nature" is to be found in the texts of premodern and early modern East Asia. The Chinese *ziran* and the Japanese *shizen* are expressions adopted in the late nineteenth century to translate the English "nature" and the German *Natur*, but in the early modern period, they were mostly utilized as adjectives or adverbs—in Japan also read as *onozukara*—meaning "in itself," "spontaneously."[55] Both Chinese and Japanese traditions, in fact, distinguished with different terms the various semantic spheres ambiguously encompassed by the English "nature." "Human nature" (*sei*) was a Confucian concept with deep social, ethical, and psychological implications that acquired metaphysical connotation only in the later tradition of Zhu Xi's thought, starting from the late twelfth century. Song period Neo-Confucianism, blending together in a novel and creative way Daoist, Buddhist, and Confucian elements, developed a complex metaphysical system of logical and material principles that provided an explanation to various physical, social, and psychological phenomena. But there was no single term like "nature" that encompassed the ordered totality of the universe. Most importantly, there was no single term like "nature" that referred to the totality of material and phenomenal reality.

In Japan, *honzōgaku* scholars and Neo-Confucian thinkers often utilized the term *tenchi* (*tiandi* in Chinese)—sometimes pronounced as *ametsuchi* and literally meaning "sky (or heaven) and earth"—to indicate the whole material world of natural phenomena. More precisely, however, *tenchi* did not encompass a generative force moving and regulating the various phenomena, nor did it include the various natural objects and phenomena—trees, herbs, fish, insects, stars, or rain. Instead, *tenchi* merely indicated "what was above and below the unfolding of the myriads of things," as Itō Jinsai put it.[56] It was metaphorically associated with the image of a "vessel" or "receptacle" (*utsuwa*) for all natural phenomena. Itō Jinsai gave one of the best definitions of *tenchi* in his *Gomō jigi*:

> When a person builds a box putting together six pieces of wood closing it with a cover, almost immediately the generative force [*ki*] spontaneously fills up the box. As soon as the generative force has filled up the inner space of the box, white mold is spontaneously produced. And as white mold is produced, then termites will be born. This is the way in which principle [*ri*] acts spontaneously. The material universe [*tenchi*] is just like a big box: *yin* and *yang* operate just like the generative force within the box; and the entire natural phenomena [*banbutsu*] are the white mold and the termites.[57]

Jinsai described heaven and earth as the boundaries within which an immanent but distinct generative force, *ki*, acted as the enzyme moving matter in a logical and coherent way (*ri*) to produce all things in the universe. Therefore, while *tenchi* indicated the boundary that human cognition could not cross, it contained but not included all physical and metaphysical things that enabled human beings to grasp the logical working of the inner forces of *tenchi*—that is, *ki*, *ri*, *yin*, *yang*, the Five Phases (*wu xing* in Chinese, *gogyō* in Japanese: wood, fire, earth, metal, and water) constituting the building blocks of material reality, and the like—as they unfold in the concrete materiality of natural objects and phenomena.

THE NAMES OF NATURE

It is as if nature had no name. It had instead aplenty.[58] To name just a few, *kenkon* (*qiankun* in Chinese) had a meaning similar to *tenchi*, that of a container, and did not include phenomena or natural species. It was used almost exclusively in the context of divination and in the tradition of the *Yijing*. *Uchū* (*yuzhou*)—today's astronomical term for the universe—appeared in early Daoist texts to emphasize the spatial and temporal infinity of the universe, but in premodern Japanese texts, it was only sparsely used as a synonym of *tenka* (*tianxia*), literally "all under heaven," often in the sense of political "realm."[59]

Since the Meiji period, the modern Japanese word for "world" is *sekai*, but the term was originally a translation of the Buddhist concept of *loka-dhātu*, or the phenomenal universe in a Buddhist sense. In Japan, early texts like the *Taketori monogatari* and the *Genji monogatari* used it as a synonym of *uchū*, while expressions like *kono yo* referred to "this world" in its concrete materiality and inclusive of human society.

Sansui or *senzui* (*sanshui*) and *sanka* (*shanhe*), respectively "mountains and waters" and "mountains and rivers," were two terms that appeared frequently in Chinese poetry to point to a landscape, a scenery, or the natural environment in general; in both China and Japan they were often associated to landscape painting. Another interesting term was *zōke* (*zaohua*): it appeared in early Daoist texts to mean the generating power of natural things, in the sense of nature's power to generate plants, animals, and everything existing and to continuously cause things to change, to transform, and to diversify. The *Kojiki* (compiled in 712) utilized *zōke* to refer to the creative power of the divinities, but the *haikai* poet Matsuo Bashō regularly used it to exalt the diversity of nature. More difficult to define is *fūdo* or *fudo* (*fengtu*), a term that was used in Chinese local gazetteers to refer to the climate, flora and fauna, and geographical conformations of particular regions.[60]

None of these terms, however, had a semantic universe as wide and all encompassing as the English "nature," nor were they used as consistently. Also, they did not include the myriads of things and phenomena—natural, supernatural, and artificial—that populated the universe. A whole set of terms had that function: *banbutsu, banji, banyū, banshō, shobutsu,* and others were all terms that, with only slight variations, represented the "myriads of things" that *tenchi* contained.

It is therefore not surprising that *honzōgaku* scholars, without a term encompassing both "nature" and the various objects it contained, generically conceptualized the minerals, plants, and animals that constituted the objects of their research in terms of the social function that their intellectual and manual labor performed. In other words, the generic names of rocks, plants, and animals depended on their instrumental utility. Or, more precisely, plants, animals, and all natural phenomena were the noematic forms they assumed in accordance with the intellectual activity of human beings. That is to say, they changed their names in accordance with the noetic stances and interventions of different scholars for different social function. Plants and animals for physicians, apothecaries, and orthodox *honzōgaku* scholars were therefore *honzō* or *yakusō* (medicinal herbs). For encyclopedists and lexicographers, they were *meibutsu* (names of things). For agronomists and naturalists engaged in survey projects, they were *sanbutsu* (products). Often, *honzōgaku* scholars utilized the clumsy *sōmokukinjūchūgyokingyokudoseki* (herbs-trees-birds-beasts-insects-fish-metals-jewels-grounds-stones). And when plants and animals became the focus of popular entertainments and spectacles, they could be referred to as *misemono* (stuff for exhibitions and sideshows) or *kōbutsu* (curiosities).

Three of the five parts in which this book is divided follow this pattern. After part I, where, in chapter 2, I offer a brief historical survey of the field of pharmacology in China and Japan up to 1600, part II focuses on the production and circulation of encyclopedias in the seventeenth century and the function of lexicographical research in accurately determining natural species. Part III reconstructs the organization of the 1736 nationwide survey of natural species under Tokugawa Yoshimune and the recruitment of *honzōgaku* specialists in the state apparatus. Part IV maps the popularization of natural history in eighteenth-century cultural circles, popular entertainments, exhibitions, and collections. Lastly, part V suggests how in the latter part of the Tokugawa period, with a wider circulation and acceptance of Western knowledge and the growing intervention of bakufu and domainal administration in matters of political economy, the eclectic field of *honzōgaku* became increasingly disciplined to the necessities of economic growth and its specialists involved in economic reforms. The epilogue will sketch the double destiny of *honzōgaku* in the early Meiji period, when it lost its name but retained much of its accumulated knowledge, recodified in the language of the newly institutionalized Western sciences, at the same time that it kept its name but erased two centuries of research in the name of a tradition buried in a distant past. The invention of *shizen* as "nature" is thus a modern story, part of the revolutionary transformations of the 1870s and 1880s, but its roots are deeply grounded in the philosophical debate of the eighteenth and nineteenth centuries.

TELEOLOGICAL SINS AND TOKUGAWA CONTINGENCIES

It would appear that this book offers yet another version of the story of how we came to experience the world as we do today—a local version of the global story of how we all became modern. Its chronological structure and its focus on social and cultural developments spanning through two and a half centuries of Japanese history certainly reinforces the impression of a necessary progression from an archaic world of mystical correspondences to an instrumentalist world of scientific exactness. The attention I give to the gradual unfolding of a series of intertwined social, intellectual, economic, and political transformations cannot help but strengthen the impression of inevitability of the modernizing process. I obviously reject any functionalistic or intentionalistic explanation of this kind for *longue durée* historical processes. I especially refuse to interpret the efforts, negotiations, practices, and constraints of *honzōgaku* scholars as either necessary or sufficient *causes* for the emergence and success of Western biological sciences in late nineteenth century Japan. I do, however, believe in the value and possibility of reconstructing the unfolding of historical processes that does not necessarily surrender to teleology. In this book, I try to avoid any form of schematic determinism and prefer to look at the interactions between the intellectual practices of Tokugawa naturalists, the sociopolitical conditions in which they operated, the ideas and beliefs they inherited, and the material objects they manipulated. In the background of the processes and events I describe, the prevailing attitudes toward natural objects and the environment that contained them indeed changed, on a macrosociological level and with different rhythms and reaches, from one of containment and exclusion to one of discovery and inclusion, followed, after the Meiji Restoration of 1868, by exploitation and nationalistic ideologicization of a "Japanese" nature.[61] I do not have any problems accepting this narrative as long as it is freed of evolutionary necessity.

I consciously tried to avoid any attempt to connect *honzōgaku* and modern science in any causal sense. Instead, I preferred to reconstruct this field of study in its own terms, as it located itself vis-à-vis its pharmacological tradition and vis-à-vis other contemporary fields of knowledge. I therefore refuse, for example, to explain the development of *honzōgaku* as natural history solely in terms of its applied instrumentality (*jitsugaku*).[62] Rather, I qualify its recruitment by the state for economic advantages in the second half of the Tokugawa period as only one among many functions (cognitive, moral, educational, aesthetic, recreational, etc.) it had. The fact that the economic utility of *honzōgaku* became hegemonic after the 1830s is a historical development connected to the social, political, and economic situation of Japan in the nineteenth-century world of imperialistic global markets rather than a logical development immanent to *honzōgaku* itself.

The only necessity in my history of nature studies in Tokugawa Japan is the contingency of the events and processes punctuating that history. Modern Japan's scientific advances and technological successes were not prefigured in Ekiken's study of plants and animals, nor did Yoshimune's patronage of *honzōgaku* cause its practitioners to search for a scientific method. The fact that natural history enjoyed great popularity did not cause Japan to eagerly adopt Western science in the nineteenth century nor does the large number of professional and amateur naturalists that were active in the field exclusively explain the rapidity of the Japanese turn to science in Meiji Japan. *Honzōgaku* developed standardized protocols of observation and description of natural species, but these did not cause scholars to convert to or accept Linnaeus's methods. In other words, I do not see in the sophistication of nature studies in early modern Japan a sign of the development of an autochthonous protoscientific attitude that *predisposed* Japanese scholars to welcome Western science. Nor do I share the view that *honzōgaku* developed an alternative conception of nature opposed to the "alienating" epistemologies and environmental destructions that modern science directly or indirectly produced. We cannot find in early modern Japan the seeds of an alternative and "more human" science, just as we cannot locate there the possibilities for an alternative "East Asian" modernity. Rather, it was the achieved modernization of Japan, in the early twentieth century, that retroactively called for the reconfiguration of the Japanese past in that light, and it is therefore not surprising that the first historian of *honzōgaku* who did precisely that was also one of the first biologists of modern Japan, Shirai Mitsutarō.[63]

WHAT DO WE MEAN BY EARLY MODERN SCIENCE?

I therefore believe that there was no such thing as a Japanese *scientific revolution* in the Tokugawa period—for the same reasons given by Steven Shapin in the case of early modern Europe.[64] This does not necessarily imply that the descriptive and observational efforts of *honzōgaku* scholars were futile or wrong. Products of society, they shared the various contradictions characterizing and moving Tokugawa society and ideas. *Honzōgaku* scholars were concerned with truthful and accurate descriptions of plants and animals and their classification in a system that was believed to reveal their true essences as much as European natural philosophers of that period. Furthermore, *honzōgaku* scholars were as much influenced by metaphysical preconceptions and the authority of old canonical sources as their European colleagues. For example, they both conceived of natural species as "natural kinds"—that is, taxonomical groupings or orderings of plants and animals independent of human interventions. From the point of view of modern biology, they both were wrong in different ways. Biological species are today far from being conceived of as natural kinds, but comparing the results of *honzōgaku* scholars' inquiries to modern scientific knowledge is not only an exercise in anachronism but also an utterly sterile enterprise.

It would be even worse to compare and judge Tokugawa scholars' nature knowledge with what eighteenth-century European natural philosophers knew about plants and animals, ascribing to the latter the attributes of "science" and arguing about Japanese scholars' failures or successes to accept the "correct" Western paradigm. Despite the fact that early modern European naturalists could be chronologically and genealogically counted as the ancestors of modern scientists, from the point of view of today's research in genetics and biology, their natural history was not qualitatively much different from *honzōgaku*'s— sustaining the contrary and attributing to early modern European naturalists methodological and empirical positions akin to contemporary science would be committing another kind of anachronism. Moreover, as scholars like Sujit Sivasundaram, Marwa Elshakry, and Simon Schaffer have recently demonstrated, Western modern sciences have been deeply affected during their worldwide expansion in the eighteenth and nineteenth centuries by their encounters with non-European cultures.[65]

In short, what I want to avoid is a conception of science as an ahistorical and neutral meter of judgment. This decision inevitably leads to the question of what science is, which goes well beyond the scopes and possibilities of this study. Modern science is the cultural product of a particular historical time, fashioned under particular social conditions and within the framework of the particular conceptual constellations that conceived it. Like any other form of knowledge, today's science is also "situated" knowledge, and as such it reflects the position of its producers in their historical, cultural, social, and material context. Science, in other words, is not an ahistorical form of knowledge that transparently reflects an ordered reality but a discipline encompassing a variety of fields of study that emerged in a particular historical moment and context under particular sociointellectual conditions. As Peter Dear put it, "The cultural activity called 'science' as it developed during the nineteenth and twentieth centuries is not the same as the old natural philosophy. The changes that the latter label had undergone during the seventeenth and eighteenth centuries resulted in the establishment of a new enterprise that took the old 'natural philosophy' and articulated it in the quite alien terms of instrumentality—science was born a hybrid of two formerly distinct endeavors."[66]

This is not equivalent to maintaining that science, because of its historical situatedness, is just a form of knowledge like any other, qualitatively not different from religious, superstitious, traditional, or folk beliefs. If in the course of the modern period science has parted ways from the absolute truth-claims of metaphysics to embrace an epistemology of empirical *accuracy* and *certainty*—that is, truth-claims that are intersubjectively determined by the justified consensus of an epistemological community sharing protocols of observation, measurement, and symbolic representation—it still shares with philosophy a dismissal for unwarranted opinions and beliefs. Affirming the historicity and social situatedness of scientific knowledge does not necessarily imply questioning the validity of its cognitive claims.[67]

Furthermore, modern science as it emerged in the nineteenth century was much more a global event than it has been previously thought. Even if we accept the conventional narrative whereby modern science is a product of nineteenth-century Europe, the various scientific disciplines it contained bore witness of the global nature of the age of empires: European culture was influenced by local knowledges as much as it influenced the development of new ideas and practices worldwide.[68] Just like the modern sciences in Europe sublated much of the content, practices, and institutions of early modern natural philosophy, many elements of *honzōgaku* were transubstantiated into the new disciplines of biology (*seibutsugaku*), botany (*shokubutsugaku*), and zoology (*dōbutsugaku*) in Meiji Japan. Itō Keisuke—formed as a *honzōgaku* scholar and later celebrated as the first Japanese scientist at Tokyo Imperial University—symbolically embodied this metamorphosis. The challenge this book embraces is to reconstruct this story without surrendering to a teleology of modernization but conceiving of it as a reflection of the social, political, economic, and cultural changes in nineteenth century Japan.

An effect of these developments in rapidly industrializing Japan was the disappearance of traditional conceptions of the natural world. This disenchantment, in turn, had two further consequences: on the one hand, it transformed the environment into a reservoir of resources exploitable for the needs of eco-

nomic growth;[69] on the other, it called for the development of new conceptualizations of the natural environment with the opposite ideological aims of
sustaining that exploitation or condemning its cost in terms of pollution, as well
as projecting into it connotations of nationalistic uniqueness.[70]

2   *The* Bencao gangmu *and the World It Created*

FUNDAMENTAL HERBS

It is hardly disputable that many elements of Japanese culture originated in China and were introduced in Japan either directly or through the
mediation of Korean kingdoms.[1] Japan adopted from China and then radically
transformed its writing system; religious traditions like Buddhism and Daoism
that deeply shaped its social, cultural, and spiritual history; schools of thought;
literary and artistic forms; political, legal, and administrative systems; technological know-how; silver and copper coins; tools, artifacts, commodities, and
natural resources of various kinds; and skilled laborers and scholars. In this respect, *honzōgaku* was not an exception. Knowledge of plants and animals may be
as old as the human species, but in premodern Japan, the systematic and institutionalized knowledge of nature depended entirely on canonical texts produced
in China.[2] In the Tokugawa period, the influence of Ming and Qing culture was
particularly strong.[3] It should not therefore come as a surprise that the entire
field of nature studies in early modern Japan began with the introduction of a
Chinese encyclopedia, Li Shizhen's *Bencao gangmu* (Systematic Materia Medica), published in Nanjing in 1596.

In China, the study of medicinal herbs—*bencaoxue*—was a "pragmatic"
branch of medicine.[4] As such, it was closely related to foodstuff preparation,
magic, and demonology. *Bencao*—literally, "the fundamental herbs"—appears
in the *Hanshu* (History of the Former Han, 76 CE), which defines it as a distinct field of study.[5] In the *Huainan zi* (The Book of Huainan), a Daoist text of
the second century BCE that tried to fuse the pharmacological and therapeutic
traditions, we read the mythical origins of *bencaoxue*:

> In ancient times the people subsisted on herbs and drank water. They col-
> lected the fruits from the trees and ate the flesh of the clams. They frequently
> suffered from illnesses and poisonings. Then Shennong taught the people
> for the first time to sow the five kinds of grain, to observe whether the land
> was dry or moist, fertile or stony, whether the land lay high or low. He tried
> the tastes of all herbs and [investigated] the water sources to see if they were
> sweet or bitter. In this way he taught the people what they should avoid
> and where they could seek help. At that time [Shennong] found on a single
> day 70 substances that were medically effective.[6]

The tradition that saw Shennong—the "divine husbandman," a mythical emperor who taught the ancient Chinese agriculture and materia medica—as the founder of Chinese pharmacology was established by the *Shennong bencao jing* (Shennong's Canon of Materia Medica), a text probably compiled between the second and the first centuries BCE and now lost.[7] It listed 365 drugs obtained by the "fundamental herbs" (*bencao*) that Shennong had tasted, dividing them in the three categories of upper, intermediate, and lower in accordance with their toxicity. Later generations of pharmacopoeias adopted its structure and expanded it with new medicinal substances. Tao Hongjing's *Bencao jing jizhu* (Notes on the Canon of Materia Medica, circa 492 CE) not only increased the number of substances treated but gave the field of *bencaoxue* its established form and metaphysical foundations until the Ming period, when works like Wang Lu's *Bencao jiyao* (Essentials Materia Medica, 1496), Chen Jiamo's *Bencao mengquan* (Enlightenment on Materia Medica, 1565), and, especially, Li Shizhen's *Bencao gangmu* radically reshaped the field of pharmacology.

In ancient Japan, the formal study and practice of medicine was restricted to the Ten'yakuryō (the Institute of Medicine), which was a department of the Imperial University (Daigakuryō) established in the late seventh century. The medical theories and practices of the Ten'yakuryō were strictly based on the Chinese model but differed in that medicine in Japan was "largely separated from society in general." As Sugimoto and Swain explain, "The *ritsuryō* Institute of Medicine was intended to serve only the imperial household and officials in the capital cities, and the medical courses in the provincial colleges likewise were for the benefit of provincial officials."[8]

The Ten'yakuryō controlled the production and usage of manuals on materia medica, which gave "theoretical classifications, descriptions, places of production, and therapeutic effects of an increasingly wide range of herbal, animal, and mineral substances."[9] Court physicians depended entirely on books imported from the continent, and the only Japanese reference texts were digests or simplified editions of Chinese encyclopedias.[10] The first pharmacological textbook adopted by the Ten'yakuryō was *Shennong bencao jing jizhu* (Collected Commentaries of Shennong's Classical Materia Medica, *Shinnō honzōkyō shūchū* in Japanese), a revised and expanded edition of Tao Hongjing's *Shennong bencao jing* (*Shinnō honzōkyō* in Japanese).[11] It contained explanations of 730 kinds of medicinal substances, divided into seven categories (*bu*) of minerals (*yushi, gyokuseki* in Japanese), plants (*caomu, sōmoku*), animals (*chongshou, chūjū*), fruits (*guo, ka*), vegetables (*cai, sai*), staples (*mishi, beishoku*), and "things with name but without use" (*youming weiyong, yūmei miyō*). It arranged all species in each category into three hierarchical grades (*pin, hin*) of medicinal substance, following *Shennong bencao jing*'s tripartite division. As Needham explained, "The naming of these [grades] was inspired by the bureaucratic order of society, for those in the first grade were known as 'princely' [*jun, kun*], those in the second were termed 'ministerial' [*chen, shin*], while those in the third and lowest were defined as 'adjutant' [*zuoshi, sashi*]."[12] The "princely" drugs were used to maintain the general health of the body and contained no dangerous substances. The "adjutant" drugs cured more severe infections but could become poisonous if taken in the wrong dosage. The "ministerial" drugs were somewhere in the middle.[13] This tripartite classification survived until the publication of *Bencao gangmu* and was connected to a cosmological order of Daoist inspiration that saw everything in the universe ruled by three powers (*sancai*): Heaven, Earth, and Man. Princely drugs, which could be consumed in large quantities because they did not contain poisonous substances, received their healing properties from Heaven (*tian*) and as such nourished life itself (*yangming*). Ministerial drugs depended on Man (*ren*) and nourished human life (*yangxing*) if taken in moderate dosage. Adjutant drugs, potentially lethal and therefore to be taken in small doses and for short periods of time, depended on Earth (*di*).[14]

In 787, the *Xinxiu bencao* (Revised Materia Medica, *Shinshū honzō* in Japanese) replaced the *Shennong bencao jing jizhu* as Ten'yakuryō's main textbook. Its fifty-three volumes, which contained information about 850 substances, had been compiled in 659 by a team of twenty-two scholars under the supervision of the Chinese imperial physician Su Jing. A "landmark of natural history at least as much as a treatise on materia medica," *Xinxiu bencao* provided also illustrations (*tu*) that accompanied a verbal descriptions of the mineral, vegetal, and animal sources of medicinal substances.[15] A millennium later, Engelbert Kaempfer, physician for the Dutch East India Company (Vereenigde Oostindische Compagnie, or VOC), quoted it among the sources in his *Amoenitas Exoticae* (1712). In 918, Fukane no Sukehito, personal physician of the imperial family, translated the entries of the *Xinxiu bencao* into Japanese, adding a brief note to its entries specifying whether the plants or animals existed in Japan or not. Titled *Honzō wamyō* (Materia Medica in Japanese), Sukehito's glossary listed 1,025 entries, 550 of which were provided with a Japanese name.[16]

Outside the closed world of the imperial academy, Buddhist temples were also accumulating pharmacological knowledge. The monk Ganjin, the eighth-century founder of the Risshū school of Japanese Buddhism, researched the pharmacological properties of plants. Before the sixteenth century, Buddhist temples were the only institutions providing medical assistance to the general population. Some of the medicaments developed in temples are recorded in the *Engi shiki* (Rules of the Engi Era, 901–23), a tenth-century collection of governmental regulations, and *Izumo fudoki* (Gazetteer of the Izumo Region), an eighth-century report on the natural resources, geophysical conditions, and oral traditions of Izumo Province (today's Shimane Prefecture).[17]

Commercial exchanges between Japan and China increased dramatically during the Song period (960–1279). Many books were imported from China, thanks in part to the frequent trips of Japanese monks to the continent. With the establishment of the Kamakura shogunate in 1192, Buddhist monks, especially of the esoteric and Zen traditions, produced the majority of medicinal and pharmacological treatises, which were still largely indebted to Chinese encyclopedias. The Chinese *Daguan jingshi zhenglei beiyong bencao* (Classified and Consolidated Materia Medica of the [Da guan] Reign-Period, compiled in 1108), known in Japan in the abbreviated form *Daikan honzō*, became the fundamental textbook of pharmacological training. In 1284, Koremune Tomotoshi, a scholar of the Imperial University, compiled a digest version in Japanese titled *Honzō iroha shō* (Materia Medica in Alphabetical [*iroha*] Order). It rearranged a selection of 590 entries from *Daikan honzō* into the *iroha* syllabic order to make it easier to use.[18]

From the thirteenth century until the introduction of *Bencao gangmu* in the early seventeenth century, there was no major change. The social disruption produced by the continuous state of internal warfare that punctuated Japanese history in this period was undoubtedly a major obstacle to the elaboration of new forms of systematic and institutionalized learning. The Ten'yakuryō shrank to a mere symbolic existence in an impoverished imperial court: it did not produce any new research on pharmacology and retained *Daikan honzō* as its only textbook. Buddhist monks of the esoteric tradition wrote a series of herbal treatises, but they circulated in temple networks and mostly concerned perfumed plants, incense, and hallucinogens to assist monks in their meditative practices.[19]

Li Shizhen's *Bencao gangmu* (*Honzō kōmoku*) entered Japan in the early 1600s as the most authoritative encyclopedia of pharmacological substances at a time of great medical innovation. In the previous century, Tashiro Sanki and Manase Dōsan had introduced the medical theories of the thirteenth- and fourteenth-century Chinese physicians Li Ai and Zhu Zhenxiang to Japan and reinterpreted them in an original fashion.[20] Tashiro Sanki, after a period of study in Ming China, discarded his Buddhist robes in 1509 and started his own medical practice at Koga, formally the Kantō headquarter of the Ashikaga shogunate until 1573. His student Manase Dōsan also "dissociated himself from

Buddhist orders and became a secular practitioner."[21] Their school assumed the name of Goseihō, the "later method." By the seventeenth century, the Goseihō became the mainstream medical school, thanks in large part to the patronage of the Ashikaga and, later, the Tokugawa shogunates. Two of the most important Goseihō schools of the period were the Keitekiin, founded by Manase Dōsan, and the Hori school founded by Hori Kyōan, who was a disciple of Fujiwara Seika and Manase Shōjun and a close friend of Hayashi Razan.[22] Influenced by Zhu Xi's cosmology, the Goseihō identified the origins of diseases in the disharmonious movements of *ri* and *ki* that caused imbalances in the metaphysical forces of *yin, yang,* and the Five Phases in the body. Furthermore, it claimed that the cause of illnesses was not external pathological agents but rather the weakness of the body, which permitted entry to those agents. Strengthening the body's resistance was thus the main target of Goseihō therapy, which aimed at raising the level of *yin* energy top set against the overabundance of the *yang* energy that was considered the origin of all disease—via therapeutic protocols including acupuncture, moxibustion (*jiu*; in Japanese *kyūji*), and the regular consumption of ginseng and other medicinal roots.

The medical innovations in the Goseihō school in the sixteenth century anticipated the arrival in Japan of Li Shizhen's work and opened the way for its success. The *Bencao gangmu* had itself been conceived in Ming China as an indispensable pharmacological companion to Neo-Confucian medicinal theories and practices.

THE GREATEST NATURALIST IN CHINESE HISTORY

Li Shizhen was the son of a physician of the Imperial Medical Academy (Tai Yiyuan). The "prince of pharmacists," as Needham defined him, "was probably the greatest naturalist in Chinese history, and worthy of comparison with the best of the scientific men contemporary with him in Renaissance Europe."[23] In the last fifteen years, the government of the People's Republic of China has celebrated Li Shizhen as a scientist and his pioneering enterprise in scientific empiricism.[24]

His father wished him to have a career in the civil service, but after failing the highest level of imperial examinations three times, Li Shizhen decided to dedicate his life to the study of medicine.[25] The encyclopedic knowledge he accumulated and his reputation as a skillful physician gained him the patronage of various patrician families. In 1549, the princely family of Chu nominated him Superintendent of Sacrifices, "with charge of their medical administration."[26] His career skyrocketed until he reached the Fifth Rank of Assistant Academician at the Imperial Medical Academy. The patronage of aristocratic families and an impressive network of fellow scholars, with whom he conducted a massive correspondence, granted him access to the sources and the financial means he needed to complete his encyclopedia.

Li Shizhen wanted his *Bencao gangmu* to prevail against "the confusion which persisted in the pandects of pharmaceutical natural history."[27] From 1556 he began traveling to the most important drug-producing provinces to collect and study specimens and medicaments. *Bencao gangmu* was completed in 1587 with a total of 1,895 entries (275 minerals, 446 animals, and 1,094 plants).[28] Li died before he could see his work in print, which appeared in 1596 in Nanjing, thanks to the intercession of Ming poet Wang Shizhen.[29] Wang wrote a preface for the text, where he praised Li's work for covering "everything from the most ancient records to vernacular stories, with nothing relevant left out."[30] After Li's son Jianyuan presented a copy to the Ming emperor, the encyclopedia was sold to the larger public.

The importance of Li Shizhen's work and its impact on East Asian cultural history was enormous. The original developments of early modern Japanese *honzōgaku* were rooted on this text. Its title associated it with pharmacopoeias of the past centuries, but its breadth of scope and method made it a landmark achievement.[31] It claimed to cover "all that is known," including all plants, animals, minerals, and monsters, whether or not they had practical value as pharmacological substances. "The book," wrote Needham, "is a pandectal treatise on mineralogy, metallurgy, mycology, botany, zoology, physiology and other sciences in its own right."[32] Wang Shizhen commemorated it with these words:

Like entering the Golden Grain Garden, the varieties and colors dazzle the eyes; like ascending to the Dragon Lord's Palace, the treasures are all on display; as if facing a (clear) pot of ice or a jade mirror, one can discern the minutest details. Abundant but not superfluous, detailed but essential, comprehensive and thorough, penetrating the deep and vast. How could we see this simply as a medical book? What diligence and grace Master Li has devoted to the intricacies of principle (*xingli*), encyclopedic works of the investigation of things (*gewu*), the secret esoterica of kings, and the valued knowledge of common people. What earnest labor on the part of Master Li![33]

As the celebratory words of Wang suggest, it was not only the exhaustiveness of its coverage that distinguished *Bencao gangmu* from its predecessors. Li's contribution to the Neo-Confucian tradition of *gewu* (the investigation of things) went well beyond the thoroughness of information gathering and encompassed a much-needed systematization of the received corpus of natural knowledge. It was a monumental project that aimed at a complete "reorganization of knowledge," as Georges Métailié put it.[34] Li achieved this goal through two strategies: the standardization of species names and the development of a hierarchical taxonomy of species.

One of the difficulties facing earlier pharmacopoeias was the choice of standard names for specimens, since the names of plants, animals, and minerals changed from region to region and from one historical period to another. Li Shizhen solved the problem through careful philological research by establishing a standard terminology (*zhengming, seimei* in Japanese). "In nomenclature," Needham explained, "he adopted a system of priority, so that the name which had first been given historically to any particular plant or animal was taken as the standard term."[35]

Li Shizhen's second contribution to the discipline was a classificatory system. First, following Tao Hongjing's *Bencao jing jizhu*, he separated natural kinds into minerals, plants, and animals and treated them in a section of the encyclopedia distinct from the pharmacological substances derived from them. He ordered the species into a systematic taxonomy and categorized the various medicinal substances according to the disease they were thought to cure.[36] All entries or "species" (*zhong, shu* in Japanese) were divided into sixteen major headings or groups, called *gang* (*kō* in Japanese). *Gang* were in turn subdivided into smaller categories called *mu* (*moku*). *Gang* thus consisted of sixteen sections (*bu*), which were further divided into "categories" (*lei*) and each category into "species" (*zhong*). In other words, *gang* and *mu* were the two organizational principles that structured the presentation of natural knowledge in *bu, lei,* and *zhong.* This taxonomy gave the work its title, *Bencao gangmu,* or pharmacopoeia classified into sixteen *gang* and sixty *mu. Gang* and *mu* were two terms referring the main rope and the meshes of a net. As a single term, *gangmu* denoted a hierarchical organization.[37]

In the prefatory note (*fanli*), Li Shizhen explained the historical origin of its classificatory system:

> The *Divine Husbandman's Materia Medica* [*Shennong Bencao jing*] in three chapters, with three hundred and sixty kinds, is divided into three grades: upper, intermediate, and lower. Tao Hongjing of the Liang dynasty added drugs so that their number was doubled, adding them following their grade. During the Tang and Song dynasties, there were duplications and revisions, each author made additions, including some, omitting others; doing so, the levels of

grading (*pinmu*) still existed but the ancient headings were all mixed up and the true meaning was lost. Now I have ordered all under sixteen section (*bu*) which represent the higher *gang* and sixty categories (*lei*) which represent their subordinates *mu*, each according to its own category.[38]

The complete classificatory system (tab. 2.1) thus comprehended species (*zhong*), organized into categories (*lei*), which were part of a superior order of sections (*bu*):

TABLE 2.1. *The Bencao Gangmu Classificatory System in Sixteen Sections*

1. *shui* 水, *sui*, *mizu*: 43 entries of "waters," further divided into two categories: "things that fall from the sky" (like fog, hail, ice, snow, different types of rain, etc.) and "water that spills from the ground"

2. *huo* 火, *ka*, *hi*: 11 entries of "fires," from the "fire" of the Five Phases to specific kinds of fire, like sparks (fire that originates from clashing stones), fire from wood, fire from coal, and so on

3. *tu* 土, *do*, *tsuchi*: 61 entries on "earths"—that is, the various kinds of terrains, clays, and so on; a total of 61 species (*zhong*)

4. *jinshi* 金石, *kinseki*: 161 entries of "metals and minerals," further subdivided into the 4 categories of metals (*jin* 金), jewels (*yu* 玉, *gyoku*), stones (*shi* 石), and salts (*lu* 鹵, *ro*)

5. *cao* 草, *sō*: 615 entries of "herbs," further divided into 10 categories: 70 species of mountain herbs (*shancao* 山草, *sansō*, or *yamagusa*); 56 species of fragrant herbs (*fangcao* 芳草, *hōsō*); 126 species of "marshland herbs" (*xicao* 隰草, *shūsō*); 47 species of poisonous herbs (*ducao* 毒草, *dokusō*); 91 species of creepers (*mancao* 蔓草, *mansō*); 23 species of water herbs (*shuicao* 水草, *suisō*); 19 species of "stone herbs" like sedge weeds (*shicao* 石草, *sekisō*); 16 species of mosses (*tai* 苔, *tai*); 9 species of sundry herbs (*zacao* 雜草, *zassō*); and 153 species of unclassified and unclassifiable herbs, which had been named in some regions, but had no known pharmacological use

6. *gu* 穀, *koku*: 73 entries of "grains," subdivided into 12 species of hemp-wheat-rice (*mamaidai* 麻麦稻, *mamakutō*), 18 species of millets (*jisu* 稷粟, *shokuzoku*), 14 species of soy and legumes (*shudou* 菽豆, *shukutō*), and 29 species of fermented products (*zaoniang* 造醸, *zōjō*)

7. *cai* 菜, *sai*: 105 entries of vegetables, divided into 5 categories: 32 species of seasoning vegetables (*huncai* 葷菜, *kunsai*; like onion, ginger, garlic, etc.), 41 "soft and slippery" vegetables (*rouhua* 柔滑, *jūkatsu*; like spinaches), 11 "gourd-like" juicy vegetables (*guacai* 蓏菜, *rasai*; acorns), 6 "acquatic" vegetables (*shuicai* 水菜, *suisai*; like lettuce), and 15 species of mushrooms (*zhiji* 芝栭, *shiji*)

TABLE 2.1. *continued*

8. *guo* 果, *ka*: 128 entries of fruits; of these, 11 species of *wuguo* 五果, *goka* (prune, apricot, jujube, plum, etc.), 34 of mountain fruits (*shangguo* 山果, *sanka*, pear, kaki, mandarin, etc.), 31 of exotic fruits (*yiguo* 夷果, *ika*, litchi, palm, etc.), 13 of spices (*weiguo* 味果, *mika*, pepper), 9 "watermelons" (*guaguo* 蓏果, *raka*), and 29 aquatic fruits (*shuiguo* 水果, *suika*, like the lotus root, the Chinese water chestnut, etc.)

9. *mu* 木, *moku*: 180 entries of trees, of which there are 35 species of fragrant trees (*xiangmu* 香木, *kōboku*), 52 species of tall trees (*qiaomu* 喬木, *kyōboku*), and 51 of small trees (*guanmu* 灌木, *kanboku*), 12 species of parasitic trees (*yumu* 寓木, *gūboku*, mistletoes and funguses), 4 species of trees growing in clusters (*baomu* 苞木, *hōboku*, bamboos), and 26 species of miscellaneous trees (*zamu* 雜木, *zatsuboku*)

10. *fuqi* 服器, *fukki*: 79 entries of objects like clothes and utensils obtained from plants, 25 types of clothes (*fubo* 服帛, *fukugei*, cotton, silk, rags, etc.), and 54 types of artifacts (*qiwu* 器物, *kibutsu*, from paper to calendar, from fireworks to lacquerwork)

11. *chong* 蟲, *chū*: 107 entries of insects, divided into three categories of 45 species of oviparous insects (*luansheng* 卵生, *ransei*, among which are classified insects like ants, bees and so on but also spiders and leeches), 31 species of insects produced by metamorphosis (*huasheng* 化生, *kasei*, like grubworms, cicadas, longicorns, fireflies, and locusts), and 30 species of insects generated from moisture (*shisheng* 湿生, *shissei*, species from the Anura order—that is, frogs and toads—Chilopoda and Diplopoda classes, which are centipedes, and all earthworms)

12. *lin* 鱗, *rin*: 94 entries of scaly animals divided into four categories, classifying 9 species of "dragons" (*long* 龍, *ryū*, lizards and mythological dragons), 17 species of snakes (*tuo* 蛇, *da*), 31 species of fish (*yu* 魚, *gyo*), and 37 species of scaleless fish (*wulinyu* 無鱗魚, *muringyo*, squids, eels, congers, catfish, globefish, sharks, rays, dolphins, mythical sirens, octopuses, shrimps, jellyfish, etc.)

13. *jie* 介, *kai*: 46 entries of shellfish, divided into 17 shelled species (*guibie* 龜鱉, *kibetsu*) of tortoises and crabs and 29 species of bivalves and gastropoda (*bangge* 蚌蛤, *boko*)

14. *qin* 禽, *kin*: 77 entries of birds, divided into 23 species of aquatic birds (*shuiqin* 水禽, *suikin*), 23 species of grasslands birds (*yuanqin* 原禽, *genkin*), 17 species of forest birds (*linqin* 林禽, *rinkin*), and 14 species of mountains birds (*shanqin* 山禽, *sankin*, including phoenixes)

15. *shou* 獸, *jū*: 86 entries of quadrupeds, divided into 28 species of domestic animals (*chu* 畜, *chiku*), 38 wild animals (*shou* 獸, *jū*, all big animals that cannot be used as livestock), 12 species of rodents (*shu* 鼠, *so*), and 8 species of wanderers (*yu* 寓, *gū*) and monsters (*guai* 怪, *kai*)

TABLE 2.1. *continued*

16. *ren* 人, *jin*: 37 entries of humans, including the different ethnic groups (tribes and people with which the Chinese empire had come into contact), mummies (*munaiyi* 木乃伊, *miira*), and humanlike monsters

The criteria that Li Shizhen followed to classify what amounted to "all things in the world" depended on "visible morphology" (*xingzhi, keishitsu*), which was a characteristic determined by *yin* and on invisible properties (*qizhi, kishitsu*), a feature of *yang*. He therefore ordered all entries in accordance with the physical pattern that *qi* (*ki*) formed by interacting with the Five Phases. For plants and animals he proceeded from small to big, but in general he determined the sixteen categories (*lei*) in accordance with a mix of morphological and ecological criteria or if a plant or an animal lives in the wild or is cultivated or domesticated. Alternatively, he took into consideration their toxicity and their flavor or taste or he followed conventional agronomical categories. The complexity of his organization of nature is evident in this prefatory note:

> In the writing of old, gems, minerals, waters, and earths were all inextricably confused. Insects were not distinguished from fishes, nor fishes from shellfishes. Indeed some insects were placed in the section on trees, and some trees were placed in that on herbs. But now every group has its own Section. At the head some waters and fires, then come earths; for water and fire existed before the myriad (inanimate and animate) things, and earth is the mother of all the myriad things. Next come metals and minerals, arising naturally out of earth, and then in order herbs, cereals, (edible) vegetable plants, fruit bearing trees, and all woody trees. These are arranged following their sizes in an ascending order, starting with the smallest and ending with the largest. A section on objects that can be worn by human beings follows (this is logical since most of them come from the plant world). Then the tale continues with insects, fishes, shellfishes, birds and beasts, with mankind bringing up the rear. Such is the ladder of beings, from the lowliest to the highest.[39]

Li Shizhen arranged the entry (*xiaogang, shōkō*) of each species into a structure of eight subchapters (*xianmu, shōmoku*)—although not all entries have eight sections. After the standard name (*zhengming*) of the species, he listed all names under which the species under consideration was known in the past or in different regions (*shiming, shakumei*). The next subchapter, *jijie* (*shūkai*), consisted of information on habitat, seasonality, morphology, properties, and other attributes of the species under consideration. *Zhengwu* (*seigo*), the following subchapter, highlighted the mistakes in the description of the species in previous sources and rectified them. Then in *xiuzhi* (*shūji*), Li provided details on how to process the mineral, vegetal, or animal species to extract and preserve the pharmacological substances. *Qiwei* (*kimi*) specified the dosage, taste, and toxicity of the drug obtained from that species in accordance with the traditional tripartite division. An account of its most important therapeutic usage (*zhuzhi*, *shuji*) and a historical reconstruction of the discovery of its medicinal properties (*faming*, *hatsumei*) followed. A collection of its prescriptions (*fufang*, *fuhō*) concluded the entry.[40]

Li Shizhen probably did not intend to add illustrations to his encyclopedia. His brother Jianyuan, however, decided to include them for the Nanjing edition. We do not have any information about the authorship of the majority of these pictures. Some were taken from Tang Shenwei's *Jingshi zhenglei beiji bencao* (Classified and Consolidated Armamentarium of Pharmaceutical Natural History) of 1082, others illustrations he may have drawn himself.[41]

SORTING THINGS OUT

Bizarre as it might appear to us, Li Shizhen's classification of species presented a coherent system that was grounded in the discursive configurations of material reality of his time and framed in Neo-Confucian metaphysics. Taxonomies—division of plants and animals into discriminated taxa, or groups—are a cultural aspect of all societies around the world and in different historical times. They are the object of study of a branch of anthropology called folkbiology or ethnobiology. Every sociocultural community, ethnobiologists argue, has divided the natural world into groups and classes of vegetal and animal species. These are often hierarchically displayed in levels, with specific taxa divided into more generic taxa and the like.[42]

Ernst Mayr has stressed the importance of distinguishing, from the perspective of historical epistemology, schemes intended for classifications from those used for identification.[43] Mayr pointed out that all classifications have two main purposes, which are often interconnected. One is the development of a schema that enables a precise identification of a species of plant or animal for practical purposes (inventorial, pharmaceutical, agronomical, encyclopedic, etc.). "Ease of information retrieval is generally the principal or exclusive objective of the classification of items."[44] The second purpose is a "classification of items that are connected by causation (as in a classification of diseases) or by origin (as in biological classification)."[45] In contrast with identification schemes, this second approach is characterized by hierarchical grouping of species into genera, orders, and classes—as does the *Bencao gangmu* with its hierarchical system of *zhong-lei-bu-gang-mu*. The identification of a species in such an order follows a law of causation or a metaphysical order of nature. In other words, often deliberately anthropocentric, identification schemes, for Mayr, are far more casual and prone to changes, as was indeed the case in medieval *herbaria* in Europe and in pre–*Bencao gangmu* pharmacopoeias in China, which arranged species in alphabetical order, according to the substances they produced, or depending on the healing properties they had. Taxonomies, in contrast, are subjected to a superimposed order that renders them more resistant to change.[46] The dividing line between identification schemes and taxonomies is, however, difficult to draw. In the premodern and early modern periods, naturalists in Europe, China, and Japan struggled over groups and species that eluded obvious classification. This was often the case for invertebrates (commonly associated with insects), amphibians, reptiles, and cetaceans.[47] For example, Li Shizhen organized his *Bencao gangmu* in a classificatory system that aimed to reproduce a sort of *scala naturae* based on physical and metaphysical principles, which classified snakes, squids, eels, congers, catfish, globefish, sharks, rays, dolphins, mythical sirens, octopuses, shrimps, and jellyfish under the rubric of scaled fish (*lin, rin* in Japanese) and toads and frogs among insects (*chong, chū*).[48] Today, cladistic systematics, developed in the 1960s by the biologist Willi Henning, arranges organisms in accordance with a historical order of branching out in an evolutionary tree rather than by their morphological similarities.[49] DNA analysis and the consequent development of cladistic taxonomy have given scientists a genetic foundation to their classificatory practices, but it has not solved the epistemological issues at the core of classifications.[50]

The fact remains that different classifications often produce a sense of vertiginous oddity. Jorge Luis Borges saw in the madness of "a certain Chinese encyclopedia called the Heavenly Emporium of Benevolent Knowledge" the clear sign of the ultimate arbitrariness of all classificatory systems: each culture has its own incommunicable, incommensurable, and subjective understanding of the world.[51] The questions then are, how can the material world express such a vertiginous diversity? Is order imposed by the human mind upon the world so that searching for an immanent organization in nature is an enterprise doomed to failure? Or if an immanent order does exist, will it ever be possible for us to access it, or are we condemned to experiencing the world only through the ordering filter of culture? These fundamental questions inform much of both Western and Asian thought and are at the core of the issue of classification.[52]

We know the universe we live in by classifying its parts. The ordering and reordering of every thing existing is a constant feature in the history of human societies and their metabolic relations with the material world. As Ian Hacking put it, "Enumeration demands *kinds* of things or people to count. Counting is hungry for categories. Many of the categories we now use to describe people are byproducts of the needs of enumeration."[53] Classifications vary in time and space: different sociocultural communities have continuously created, adopted, modified, and abandoned more or less systematic orderings of vegetal and animal species as well as of a vast array of material, artificial, or transcendental phenomena. Each classifying system is alien to the other ones, an "exotic charm of another system of thought," as Michel Foucault put it.[54] As the structural core of every human culture, it internalizes the astonishing paradox of the existence of a heterotypic multiplicity of taxonomies of the same material reality. The world Eskimo communities inhabit is different from the world of a Tuareg or a Swiss banker: what the latter simply calls "snow" or "sand" for the formers are universes of subtle differences that constitute a matter of life or death; what for an Eskimo hunter or a Tuareg merchant is a simple form of direct social relation of exchange of goods or services, for the Swiss banker becomes a highly complex and impersonal system of investments, interests, hedge funds, derivatives, securities, bonds, and so forth. Hence we might say that human communities create the world they live in by dividing and ordering its components to satisfy some practical necessities, in a strictly utilitarian, cognitive, or hedonistically trivial way. For the historian, the interest of classifications and their legitimation lies in what they reveal of the society that engendered them, as Harriet Ritvo argued about Victorian Britain.[55] The truth-value of a specific classificatory system thus depends on the plausibility of its justifications. In turn, the plausibility of a symbolic representation of nature is connected to—without being reducible to it—the social status of the producers of that discourse, their position vis-à-vis the political and economic power, their dominant beliefs, and their intellectual and manual practices certified by established institutions. That is, the authoritative status of a classificatory system is directly proportional to the epistemological authority given by the larger society to the community of specialists that developed that system.

We know things in the world by naming them. Language thus performs a primeval foundational ordering of reality into categories, as *Bencao gangmu*'s effort to standardize names clearly shows. "Words," Marcel Proust reflected in the first volume of his *Recherche*, "present us with little pictures of things, clear and familiar, like those that are hung on the walls of schools to give children an example of what a workbench is, a bird, an anthill, things conceived of as similar to all others of the same sort."[56] Language, Proust suggested, forces reality into its own grid, segments the continuum of meaning in discrete semantic units, and fills reality with — or rather flattens reality into — a multiplicity of sets, each with its own label and its properly arranged content. Li Shizhen's quest to rectify the names of species had the same function: ordering nature began with an ordering of its names. When I say, "There is a tree," I actually group together into one big set all those myriads of things I see that are reducible to a cognitive model I inherited from my culture, a cognitive model inherent to the "encyclopedia" of the language that I learned from metaphorical "pictures hung on the walls of schoolrooms" as much as from every speech act I exchanged.[57]

Things are however more complicated than that. Despite the firm belief that our usage of words like "tree" or "bird" or "anthill" is straightforward and perfectly natural, words are by no means simply mimetic of the reality they describe. As Edward Sapir once put it, "No two languages are ever sufficiently similar to be considered as representing the same social reality," because "the world in which different societies live are different worlds, not the same world with different labels attached."[58] Not only are the worlds Eskimo, Tuareg, and Swiss bankers live in different, but there seem to be as many worlds as languages are spoken, or, more radically, as many as there are speakers. The lexicographical investigations of Li Shizhen seem to conform to this view.

This statement has become almost commonsensical, but it opens up two further sets of problems. On the one hand, there is the issue of what I would like to call "linguistic determinism," which regards the role of language in shaping individuals and societies as well as the world they live in. Heidegger believed that "man acts as though *he* were the shaper and master of language while in fact *language* remains the master of man."[59] According to this view, human beings are embedded in the symbolic universe that language constantly builds for them, and it is only because of this function of language that we can have and accumulate knowledge at all. In other words, it is because of language that there is such a thing as "the world" and not just an incoherent series of neuronal stimuli.

On the other hand, if indeed we can never escape from language, then the question that needs to be addressed concerns the status of our cognitive claims about the world, of which biological taxonomies are but one example: is their order a functional or utilitarian fabrication of human societies or do they in some sense represent an inherent order of the natural world in itself? In other words, if language—or any other symbolic system—is the necessary and inescapable means through which we make sense of reality, then any epistemological statement we utter about the world will remain necessarily a claim *for us*: we would never be able to talk about objective reality in itself, unless in the approximate terms of truthlikeness to the best explanations. The gap between our historically situated cognitive models—like, for example, biological taxonomies—and material reality would remain insuperable, the questions of being inseparable from the question of thinking. Immanuel Kant's transcendental epistemology—the philosophical roots of this correlationist view—seems to be the insuperable horizon of human cognition.[60] But what about matter? Does the "myriad of things" populating the universe that Li Shizhen tried to catalog in his encyclopedia merely exist only in language?

One way out from linguistic determinism that avoids a return to the naïve realism of pre-Kantian dogmatism or modern positivism was offered by Theodor W. Adorno in his *Negative Dialektik*. Stressing the nonidentity of matter and concepts, Adorno postulated the existence of an excess, or preponderance of matter over thought. This strategic twist allows Adorno to preserve a fundamentally materialist approach without renouncing the historicity of our knowledge of the world. On the one hand, Adorno conceded that we are captive "in the prison of language."[61] On the other, by conceiving of reality and thought as imbricated, he developed an oppositional dialectic that historicized matter without falling back into idealism. In the various historical epochs, thought always failed in its attempts to subsume matter into a conceptual constellation.[62] It is out of this struggle between nature and history—two concepts that cannot be separated without falling back into ontological dogmatism or postmodern relativism—that comes the metabolic relation of the material world and human societies. For Adorno, "concepts not only refer to non-conceptual, material particulars, but also emerge in historically situated and conditioned encounters with them."[63] What survives the changes in the conceptual understanding of the natural world in different historical situations is precisely the nonconceptual, the excesses of material reality vis-à-vis thought.

Concepts imprison material objects and render them manipulable for human needs, whether intellectual (knowledge, aesthetics) or material (agricultural, pharmaceutical, gastronomic, etc.). However, concepts fail to completely subsume matter under their meanings. There is always something in excess of the various conceptualizations of the world, something that escapes its reduction to concepts. By dint of this excess, we can understand the sociohistorical situatedness of knowledge without conceiving of its conceptual apparatus as asymptotically approaching an immutable reality (as in positivism) and without reducing reality itself to the expressions we use to understand it (as in postmodern relativism or in social constructivism). Rather, the continuous creative adjustments of the *Bencao gangmu* in Tokugawa Japan resulted both from the different sociohistorical contexts in which Japanese naturalists operated and from the different objects that failed to be captured by its taxonomical and conceptual grid. In fact, a constant element in the naturalist research of Japanese *honzōgakusha* was the adjustment of the information they retrieved from textual sources to the plants and animals living and growing in Japan. Scholars at first attempted to solve the incongruences of what they read in canonical encyclopedias about Chinese plants and animals with actual plants and animals growing and living in Japan through lexicographical and philological research and through the juxtaposition of different texts (the focus of part II). But by the eighteenth century, new observational, descriptive, and representational practices developed in the attempt to solve these incongruities (parts III and IV). Taxonomical practices too would later be affected (part V). Furthermore, the changes in the cognitive practices of *honzōgaku* scholars also affected the objects of their investigations through experiments in selections, interbreeding, and cultivation.

THE INVESTIGATION OF THINGS

Historians of science often celebrate *Bencao gangmu*'s long and precise descriptions of vegetal and animal species from a natural historical perspective. In fact, it was the Neo-Confucian metaphysical framework of the encyclopedia that justified Li Shizhen's emphasis of each subchapter.[64] As he argued in his memorial, which his brother Jianyuan read posthumously to the Chinese emperor at the official presentation of the first edition, Li Shizhen intended to illustrate the fundamental principles that moved all things in the universe (*shigai wuli*). He explained the Neo-Confucian philosophical inspiration in the prefatory note to the work: "Although the natural objects spoken of in this work provide the drugs of varying value used by physicians, yet the researches needed to explain (their identifications and properties) and (to elucidate) the truth about their natures and patterns [*qi kaoshi xingli*] are (part of) what our (Neo-) Confucian philosophers have called the 'science of the investigation of things' [*shiwu ru gewu zhi xue*]; and they can complement the deficiencies of the ancient books."[65] The "science of the investigation of things" (*gewu zhi xue, kakubutsu no gaku*) was a fundamental aspect of Zhu Xi's thought. Expressed differently as *kakubutsu chichi* (the knowledge of all things), it came to play an important role in the philosophical speculations of later Japanese scholars.

The teachings of Zhu Xi—what Tokugawa scholars called *shushigaku* and is usually, but improperly, translated in English as Neo-Confucianism—revolutionized the interpretation of the Confucian classics by adding metaphysical elements of Buddhist and Daoist inspiration to their norms of practical ethics.[66] According to Zhu Xi, the universe originated in and its existence was supported by the activity of two principles: *li* (*ri* in Japanese) and *qi* (*ki*). *Qi* was the material energetic principle through which the dynamics and movements of all things originated. The movement of *qi* alone, however, was not sufficient to create things as we can see them, because *qi*, in itself, had neither logic nor reason. *Li* is what gave order and coherency to the things formed and animated by *qi*. The immanent forces of *qi* and *li* generated and ruled everything in the universe. It was impossible for *qi* and *li* to exist without the other, but since *li* gave order to all things, Chinese thinkers often practiced the a priori study of its logic over the empirical research on its material manifestation in *qi*. As *qi* acts coherently (*li*) upon inert matter, two other forces appear—*yin* (*in* in Japanese) and *yang* (*yō*)—whose interactions in turn generate the Five Phases (*wuxing* [*gogyō*], sometimes translated also as Five Elements): wood, fire, earth, metal, and water. The interactive permutations of *yin*, *yang*, and the Five Phases form all things in the universe.

Everything in nature has both *qi* and *li* in itself, but depending on the kind of *qi* (i.e., the kind of mixture of *yin*, *yang*, and Five Phases), different material things responded differently to *li*. Like all material things, human beings also have both *qi* and *li* within them, but, because of their capacity to discriminate the workings of *li* in the universe, they occupy a superior position over animals. Li Shizhen applied this logic of *li* and *qi* in the classificatory system of his *Bencao gangmu*.

Zhu Xi identified "human nature" (*xing*, *sei*) as the *li* of human beings, so that human nature, which is identical to cosmological *li*, is inherently true and good. Acting badly, therefore, meant acting against one's true nature. That could happen because the human mind (*xin*, *shin*) is constituted not only of its true nature (*xing*, *sei*), which is identical to *li*, but also of passions or emotions (*qing*, *jō*), which are derived from *qi* and hinder the actualization of *li* in human nature. Together, *qi* and *li* provided powerful conceptual instruments with which to investigate nature. On the one hand, everything was theoretically understandable by principle (*li*), which regulated all things in the world. On the other, since the structure of the universe was reflected in human beings, the universe was also subjectively comprehensible through introspection. By mastering his interiority, the sage could comprehend the universe.[67] Conversely, by investigating external phenomena, the sage could discover himself in the workings of nature.[68]

In the *Bencao gangmu*, Neo-Confucian influence was most evident in the assumption that entries should reflect the metaphysical order of things in a "ladder of beings, from the lowliest to the highest."[69] This order evoked Zhu Xi's treatment of both the incarnation of *li* in living creatures and the obfuscation of *li* caused by the perturbing presence of *qi*. Human beings were at the top of the ladder because their nature (*xing*) was such that *li* could impose itself upon *qi* in the working of *xin* (human mind-heart, consciousness) in a far purer manner than in beasts or monsters. At the top, one finds, of course, the Neo-Confucian sage, the literati dominating the social hierarchy of Imperial China. Through inner meditation and through the investigation of things, official literati were able to discern the identity of human nature (*xing*) and the principles regulating the universe (*li*). Because of their superior understanding of the logic of *li*, they were able to operate and correct any material disruption to its smooth flowing. These disruptions occurred in the natural world in the form of natural disasters and diseases and in human society in the form of wars and crimes. As Elman put it, "'Investigating things and extending knowledge' (*gewu zhizhi*) presupposed that there was a 'principle for all things' (*wanwu zhi li*) in a real, not illusory, world."[70] And that was true of both the natural and the human world.

Li Shizhen explicitly framed his *Bencao gangmu* and its taxonomy in the Neo-Confucian logic of *qi* and *li*, but, as a matter of fact, this logic served merely as metaphysical justification of his classificatory decisions. As we have seen, Li sorted species in different categories (*lei*) on the basis of his observation and interpretation of their "visible morphology" (*xingzhi, keishitsu*), which included shape, size, color, smell, habitat, and other noticeable properties. These, in turn, were explained in terms of the "invisible properties" (*qizhi, kishitsu*) deriving from the permutation of metaphysical principles. Georges Métailié has argued that "Li Shizhen's criteria are largely based on subjective judgment and are not mutually exclusive."[71] For him, thus, the *Bencao gangmu* "in no way constitute[s] a modern taxonomy of natural kinds."[72] But if we avoid the comparison with modern botanical practices and instead consider Li's work in the light of sixteenth-century materia medica, the *Bencao gangmu* still bears witness of a new empirical approach to plants and animals. In fact, the belief that the properties of all things derived from the interactions of *yin* and *yang* elements, the infinite configurations of Five Phases, the interaction of *li* and *qi*, and so on did not offer Li a method for consistently classifying all natural species. On the contrary, Neo-Confucian metaphysics served him only as a legitimating platform for his classificatory decisions based on empirical observation, not as an axiomatic premise from which he deduced the sorting out of things. As all scholars of the period, he shared the assumption that all things in the universe had an inherent and theoretically intelligible order that did not require being proven or substantiated. But this order did not directly affect his ordering of species. In fact, in *Bencao gangmu* the Neo-Confucian metaphysical paraphernalia usually function as explicatory tool only at the category (*lei*) level. We read, for example, in the introduction to the trees section (*mu bu*), "*Mu* (tree/wood) is a plant (*zhiwu*), [it] is [also] one of the Five Phases. Its nature (*xing*) is linked to the soil: the mountains, valleys, plains, and marshes. At the beginning it is through the transformation of the material force (*qi*) that it receives shape and quality (*xingzhi*): standing up, with drooping branches, in clusters and in thickets. Roots, leaves, flowers, and fruits; firmness, softness, beauty, and the ugly; all depend on the Supreme Pole (*Taiji*)."[73] As Li put it, "With color and fragrance, quality and flavor, one distinguishes grades and categories [*pinlei*]." In fact, he thoroughly explained the structure of his taxonomy not on the premises of Neo-Confucian metaphysics but from a purely empirical perspective.

The grouping of various plants and animals by morphological similarities in different categories (*lei*) in the *Bencao gangmu* largely coheres with modern biological families—as indeed many folktaxonomies do. As Needham commented, "The Chinese perceived very clearly relationships between plant genera, even though they were often 'submerged' within their oecological and physiological classification."[74] Métailié—more cautious than Needham to credit Li Shizhen's work with any qualification of "scientificity"—remarks that "it is not entirely surprising to observe some convergence between Li Shizhen's 'folk classification' and modern scientific taxonomy."[75]

Similarly to the premodern notions of "species" (εἶδος) and "genera" (γένος) in Europe, in *Bencao gangmu* the relation of *zhong* and *lei* is conceived as both logical and ontological.[76] As John S. Wilkins persuasively argues, the essence of Aristotelian logic of genera and species survived, in varying forms, in the early modern period in the writings of Nicholas of Cusa, Marsilio Ficino, Leonhart Fuchs, Conrad Gesner, Andreas Cisalpino, and many others.[77] In other words, before being biological categories, "species" and "genera," like *zhong* and *lei*, were and still are logical categories. *Zhong* by no means conceives of natural species as natural kinds in the same manner and methodological consistency as in Linnaeus and modern taxonomists. But as research on phylogenetic taxonomy has evidenced, neither does the modern classificatory system—as conceived by Linnaeus's descendants like Ernst Mayr, H. W. B. Joseph, George Gaylord Simpson, or even David Hull—identify "natural kinds."[78]

Li Shizhen presupposed that the diversification of natural species and phenomena in a hierarchical system of relations was determined by the action of the principles *li* and *qi* upon matter. The wondrous diversity of natural and artificial things populating the world was thus regulated by the categories of *yin*, *yang*, and the Five Phases. Needham has recognized an affinity of Li's system to the Aristotelian notion of a *scala naturae*. A conspicuous difference, however, distinguished Li Shizhen's Neo-Confucian view from the Aristotelian one. Aristotelian "species" were in essence static and eternal: their physical properties (material cause) directly derived from their species-specific composition of four elements (στοιχεῖον): earth, water, air, and fire. Change (κίνησις), therefore, was an inner condition of species, an actualization of a potential inherently present in every individuals of that species, but it could not cause a species to become something other than what its essence dictated it to be[79]—that would have contradicted the principles of identity, noncontradiction, and the excluded middle Aristotle exposed in Book IV of his *Metaphysics*.[80]

Zhu Xi's model of the Five Phases (*wuxing*) was on the contrary dynamic and transformative, as Carla Nappi has demonstrated.[81] Everything constantly transformed, not merely by decaying or dying but also by metamorphosing. As Nappi puts it, "One Heraclitean constant informed the *Bencao [gangmu]*: nothing was constant. Transformation and processes of change were ubiquitous."[82] Li Shizhen often expressed in the text how "creations were truly remarkable," how natural species continuously broke through all epistemological barriers. The state of constant flux characterizing the existence of natural species deeply affected the systematicity of the *Bencao gangmu*'s taxonomy: how was it possible to contain ever-metamorphosing plants and animals in a rigid system of correspondences, in a kind of Porphyrian tree? Indeed, the transformative power inherent in all things in the universe was the reason minerals, plants, and animals could be transformed and used as medicinal substances.

WHAT'S IN A NAME?

The pervading "literati theory of knowledge" characterizing the *Bencao gangmu* is evident in the importance Li Shizhen attributed to the philological analysis of species names.[83] While reorganizing the received knowledge of pharmacological substances, Li compiled a massive encyclopedic work in the best literati tradition: "It is primarily as cultural objects that plants interests Li Shizhen."[84] Of the 932 titles he quoted as sources for his encyclopedia, only one-third were pharmacological and medicinal works. The remaining titles included a heterogeneous ensemble of encyclopedias, gazetteers, Confucian classics, poetic anthologies, and so on that was unprecedented in the canon of materia medica. Because of its encyclopedic reach, after its publication, the *Bencao gangmu* became an indispensable source not only for Ming and Qing apothecaries and physicians but also for literati, painters, poets, and scholars.

One of the most striking characteristics of the *Bencao gangmu* is its emphasis on accurate philological analysis of species names. In particular, Li Shizhen's preoccupation with finding the correct name (*zhengming*) of each species and clarifying the origins and meanings of the other names under which a species was known in different geographical areas or in different historical period bears witness to his compliance with the Confucian doctrine of the "rectification of names" (*zhengming*, in Japanese *seimei*)—"a perennial literati concern" and "passionate aim" of all scholars.[85]

Confucius himself in the *Luyun* (Analects) first emphasized the importance of correcting the correspondence of things and names. In a passage of the *Luyun*, we read that "Xilu asked: 'The Prince of Wei is waiting for you to take control of the administration: what would you do first?' The Master [Confucius] replied: 'What must be done first of all is the rectification of names.'"[86] Rectifying the name originally meant giving political titles their proper function, which implied that "to rule (*zheng*) means to rectify (*zheng*)."[87] In other words, the "rectification of names" was strictly connected to the political doctrine of maintaining the proper hierarchical order of society. Elsewhere, though, Confucius used *zhengming* in a much wider, almost nominalist sense. In *Luyun*, *pian* 17, *zhang* 9, for example, he urged his students to study the *Book of Song* in order to "learn as many names of birds, beasts, and plants as possible."

Xunzi in the third century BCE gave the doctrine of "rectification of names" greater sophistication. "Names," he wrote, "were made to denote actualities [*shi*]" and to "distinguish things that are the same from things that are different."[88] By denoting differences and similarities, names indicate the proper nature and function of things. But in order to do that, names must have precise meanings. Xunzi's philosophy of language conceived the "rectification of names" as both a way of maintaining the proper social order and a logical-epistemological method to unveil the inherent order of things: "In order to separate different kinds of things, we use 'separating names' such as 'bird' and 'beast', stopping only when we cannot further separate different kinds."[89]

Zhu Xi reconfigured the notion of the "rectification of names" in accordance with its metaphysical system: if *li* was the organizing principle of all things in the universe, then it regulated not only things but also the signs used to express those things. In other words, things and their names were subjected to the same laws of *li* and *qi*. It followed that the order of the names of natural objects matched the inherent order of those natural objects. Qing scholar Li Gong, for example, in his analysis of the classic *Daxue* (The Great Learning), argued about the necessity of coordinating actions, things, and names: for him, rectifying one's behavior or rectifying things so that they fit their name was a means of reaching a deeper understanding of the proper order of things.[90]

The structural homology of words and things in Zhu Xi's philosophy, in Li Shizhen's materia medica, and in other literati encyclopedias profoundly influenced the study of nature from both theoretical and methodological perspectives. The metaphysical identity of words and things determined the ways in which the early generations of *honzōgaku* scholars studied nature: observation and experimentation with plants and animals was never distinct from lexicographical and philological research into the names of plants and animals. Philology thus played as important a role in the study of nature as observation. Métailié argues that "Li Shizhen's field work is of an anthropological rather than a field-botanical kind," and it only complemented his philological research.[91] On the other hand, Li Shizhen was forced to constantly negotiate between Zhu Xi's rationalist determinism and the inherent dynamical conception of reality engendered by the incessant metamorphoses of natural things due to the active force of the Five Phases. As Nappi points out, "Observation was a primary and reliable way of knowing about the natural world and confirming textual claim"—precisely because of the inherent difficulty to systematically describe a reality that is always in motion.[92]

The role and importance of empirical observation in *Bencao gangmu* remains a contentious matter that goes beyond the scopes of this book. It is however unquestionable that names and correct naming were central concerns for Li Shizhen. In *Bencao gangmu*, a species was first identified with its correct name (*zhengming*)—that is, the earliest name under which a species was known; second, the sections on *shiming* provided an explanation of the variant names of mineral, vegetable, and animal species. *Zhengming* and *shiming* preceded the subsections exploring their morphological, medicinal, and pharmacological properties: naming a thing correctly determined its correct positioning in the order of things. Errors, therefore, were not so much a matter of misplaced observation or misconceived experiments but a problem of incorrect naming.[93] If names and things were epistemologically analogous, then the study of plants and animals involved the study of the linguistic uses of their names. The "ladder of beings" described in *Bengcao gangmu* consisted of a semantic network of verbal definitions and descriptions that were conceived to be identical to the things they defined.[94] In other words, in Ming China, much as in medieval Europe, the book of nature was, first of all, a book of the names of nature. In a Neo-Confucian context, names, more than modes of *representation* of things, were different modes of *being* of things.

Li Shizhen would have probably dismissed Juliet Capulet's cry: "What's in a name? That which we call a rose / By any other name would smell as sweet."[95] He would have rather agreed with Bernard of Cluny that "*stat Roma pristina no-mine, nomina nuda tenemus*" — "Of ancient Rome only its name endures, names are all that we have."[96] To him, the names of things internalized and revealed their nature, their essential properties, the logic of their metamorphoses, their relationship with the "myriads of things," and their proper place in the order of things. And yet, as Li explicitly explained, he accumulated decades of traveling in different provinces of the Ming Empire, he interviewed apothecaries and agronomists, he performed experiments with various substances, he tested the efficacy of the received knowledge, he dramatically expanded the number of known species, and he excavated the palimpsest of meaning accumulated in centuries of materia medica to retrieve the foundational meaning of species names.

It was as if the *Bencao gangmu* had two souls, one logical the other empirical, working in synergy. Its introduction to Japan contributed to create a vibrant field of study that expressed these two approaches throughout its two and a half centuries of development. The *Bencao gangmu* achieved unprecedented success and was reprinted continuously until the twentieth century. It was the standard manual for the pharmacological training of physicians in Ming and Qing China, as well as a required textbook for the study of nature. A copy of *Bencao gangmu* was an essential presence on the bookshelves of educated men and literati, who appreciated it not only for its practical value but also as aesthetic inspiration for poetic compositions and landscape paintings.

Modern scholars give varied explanations for this success. First, *Bencao gangmu* was the first completely original work on materia medica produced since the eleventh century. Furthermore, even though many other encyclopedias were compiled in the centuries following its first edition, it remained the foundational and authoritative source for them all. Second, it dramatically increased the number of known species. Last and most important, the order in which it classified natural species was not based on an artificial division of substances according to their utilization as drugs but on the perceived order of nature.

PART II *Ordering Names: 1607–1715*

*[The sages] instituted names to refer to objects, making
distinctions in order to make clear what is noble and what
base and separations in order to discriminate between
things that are the same and those that are different.*
— Xunzi

*One must apply truth and energy in naming things.
It elevates and intensifies life.*
— *Thomas Mann,* The Magic Mountain

The arrival of the *Bencao gangmu* in Japan in the early years of the seventeenth century marked the beginning of a new phase in the development of Japanese materia medica. The introduction of Li Shizhen's encyclopedia was part of a process that saw in the decades before and after the year 1600 a massive rise in the quantity of Chinese texts introduced in Japan from China and Chosŏn Korea.[1] Simultaneously, a process of secularization of intellectual production began in different regions—in particular in the provinces of Harima and Tosa and in the Osaka-Kyoto metropolitan areas. Buddhist monks broke ties with monastic institutions, renounced vows, and put their intellectual expertise into practice as advisors, tutors, or administrators of samurai elites. After the battle of Sekigahara in the autumn of 1600, these "free-floating" scholars moved to the main cities and castle towns of the archipelago to earn a living as tutors, librarians, scribes, and councilors. In time, they played a valuable role in legitimizing the new Tokugawa regime. The first steps in the formation of a field of nature studies in early modern Japan was coeval to these large-scale processes: an informational surge, the secularization of cultural production, and the consolidation of a new regime.

To understand the explosion of cultural production in various fields of knowledge in early modern Japan, we should therefore reconstruct the dynamic processes that gave birth to a new socioprofessional identity of scholars (*jusha*). In the premodern era, in particular between the ninth and fifteenth centuries, intellectual production—comprising Buddhist and Confucian commentaries, official chronologies, miscellaneous historiographical essays, and aesthetic treaties—was virtually confined to the imperial court nobility and the Buddhist clergy, to such an extent that it constituted an exclusive expression of their social position. That is, culture as such was a social expression of the civil and, later, military aristocracy and of Buddhist monastic institutions. Starting from the second half of the fifteenth century, as the Ashikaga's grip of actual political power considerably weakened, Buddhist monks and impoverished members of the imperial aristocracy found patronage among local domainal lords (*daimyō*). Increasingly in control of their land as feudal lords, these *daimyō* sought to change their coercive military might into more symbolic forms of domination and to compete for distinctions with their peers. The celebrated story of the tea master Sen no Rikyū or the glittering court of the Ōuchi in Yamaguchi, the so-called Little Kyoto, are perfect illustrations of this process.[2]

By the end of the sixteenth century, the phenomenon reached larger proportions when the first generation of scholars—whom twentieth-century historians would call "Neo-Confucian"—began a struggle for acknowledgment of their scholarly labor and for a new social role for intellectuals outside of Buddhist and Imperial institutions. Fujiwara Seika, Hayashi Razan, Matsunaga Sekigo, Nawa Kassho, Hori Kyōan, and Manase Dōsan, among others, were all former monks who invented a new way of being a scholar in Japan by negotiating their aspirations to import the Ming ideal of the scholar-official (*shi*) and the reality of the feudal society ruled by Toyotomi Hideyoshi and, after 1600, Tokugawa Ieyasu. The events of their lives and the compromises they made in order to earn a living by their intellectual expertise influenced the style, the language, the problems, the topics, and the compartmentalization of knowledge characterizing their scholarly production.

The first generations of Tokugawa scholars heavily relied on imported commentaries in the Neo-Confucian tradition. Its language, style, and metaphysical premises ended up influencing their cultural production in different fields. Neo-Confucianism—which, since the thirteenth century, was the hegemonic school of thought in East Asia—was the only available alternative to Buddhism, Daoism, or even classical Confucianism, which ruled court scholars' (*hakase*) practices at the time. Not only did it offer a conceptual system connecting politics and ethics to the metaphysical order of the universe, but it also placed the literati at the top of the ideal social hierarchy. In addition, Neo-Confucianism comprised a corpus of technical manuals on medicine, pharmacology, botany, fiscal policies, administration, and military techniques that had practical utility for the new authorities in Edo. *Honzōgaku* was not different: *Honzō kōmoku* (the Japanese pronunciation of *Bencao gangmu*), the fruit of years of empirical research by Li Shizheng, was, as we saw, inserted in a Neo-Confucian framework. Its terminology, style, and aims—both practical and theoretical (the "investigation of things")—influenced similar works produced in Japan throughout the early modern period.

One of many encyclopedias imported (or republished) in Japan in large quantities in the seventeenth century, *Honzō kōmoku* responded to the needs of early scholars like Fujiwara Seika and Hayashi Razan, who were engaged in an effort to understanding, digesting, interpreting, and popularizing the corpus of Neo-Confucian commentaries. Encyclopedias epitomized the concrete realization of the ideal of the "investigation of things" (*gewu zhi xue, kakubutsu no gaku*), a fundamental aspect of Zhu Xi's philosophy. Chapters 3 through 5 survey the early history of Japanese encyclopedic production in the field of nature studies, from Razan's glossary of *Honzō kōmoku*, *Tashikihen*, to Inō Jakusui's *Shobutsu ruisan*. Hayashi Razan, Kaibara Ekiken, and even Inō Jakusui did not define themselves as *honzōgaku* specialists but considered themselves and their intellectual pursuits in the broadest terms. Later generations of naturalists would however retroactively recognize in their works the foundation of the field of nature studies in Japan. As Hiraga Gennai wrote in 1762, in the manifesto advertising his great exhibition of medicinal substances, "the discipline of *honzōgaku* was first studied in Kyoto by Master Inō during the Genroku period. After him, Master Kaibara and Master Matsuoka continued his studies. Their studies have proved useful and have produced a large number of texts that we are now benefiting from. It is thanks to their activities that the number of persons relying on foreign medicines has decreased considerably. Moreover, our knowledge of indigenous species of medicinal plants and animals has greatly improved, thanks especially to the great achievements of these three masters."[3]

The following chapters reconstruct the struggle for the intellectual recognition of Hayashi Razan (chapter 3) and Kaibara Ekiken and Inō Jakusui (chapter 5) and survey the production of the first *honzōgaku* encyclopedias in the first century of the Tokugawa period. They focus in particular on Razan's *Tashikihen*, Ekiken's *Yamato honzō*, and Jakusui's *Shobutsu ruisan*, but they also map the circulation of other Chinese and Japanese encyclopedias and their role in shaping natural knowledge in seventeenth-century Japan (chapter 4).

3   Knowledge in Translation
Hayashi Razan and the Glossing of *Bencao gangmu*

. . . .

THE MAKING OF HAYASHI RAZAN

We could do worse than to assume that it all began in 1607, the year Hayashi Razan presented the retired shogun Tokugawa Ieyasu with a printed copy of *Bencao gangmu*.[1] Razan's originality in the intellectual panorama of early Tokugawa Japan has been repeatedly questioned and his importance downsized since Maruyama Masao's portrayal of Razan as the architect of the ideological apparatus of the Tokugawa regime.[2] Yet to begin with Razan is justifiable for a number of reasons. Razan's promotion of Li Shizhen's encyclopedia popularized the discipline of *honzōgaku* to a wider public of scholars than the cloistered world of court scholars and Buddhist monks. He edited and published an annotated version of the introduction to the *Honzō kōmoku* and compiled a slim digest of the mammoth encyclopedia, which was titled *Tashikihen* (The Explanation of Many Things) and consisted of a Japanese translation of the encyclopedia's entries.[3] This glossary, inaccurate and incomplete as it was, served nonetheless as one of the bases for a number of subsequent encyclopedias, from *Kinmō zui* and *Wakan sansai zue* to Kaibara Ekiken's celebrated *Yamato honzō* and Inō Jakusui's *Shobutsu ruisan*, all of which cited it in their bibliographical notes. Razan might not have been particularly original and all his writings might be vulgar oversimplifications of Chinese texts, but we cannot ignore his painstaking adoption and adaptation of Chinese knowledge in various fields, which set a standard for transliteration, translation, and systematization of Chinese names for much of Tokugawa's intellectual production. Most importantly, Razan might not have been the ideological éminence grise that Maruyama Masao imagined him to be, but he contributed more than any of his contemporaries to the creation of a social niche for a new figure of scholar, a function confined in the previous centuries to Buddhist institutions and the Imperial court.

Much of the misconceptions and idealizations of Hayashi Razan comes from the *Tokugawa jikki* (True Records of the Tokugawa), a text produced in the nineteenth century under shogunate sponsorship as part of a campaign to legitimate Tokugawa hegemony. It was *Tokugawa jikki* that retroactively elected Razan as the founding father of Japanese orthodox Neo-Confucianism, projecting into the past ideological conceptions that were proper of late eighteenth-century Japan. Its compilation began in 1809 under the supervision of Hayashi Jussai, a descendant of Razan, who in 1793 became the head of the state-sponsored Hayashi academy of Zhu Xi studies, the Shōheikō. Completed after Jussai's death in 1849, the *Tokugawa jikki* recounted the history of the Tokugawa regime according to a Neo-Confucian framework in line with the doctrine of the Hayashi school, which a series of political reforms in 1790 had elevated to official state ideology and philosophical "orthodoxy."[4] In the early twentieth century, intellectual historians accepted *Tokugawa jikki*'s narrative of the establishment of the Tokugawa shogunate as factual, and the narrative acquired further authority with the publication of Maruyama Masao's *Studies in the Intellectual History of Tokugawa Japan*.[5] Tokugawa Ieyasu, Maruyama argued, "made the Hayashi family the custodian of the official philosophy of the regime."[6]

Counter to this interpretation, Herman Ooms has demonstrated how the first shoguns, displaying no exclusive commitment to Hayashi Razan or his ideas, adopted diverse strategies of ideological legitimation. Other historians have also argued that the diversity of Neo-Confucianism in the early years of the seventeenth century raises the question of whether ideology is the proper term for the intellectual hegemony of Neo-Confucianism in early Tokugawa Japan.[7] Ooms explained that in 1605, Razan "entered Ieyasu's service, not as a Confucian scholar or Buddhist priest but as an exceptionally learned young man who impressed his interviewers with his broad knowledge of Chinese scholarship."[8] Indeed, Razan was at the center of a nascent network of learning among samurai that were soon to be employed either as teachers and lecturers or as officials hired by the shogunate or by local domains for their intellectual skills.

RAZAN, THE BOOK FINDER, AND *HONZŌGAKU*

Razan did not intend to found a new discipline of nature studies, but his work provided a foundation for future generations of *honzōgaku* scholars to do just that. Razan's production covered a broad range of fields, including exegesis of Confucian texts, Shintoism, Buddhism, poetic composition, Japanese history, medicine, and *honzōgaku*, among others. The polymathy of early Tokugawa scholars like Razan was a response to the opening of a new space of intellectual production that had not yet diversified into specialist divisions of labor. Scholars were able to read nearly all the texts that were in circulation, and their number was manageable enough that they did not feel the pressure to specialize. His work facilitated the development of other disciplines, including Neo-Confucian philosophy proper (*shushigaku*), lexicography (*meibutsu-*

gaku), administration (*keizaigaku*), and historical genealogy (*keizugaku*), among others. Razan helped to invent the Neo-Confucian scholar (*jusha*), a profession that did not exist during his lifetime but that became a reality for future generations of scholars in part because of his activities. Although Ooms's characterization of Hayashi Razan as a bookish young man is supported by the fact that Ieyasu hired him as book purchaser, it does not say much about the nature of his intellectual labor, the degree of distinction that this labor provided him within the larger field of social positions, and the impact of his intellectual production. To assess Hayashi's role as a pioneer of Neo-Confucianism, *honzōgaku*, or other fields of knowledge, we should describe the strategies he used to fashion himself as a scholar, the pressures that constrained him, and the recognition that this socioprofessional identity gained among his contemporaries.[9] In other words, what kind of social role did Razan play—A librarian? A scholar? A ba-kufu retainer? Ieyasu's servant?—when he presented Tokugawa Ieyasu with the fifty-two volumes of Li Shizhen's *Bencao gangmu*, which he had purchased in Nagasaki in 1607, carried to the Sunpu retreat of the retired shogun, and later helped archive in the shogunal library of Momijiyama Bunko?

This question is even more relevant if we consider the fact that recent studies have suggested that *Honzō kōmoku* was probably being read in Japan prior to 1607.[10] In the entry in *Razan sensei nenpu* covering the events in Razan's life in 1604, there is a list of all books that he had read up to that year. The list was taken from *Kidoku shomoku* (Index of the Books I Read), a manuscript in which Razan recorded more than 440 titles, thematically arranged. Li Shizhen's *Honzō kōmoku* appeared in the impressive list of Chinese, Korean, and Japanese texts, which means that Razan listed it as "read" at least three years before its so-called official introduction.

Other references to *Honzō kōmoku* show that the encyclopedia was also known outside Razan's network. In 1608, Kyoto physician Manase Gensaku, attendant physician of Tokugawa Ieyasu, published the manual *Yakusei nōdoku* (On the Medicinal Potential of Herb Extracts), which had previously circulated in manuscript form and consisted of a revised and expanded edition of *Nōdoku* (Properties of Herb Extracts), a pharmacological manual written by Gensaku's adoptive father Dōsan, a figure of importance in the history of Japanese medicine. The explanatory note in *Yakusei nōdoku* listed *Honzō kōmoku* as one of the reference sources for the pharmacological treatment of plants and animals. Another reference to the *Honzō kōmoku* with no direct link to Hayashi Razan appeared in the writings of a Confucian scholar and physician named Kantokuan. Kantokuan recorded in his diary that in the early 1600s, he had shown *Honzō kōmoku* to his teacher Fujiwara Seika.[11] Kantokuan, of whom we know very little, studied medicine under Manase Dōsan and Gensaku before entering the school of Fujiwara Seika. He may have been a node in a network that linked the Manase family with Fujiwara Seika and, years later, Hayashi Razan.

The significance that later generations of *honzōgaku* scholars would attribute to 1607, despite the fact that physicians and apothecaries knew of *Honzō kōmoku* before Razan's official introduction, suggests a close relationship between scholarly activity and political authority, between learning and power. Razan's struggle for recognition opened up the possibility for the existence and social acceptance of a new category of scholars, identifiable as such only for their intellectual labor, the value of which was vouchsafed by their association with the new political authorities.

A reconstruction of the events of Razan's struggle to become a scholar, with all the incidents, contradictions, negotiations, and compromises he faced, is therefore fundamental to understanding the constrains, methods, language, and attitudes characterizing the following generations of scholars in the seventeenth century.

HAYASHI RAZAN, SCHOLAR FOR HIRE

Hayashi Nobukatsu, later named Razan, was born on the eighth day of the eleventh month of the eleventh year of the Tenshō era (December 21, 1583).[12] The hagiographic account in *Razan Rin sensei gyōjō* introduces the Hayashi family as descended from a branch of the highborn Fujiwara clan, but Razan's ancestors were in fact rural warriors (*gōshi*) or low-ranking *jizamurai* from the Kaga Province.[13] Razan's father, turned *rōnin*, moved to Osaka and lived off the earnings of the rent from their lands. As is customary in such eulogies, *Razan Rin sensei gyōjō* portrayed Razan as a "very intelligent child who could speak fluently at two years of age."[14]

The two biographies by his sons do not suggest that the Hayashi family experienced any economic difficulty, but in the turbulent last decades of the sixteenth century for *gōshi* like the Hayashi, detached from their land and without connections to the new centers of power, investing in education was often the only option for social mobility apart from renouncing their samurai status and becoming merchants or artisans. Probably for this reason, Razan entered the Kenninji in 1595 as an acolyte of the Rinzai Zen temple. The Kenninji was an important part of the powerful network known as the Gozan, or "Five Mountains," perhaps the most prestigious center of cultural production in late medieval Japan.[15] It was there that Razan first came into contact with the commentaries of the Song scholar Zhu Xi.

Zhu Xi's commentaries on canonical Confucian texts played a fundamental role in the political and intellectual history of China from the Song to the Qing dynasties and the Chosŏn kingdom in the Korean peninsula. Zhu Xi revolutionized the interpretation of the Confucian classics by adding metaphysical elements of Buddhist and Daoist inspiration to their practical ethics.[16] His texts reached Japan during the Kamakura period (1192–1336), and in the course of the fourteenth century, Zhu Xi's Neo-Confucianism began to be studied intensively by Zen monks of the Gozan temples of Kyoto. In particular, the Rinzai school treated his commentaries as texts encouraging meditation. Neo-Confucian thought began to spread outside the cloistered world of Buddhist temples during the civil wars of the sixteenth century, when monks in search of refuge found asylum in various fiefdoms as advisers or educators.[17]

The role that Zhu Xi's thought played in the formation of the unified state first under Toyotomi Hideyoshi and later under the Tokugawa remains a contentious issue among contemporary scholars. There is little doubt that by the late sixteenth century, Neo-Confucianism had begun to occupy a significant part of intellectual production. Allusions to and quotations from the Confucian classics and Zhu Xi's commentaries appeared frequently in Buddhist works, in the house rules (*kakun*) of military households, and in oral storytelling (*otogizōshi*). Most importantly, by the late sixteenth and early seventeenth centuries, Neo-Confucianism had become a medium for a new philosophical language that introduced new forms of knowledge, new fields of study, and new conceptions of society. Its emphasis on order and regulations (order of nature, order of society, order of the family, order of the mind), grounded in a complex metaphysical apparatus, appealed to the social and political forces seeking to impose a new regime in Japan. Neo-Confucian texts also inspired new forms of financial and organizational administration of households and governments. Learned people acquainted with these texts were therefore of practical value as councilors and advisers for the new leaders.

The new order taking shape after the military and political unification of the Japanese archipelago under the Tokugawa created an opening for the emergence of new social groups. These included lower-ranking samurai who during the warring states period had assumed the role of local military figures as small landholders and yeomen (*gōshi* and *jizamurai*). Less important members of these groups such as the Hayashi family, cut off from direct access to governing positions by the Tokugawa authorities, often found opportunity for social advancement in cultural pursuits. Culture played an important role in the legitimating process of the new shogunate in the early seventeenth century. With political power in their hands, the new military elites headed by the Tokugawa began to adopt various strategies to enhance their status and legitimate their ruling position. Besides a sheer display of their military might in parades and in subduing revolts like the ones in Osaka (1614–15) and Shimabara (1637–38), they used sophisticated symbolic approaches. These included inventing an uninterrupted lineage with the military aristocracy of the past, patronizing Buddhist institutions, the divinization of Tokugawa Ieyasu after his death (1616), and the patronage of artists and scholars for the construction of visible signs of power, not only in the form of palaces and paintings, but also in printing facilities and libraries, schools, and scholarly academies.[18]

Buddhist temples were the only institutions that offered an education to the younger generations of low-ranking samurai who wished to dedicate their lives to learning. Through this study, young samurai students became familiar with a number of Chinese texts. Knowledge of Chinese, in turn, enhanced their value as tutors, lecturers, or advisers employed in large samurai households and later in merchant families as well. Neo-Confucianism provided a new, fresh, and systematic interpretation of the classical Confucian ethical and political rules known since the first Japanese unified state in the seventh century. Although all the social and intellectual elements necessary for making Neo-Confucianism a powerful ideological apparatus were already on hand by the early seventeenth century, it would take two centuries to transform it into the mature ideological apparatus designated as "orthodox" by the Tokugawa state in its 1790 edict.

It was in this context that at twelve years of age, young Razan entered the Kenninji at a time when "in Rinzai Zen temples, the reading and memorizing of Neo-Confucian texts had superseded the recitation of [Buddhist] scripture in the training of youngsters."[19] At the temple, he started accumulating both knowledge and books, which he carefully listed in diaries and manuscripts in order to supplement his scholarly curriculum. He soon mastered Chinese and devoured every text he could lay his hands on.[20] In 1597, after three years of training and having shaved his head and taken the vows of his order, he "furtively escaped the temple and returned home."[21] Once home, he continued reading Chinese texts on diverse topics, but it would not have been easy for young Razan to obtain books, an issue addressed in his two biographies. His hunger for knowledge was relentless. He believed, in fact, that "we are in a historical period when there is no man who is not acquainted with the five classics. It is therefore the duty of contemporary scholars to look for the largest number of books, until there is not a single one left for him to read."[22]

Razan's desire to expand his erudition led him to copy and purchase texts dealing with pharmacology and medicine as well, either from Ming China or from Chosŏn Korea. He soon started giving public lectures on important texts of the Confucian tradition, mostly the five classics, interpreting them through the Neo-Confucian perspective of Zhu Xi. His public exposure as book pursuer and lecturer in Kyoto eventually put him into contact with other young scholars, who either became his students or collaborated with him in an exchange of books and opinions.[23] Most of these scholars had to fashion their professional identity from scratch. They aspired to become scholar-gentlemen (*shi*), who in imperial China served as government officials chosen for their superior intellectual qualities and certified by a system of public examinations.[24] But Japan had no such system, nor did the scholar exist as a public figure outside the Buddhist clergy or the small community of court academics (*hakase*). Thus Ooms is right to write that "Razan's career was not that of a Confucian scholar."[25] *Jusha*, the Tokugawa term for Neo-Confucian scholars, had yet to become the designation of a socially and professionally identifiable figure. "Rather," as Ooms states, "[Razan] used his career to establish himself as one."[26]

Dreaming of China as the land of scholar-officials and combating the control of intellectual production by Buddhist and court institutions, Razan had at the time a model to emulate in Fujiwara Seika, whose disciple he eventually became.[27] Seika was a former Zen monk who also abandoned his Buddhist robes after he came into contact with a captive from the Korean peninsula, a former official named Kyokō. Kyokō lectured him in Neo-Confucianism—which Seika had already begun to study as a Zen acolyte—and explained to him the important social and political role of Confucian literati in Chinese and Korean societies. As a result of this relationship, Seika opened a small school of Neo-Confucian studies in his house and started wearing a *shin'i*, the formal costume of Confucian scholars on the continent (fig. 3.1). The dress was similar to the one officials wore in ancient China and was supposed to emphasize the active role that Neo-Confucianism and Neo-Confucian scholars played in public life, a role that Seika himself never came close to achieving.[28] Seika's activities in his little school on the periphery of Kyoto involved a large number of students and colleagues, such as Matsunaga Sekigo, Nawa Kassho, Hori Kyōan, Manase Dōsan, and Gensaku. They all soon became main actors in Kyoto intellectual life.

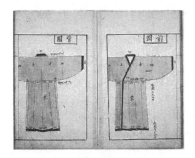

FIGURE 3.1.
A *shin'i* from Nakai Riken,
Shin'i zukai (1795). Tokyo:
National Diet Library.

After 1600, Razan began to lecture more frequently, enjoying increasingly larger audiences. His lectures focused on the ancient Chinese and Japanese classics and texts of the Confucian tradition, "but Zhu Xi's *Lunyu jizhu* (Collected Commentaries on the *Analects*) occupied a privileged position on Razan's desk."[29] It is difficult to imagine how these lecturing pseudomonks were perceived at the time. Twenty years later, at the apex of his career, Razan's rank among shogunal personnel was that of *ohanashishū*, a position not much higher in status than those of "attendants (*koshō*), pages (*konando*), doctors (*ishi*) and Buddhist monks (*bōzushū*), who were in charge of the most humble services."[30] In the early decades of the seventeenth century, people likely associated public lecturers like Razan with the array of peddlers and street performers who did everything from delivering religious sermons to storytelling and performing on stage or in acrobatic routines.[31] This lecturing practice continued throughout the period (fig. 3.2).

Other scholars in search of patronage for their intellectual expertise shared Hayashi Razan's early career as the scholarly equivalent of street performers. Yet to establish associations and educational institutions, they were disseminating books, interpretations, and ideas that had before been enclosed in temple compounds or secretly transmitted in the Daigakuryō. Razan's visibility on Kyoto's street corners (or *yose*, as they came to be called by the late seventeenth century) and his intellectual influence increased to such an extent that his lectures were often quoted in contemporary diaries.[32] His reputation grew among his fellow lecturers, and he began to benefit from the expansion of his network of fellow scholars, who introduced him to wealthy families for patronage and financial assistance. Razan's case illustrates how scholars, who in those years reached Kyoto in search of support for their study, began to form a network through which they could exchange information about new books, commentaries, and lectures and also about samurai and wealthy townsmen families in search of tutors and

FIGURE 3.2. A scene from the Ikutama shrine. Yuzuke Gansui, *Gonyūbu kyara onna* (1710). Osaka: Osaka Furitsu Toshokan.

instructors. Their association with wealthy merchants allowed them to acquire texts imported from the continent and thus increase their curriculum.

Razan's friendship with the wealthy merchant Yoshida Haruyuki, an amateur scholar who had sponsored a Japanese edition of *Shiji* (The Records of the Grand Historian, *Shiki* in Japanese), proved vital to his career. Razan first contacted Haruyuki to purchase volumes of the Chinese classic. The subsequent friendship, as an epistolary exchange reveals, had consequences Razan may well have envisioned from the start. Haruyuki was son and heir of Suminokura Ryōi, one of the most prominent merchants of the time with an official license to trade outside Japan. Born into a family of physicians and moneylenders, Ryōi favored the latter occupation. He collaborated closely with the Toyotomi family and later with Tokugawa Ieyasu and the shogunate. For Razan this friendship meant a constant flow of information and books from the Suminokura ships trading with China. Most importantly, Haruyuki was able to organize the long-awaited meeting of Razan with Fujiwara Seika in the eighth month of 1604. Razan soon became one of Seika's disciples and gained access to his master's network of scholars and officials.

The fruits of this series of encounters soon emerged. In the beginning of 1605, Razan began to correspond with members of the Korean embassy to Japan. In the fourth month, he had his first meeting with Tokugawa Ieyasu at the Nijō castle in Kyoto. By 1606 his meetings with Ieyasu became regular and he gained free access to Ieyasu's library at the Fushimi castle. By 1607 he had begun to lecture Hidetada, Ieyasu's third son who had become the new shogun in 1605. In these years, Razan darted from Kyoto to Fushimi, to Sunpu, to Nagasaki and back to Kyoto, meeting with envoys from the continent, buying books from Chinese merchants, and organizing lectures.

His contacts with the Tokugawa, however, should not be taken as a sign that the new government supported his philosophical vision of a society led by scholars. Razan had to wait until 1630 to receive from the shogunate a "gesture of gratitude for twenty-five years of service": a small plot of land in Edo and a small grant of two hundred *ryō* to build his own private academy, which he named Kōbun'in.[33] But it would be a mistake to undervalue the shogunal endorsement of Razan's scholarly labor, whether if only as a librarian and book purchaser, or as "ghostwriter" of diplomatic correspondences with the continent, because the Tokugawa's recognition of his intellectual expertise effectively legitimized his socioprofessional identity as a "scholar" independent of the traditional cultural institutions of learning.

Razan's growing visibility in Kyoto, in conjunction with his entrance into the Seika-Haruyuki network of fellow scholars, brought him some notoriety, which in turn put him into direct competition with established intellectual producers at court and in the Buddhist temples. Launching a virulent campaign against Buddhism, Razan attacked the alleged unethical and vulgar behavior of monks and denounced the economic interests of temples as landholders.[34] Razan also had to fight against court scholars protecting their control over the right to lecture on and interpret Confucian texts. His confrontation with Kiyohara Hidekata, one of the most powerful court specialists in Confucian studies, illustrates the intellectual environment at the turn of the seventeenth century.

Kiyohara Hidekata was a personal lecturer on Confucian classics (*myōgyō hakase*) to various emperors.[35] Hidekata and Razan met in 1603, thanks to the mediation of the Kenninji abbot Jikei, and at first enjoyed a friendly exchange. Hidekata's diary indicates that Razan visited Hidekata's mansion three times between the tenth month of 1603 and the ninth month of 1604.[36] The meetings are described as convivial occasions where Razan and Hidekata exchanged and discussed Chinese texts. Wajima argues that Razan was probably hoping to be accepted under Hidekata's patronage and, through him, obtain access to the highest spheres of the cultural establishment in the capital.[37] But Razan's plan failed when in the following year, 1604, he started lecturing on Zhu Xi's *Lunyu jizhu*, which broke the court's control over Confucian matters and incurred Hidekata's anger. In *Nozuchi* (The Deities of the Field, 1621), Razan explained that

> one or two years after I started lecturing on Zhu Xi's *Lunyu jizhu* and other [Neo-Confucian] texts of the Ming [period], at 21 years of age I started wearing a *shin'i* when discussing Confucian matters.[38] At the time in Japan if someone lectured on the Confucian Classics without previously applying for the permission of the court, he committed a felony. Or, at least, there were some people at court who made such accusations. But I did not care and went on lecturing even more than before.[39]

A letter from Seika to Razan dated the twenty-sixth day of the second month of 1605 corroborates this account. Seika warned his new student that he "should not dress in *shin'i* or flashy garments while lecturing," in order to avoid any reprimand from "those who occupy a higher position in society." Seika's cautiousness reveals the precarious position that the new scholars occupied in the early Tokugawa society. The philosophical justification further demonstrates his prudence: "To follow a superior's orders is, in accordance with the way of man, the righteous course to take in order to follow the way of 'principle' (*ri*)."[40] Despite his master's warning, Razan did not interrupt his lecturing practice.

Razan's decision to continue his public lectures without compromising with the court eventually earned him retaliation from Hidekata, who launched accusations against Razan throughout 1604 and 1605, filing suits against him to the imperial court and to Ieyasu, but to no avail. Ieyasu did not take action against Razan, but he eventually granted Razan—and by default, all other public lecturers—freedom of intellectual pursuit.[41] This surprising decision can be interpreted as a sign of the changing social order under the Tokugawa regime. That a townsman of low samurai status like Razan could have prevailed against a member of one of the oldest noble houses of the capital was remarkable.

Historians generally interpret Ieyasu's decision in the Hidekata-Razan dispute as part of his plan to make Neo-Confucianism the philosophical foundation of his regime; that is, Neo-Confucianism had to be freed from its association with Buddhist temples and the court.[42] Ooms, however, has persuasively shown that "Ieyasu certainly did not take either to Razan's virulent anti-Buddhism or to his love of Neo-Confucianism."[43] In fact, the printing presses funded by the shogunate published mostly texts on Buddhism, only two books of classical Confucianism, and not a single one on Zhu Xi's thought. It may have been simply the serendipitous encounter of two necessities. Ieyasu needed to transform his hegemony from one based on military force, typical of the previous century of civil disturbances, to one based on symbolic legitimation. Ieyasu's regime needed religious, legal, philosophical, and ethical bases to be in line with Ieyasu's idealized models of the Minamoto and Ashikaga shogunates.[44] As we have seen, he followed diverse strategies, all designed to provide the new government with authority. At the same time, Razan was struggling to establish himself as a Neo-Confucian scholar modeled upon the Ming scholar-gentleman. Only the patronage of the shogunate could grant intellectual authority to his Neo-Confucian thought, since he could not rely on either the court or the Buddhist temples. In sociological terms, the shogunate and the scholars each needed the legitimation of the other.

Whatever the motivation, Ieyasu's authorization of Razan's public lecturing and his later recruitment of the young scholar as private tutor, lecturer, ghost-writer, proofreader, adviser of international diplomacy, and book purchaser elevated Razan's status. The patronage of the shogunate, however minimal, gave Razan the recognition he needed to be socially accepted as a scholar despite the fact that he was neither a Buddhist monk nor a member of the aristocracy. Later generations of scholars regarded this event as the beginning of the public role of the *jusha* (Confucian scholar). Razan did not necessarily provide straightforward ideological legitimation of the new regime, but he did lend the shogunate the respectability of cultural expertise in the form of up-to-date knowledge of Chinese philosophical discourse and practices like diplomatic etiquette and matters of fiscal and administrative organization. By patronizing a new type of scholar, Ieyasu and his successors avoided excessive reliance on potentially competitive forces, thereby preventing the court and the Buddhist clergy from remaining the powerful political actors they had been in previous centuries.[45] The decline in shogunal sponsorship of Buddhist temples toward the end of the seventeenth century derived from the same motivation.

Backed by Ieyasu, Razan continued to lecture publicly while serving the shogunate in matters related to culture and international diplomacy. When Ieyasu hired Razan in 1607 as an adviser, the former shogun forced him to shave his head and dress like a monk. The order, according to *Razan sensei nenpu*, arrived unexpectedly when Razan joined Ieyasu at his Sunpu castle for a poetic meeting. Razan expressed his despair in a letter to Yoshida Haruyuki, in which he complained, "Now I am bald like a monk and always have to wear these black robes. This is not right."[46] There are many different speculations concerning Ieyasu's order, but it is plausible that Ieyasu had little choice in the matter. If he wanted the domainal lords to acknowledge and respect Razan's intellectual skills, Ieyasu had to make him physically resemble what at the time was immediately perceived as an authoritative scholar—that is to say, a monk. Razan obediently changed his name into Dōshun but never took vows or registered at any temple.[47]

Contacts with influential people in the government and in Kyoto society granted Razan some fame and public recognition of his talents. However, as Nakai suggests, "The majority of early Tokugawa Confucians were people whose status in Tokugawa society was fraught with ambiguity."[48] Hoping to resemble imperial Chinese gentlemen-officials, Japanese scholars could aspire only to employment for their linguistic skills and their knowledge of Chinese matters. Their intellectual labor had to appeal to their patrons, the new rulers of Japan, who were not interested in having Confucian scholars controlling the bureaucratic administration of their lands.

TASHIKIHEN, BOOK OF NAMES

In the tenth month of 1607, Ieyasu dispatched Razan to Nagasaki to purchase a complete edition of *Bencao gangmu*. Rumor had it that Ieyasu "loved that encyclopedia so much that he always wanted to have it at hand."[49] Razan had already visited Nagasaki in 1602. The port city attracted scholars primarily for its market of Chinese books that arrived with shipments from the continent. The purpose of Razan's 1602 visit is obscure. He probably went to purchase books or meet with scholars from the continent who gave private lectures in the city. He might well have heard of *Bencao gangmu* on that occasion.[50]

Razan had no training in medicine, but his record of books listed many medical and pharmacological titles. It is impossible to judge Razan's expertise in these fields, even though he corresponded throughout his life with eminent physicians. He certainly acknowledged the intellectual value of Li Shizhen's text as an empirical application of Zhu Xi's metaphysics. For Nishimura Saburō, Razan found in *Bencao gangmu* a useful source of information on *meibutsugaku* (semasiology), a discipline that developed in Japan as a subfield of Neo-Confucian studies. *Meibutsugaku* consisted of philological analyses intended to ascertain correspondences between Chinese "names" and Japanese "things."[51] Razan's use of the *Bencao gangmu* was essentially lexicographical and encyclopedic.

His work on *Bencao gangmu*, first in *Tashikihen* and later in preparing a Japanese edition of Li Shizhen's introduction, stimulated a wider interest in the Chinese text.[52] Razan completed *Tashikihen* in its manuscript form in 1612. It was printed in two volumes in 1630 and expanded to five volumes as *Shinkan*

Tashikihen the following year. It consisted of a digest of *Honzō kōmoku* and *Nongshu*, a 1313 agronomical encyclopedia compiled by Wang Zheng. Rather than an abridgement, it is better described as a glossary of 2,315 words that appeared as entries in these Chinese encyclopedias, with a translation in Japanese (including regional variations).[53]

Tashikihen listed most of the entries from the *Honzō kōmoku* in the original order and provided for each a Japanese translation in *man'yōgana*.[54] Sometimes the entry had a note with a reference to names the plant or animal might have been known by in other Chinese sources. Its structure suggests that it might have been used as a quick reference or a glossary to accompany the scholarly reading of Chinese sources. Razan's choice of a simplified form of *man'yōgana* corroborates this hypothesis.

For example, *Tashikihen*'s first entry—also the initial entry of the *Honzō kōmoku*—read simply,

雨水　下米　今案　阿末美豆

The Chinese word 雨水 (*yushui*) was followed by the Japanese translation *ame* (written in phonetic characters as 下米), or "rain."[55] Then the expression *ima anzuruni* (personal interpretation) indicated that the following translation was based on Razan's speculation, meaning that 雨水 "possibly" means *amamizu*, rainwater. Razan's translation of *Honzō kōmoku*'s entries into the anachronistic system of *man'yōgana* may appear strange, but he was in fact following the style in which similar glossaries had been written since antiquity. The tenth-century *Honzō wamyō*, one of the reference sources for his translations, was also glossed in *man'yōgana*.

Tashikihen followed the *Honzō kōmoku* when classifying regional variations of a species, listing the subspecies in a lower row. When he could not find any satisfying Japanese translation, or when a translation was not needed because the Chinese character was in current use, Razan just copied the *Honzō kōmoku* entry accompanied by its reading in *man'yōgana*. For example, the entry for *kaki*, or Japanese persimmon, appeared as follows:

柿 可岐

It consisted merely of the Chinese character 柿 followed by its Japanese reading in phonetic characters: "可岐," "*kaki*."[56] Razan did not add any further information from *Honzō kōmoku*—no descriptions, no morphology, no physiology, no seasonality, and no pharmacological use. All he did was translate the name into Japanese.

Razan then ventured possible Japanese translations of compound expressions with *kaki* without any further qualifications:

柿 可岐

烘柿　今案豆豆美可岐
Hongshi, ima anzuru ni tsutsumigaki.
Baked persimmons. Possibly rolled persimmons.

白柿　今案豆利可岐
Baishi, ima anzuru ni tsurigaki.
White persimmons. Possibly dried persimmons.

烏柿　今案阿末保志
Wushi, ima anzuru ni amaboshi.
Wholly flowered persimmon.[57] Possibly sour black persimmon.

醂柿　今案阿和世可岐
Lanshi,[58] ima anzuru ni awasegaki.
Persimmon jam. Possibly *awasegaki* [fermented *kaki*].

Razan did not distinguish cooked persimmons from subspecies of *kaki*. His translation of *wushi* as "sour persimmon" (*amaboshi*) omitted the identification in *Honzō kōmoku* of the *Diospyros eriantha* variant of the *Diospyros* genus. It is as if an animal encyclopedia were to list recipes for pork alongside biological information about species of hogs and boars in the Suidae family.[59] Other entries followed the same pattern. Razan sometimes noted whether earlier encyclopedias known in Japan listed that particular species under different names, but in those cases too he just copied the entry and gave its Japanese reading. For example, the entry for a species of "leech" (Clistellates order, Hirudinidae family) appeared as follows:

蛭　比流　異名　至掌　別録　馬蛭　唐本
　　馬蛭　今案牟末比流　木痴　草痴
　　石蛭　泥蛭　山蚑

After the Chinese character *zhi* (蛭), or "leech," Razan offered the Japanese reading "*hiru*." Then he faithfully copied the "different names" (*imei*) proposed in *Honzō kōmoku*. In the next two lines, Razan transcribed the six subspecies recorded in *Honzō kōmoku*, but he translated only the first, perhaps because it was the only one Razan thought of as equivalent to the Japanese *uma hiru*, the green horse leech.[60]

The following example shows the mistakes that could result from Razan's lack of expertise in materia medica. The entry for the black henbane (*Hyoscyamus niger*) in the 1612 manuscript consisted solely of its Japanese translation:

莨菪

Here Razan just attached two Japanese readings to the Chinese characters 莨菪 (*langdang*, or black henbane): *onihirukusa* and *ōmirukusa*, which he took directly from Fukane no Sukehito's *Honzō wamyō*. Razan was probably confused about what kind of plant *onihirukusa* or *ōmirukusa* was. In fact, in the subsequent printed editions of *Tashikihen*, he associated *langdang* with tobacco:

莨菪　於保美久佐今案多波゛古
Langdang [rōtō]: ohomikusa ima anzuruni tabako

Razan still translated the *Honzō kōmoku*'s entry for *langdang* as *ōmikusa*, but then he added the supposition that the plant was probably *tabako* (tobacco).[61] If Razan had known even the basics of Chinese pharmacology, he would have known that *langdang* (*H. niger*) was an herb that, chewed or inhaled, had been used since antiquity for its powerful narcotic and sedative properties. In China, *langdang* was classified in the same order of *lang*, a plant also used as a narcotic.[62] Although any pharmacologist would have known this, Razan's main interest was ordering words, not things, and his sources were books, not actual plants.[63]

At a cursory glance, the *Tashikihen* reads like a glossary of Chinese terms: a book of names rather than a book of nature. Razan did not base his Japanese translations upon observations of plants and animals but either copied, when he could, from earlier dictionaries or suggested Japanese equivalents based on speculation from the literal meaning of the Chinese characters. There were many mistakes, not surprising for the first work of its kind in Japan. That the *Tashikihen* had undergone revisions suggests that Razan tried to improve his glossary. Unfortunately, his correspondence does not reveal whether he contacted any other scholar for information about plants and animals. Only a few of the later encyclopedias of natural history would list *Tashikihen* as a source, especially the early ones. Instead, manuals of composition of Chinese poetry, manuals of painting, and Neo-Confucian commentaries often quoted it as a source for understanding the Chinese names of minerals, plants, and animals that appeared in aesthetic or philological works.

Tashikihen was not a text of natural history because no such field yet existed. Nonetheless, the *Tashikihen* was the kind of scholarly tool that allowed future generations of specialized scholars to access a large number of Chinese sources of natural historical interest. Razan took the title of his glossary from *Duoshibian* (The Explanation of Many Things), a text in seven volumes by the Ming Neo-Confucian scholar Lin Zhaoke. *Duoshibian* was also a glossary of the names of "birds, beasts, herbs, and trees" quoted in the Chinese classics that Zhaoke collected to refine his "knowledge of many things."

A precise knowledge of names constituted a fundamental step in education. Ordering nature in a taxonomy amounted to ordering the names of minerals, plants, and animal in a logical order. As the third-century BCE *Xunzi* stated, "[The sages] instituted names to refer to objects, making distinctions in order to make clear what is noble and what base and separations in order to discriminate between things that are the same and those that are different."[64] Just as Chinese civil examinations were structured to measure the ability of the candidate to quote the classics from memory and to demonstrate understanding of the meaning of their words, those who mastered the order of names were thought to master also the order of things. Furthermore, those who mastered the order of names, and thus things, were metaphysically and ethically entitled to govern. Philological exactness was required to judge intellectual ability from both philosophical and sociological points of view, and this was reflected in the structure of the learning process in imperial China.[65] In early seventeenth-century Japan, where no system of civil service examination existed and no socially recognizable Confucian scholar had yet emerged, for scholars-to-be like Hayashi Razan, glossaries like *Tashikihen* were an indispensable tool for understanding the increasingly large corpus of Chinese books that were arriving in Japan. *Tashikihen* served, in Nakai's terms, as a vehicle for the naturalization or a modification of "the more alien elements of Confucianism in its original Chinese form."[66]

A reconstruction of how contemporary readers classified and used *Tashikihen* supports this interpretation. In his *Nihon bungaku daijiten*, literary historian Kameda Jirō defined *Tashikihen* as a text widely used in the Tokugawa period to compose poetry in Chinese.[67] It was often quoted in the introductory notes of essays and anthologies of Chinese poetry, as well as in dictionaries of various kinds. In the eighteenth century, when the field of semasiology (*meibutsugaku*) became fashionable among scholars, *Tashikihen* was a reputable source on how to translate Chinese names of plants and animals into Japanese. One of the most influential catalogs of books, *(Wakan) shoseki mokuroku* (A Catalog of Japanese and Chinese Books)—and its updated editions *(Zōho) Shoseki mokuroku* (Expanded Catalog of Books)—classified *Tashikihen* under the heading *jisho* (dictionary), which is, after all, a book of names.[68]

Writing Nature's Encyclopedia

For us, an encyclopedia is a map that helps one navigate through a mystifying forest of books and information. As Diderot wrote in the preface of his *Encyclopédie*, "As long as the centuries continue to unfold, the number of books will grow continually, and one can predict that a time will come when it will be almost as difficult to learn anything from books as from the direct study of the whole universe. It will be almost as convenient to search for some bit of truth concealed in nature as it will be to find it hidden away in an immense multitude of bound volumes."[1] The encyclopedia is a solution to informational overload. In addition, an encyclopedia, as Diderot and D'Alembert argued, is a powerful educational tool, as the origin of the term "encyclopedia"—ἐγκύκλιος παιδεία, literally "all-around, general education"—reveals.

Just as a map cannot reproduce the curvature of the earth without distorting it, however, encyclopedias standardize knowledge, organize consensus around certain theoretical views, and keep in check the proliferation of heterodox ideas and interpretations.[2] In doing so, they often legitimized orthodox scholarly enterprises and recruited knowledge to the cause of the established power.[3] What was the nature of early modern Chinese and Japanese encyclopedias? Were they maps to navigate the labyrinths of knowledge? Or were they more like the work of Borges's cartographers of "Del rigor de la ciencia," who developed a map of the empire so big that it coincided point by point with its actual territory?[4]

THE REDISCOVERY OF CHINESE "ENCYCLOPEDIAS"

The introduction and reprinting of Chinese encyclopedic works and later the compilation of original ones played a fundamental role in the constitution of several disciplines in early modern Japan, *honzōgaku* included.[5] With the refashioning of the intellectual landscape under the newly established Tokugawa regime, the seventeenth century can be called an encyclopedic age.[6] Many of the specialized fields of knowledge that would characterize Tokugawa intellectual history originated in the thematic divisions of influential Chinese encyclopedias like the *Erya*. Moreover, encyclopedias and encyclopedic textbooks expressed the ideal of comprehensiveness in the "investigation of things." With their systematic organization, they reproduced the ordering activities of the metaphysical principle of *li* in material reality.

In the field of *honzōgaku*, as soon as the *Honzō kōmoku* began circulating in Japan, it became the "architext" influencing the style, terminology, structure, views, and content of original manuals and encyclopedias of materia medica compiled throughout the Tokugawa period.[7] Its classification of natural species remained unquestioned until the nineteenth century, and despite the corrections and updates, it continued to symbolize orthodoxy in the field of nature knowledge. Following Razan's official introduction of the encyclopedia, the growing demand for copies of the *Bencao gangmu* soon surpassed Nagasaki traders' ability to purchase it from Ming booksellers. In Japan, the first edition of *Honzō kōmoku* was most likely based on the first Nanjing edition and was printed in Kyoto in 1637 by Noda Yajiemon, a bookseller who specialized in trading Chinese texts. The publisher added the first *kundoku* annotations, phonetic markers that made it possible to read the Chinese text in classical Japanese.[8] In the second month of 1653, the same bookseller added illustrations taken from other Chinese pharmacopoeias and published a new edition with revised *kundoku* annotations. The third edition of 1659 consisted of a revision of the 1653 edition by Nomura Kansai but with original illustrations by Japanese artists. A fourth one, in 1669, by Matsushita Kenrin, was a revision of the 1659 edition and was soon followed by a fifth in 1672, thought to have been edited by Kaibara Ekiken.[9] An independent edition, completely revised by Inō Jakusui, was published in 1714 under the title *Shinkōsei Honzō kōmoku* (New Revised Edition of *Honzō kōmoku*). Ono Ranzan's masterpiece, *Honzō kōmoku keimō* (Clarifications on *Honzō kōmoku*), published between 1803 and 1806, was intended to be a radically new annotation of Li Shizhen's masterpiece.[10]

Following the introduction of *Honzō kōmoku*, it took more than a century before a scholar published Japan's first original encyclopedia of materia medica. Yet this work, Kaibara Ekiken's *Yamato honzō* (Japanese Materia Medica), too, was something of an anomaly when compared with the overtly lexicographical approach of contemporary *honzōgaku* works. Ekiken proposed a new way of looking at nature, which stressed empirical research and field observation to complement textual sources like *Honzō kōmoku*. Yet while later generations of *honzōgaku* scholars celebrated Ekiken as one of the founders of their discipline, his stress on empirical methodology was not adopted by subsequent generations of naturalists until the late 1700s.

Throughout the seventeenth century, a growing number of scholars studied *Honzō kōmoku* and other Chinese pharmacopoeias, many of them while employed as advisers to domainal governments. Others lectured in private schools or worked as town physicians (*machi isha*)—private physicians who, beginning in the early sixteenth century, established their practice in castle towns to serve townspeople and rely on them for an income.[11] The growing interest in Chinese texts prompted many scholars to produce abridged, simplified, and illustrated versions of Chinese materia medica for a Japanese audience. These translations were largely intended to be of practical use to physicians, peasants, and gardeners. In time, they provided also an intellectual and aesthetic reference for educators, poets, artists, and scholars.

In the seventeenth century, scholars were only just beginning to experience the information explosion that would later contribute to what has been characterized as a "quiet revolution in knowledge" in the Tokugawa period.[12] The stream of books and ideas from abroad, while feeding a desire for knowledge, also threatened to create misunderstandings, as the imported dictionaries, encyclopedias, textbooks, manuals, and monograph treaties were often contradictory and confusing.[13] To quote Brian Ogilvie, in Japan as in Europe, "the book of nature had become illegible unless it was accompanied by nature's bibliography."[14] Among the strategies scholars adopted to cope with this information overload were thematically arranged bibliographies, indexes, and tables of contents, which began circulating in manuscript form starting in the early years of the seventeenth century.[15] Razan's *Tashikihen* was one such tool. Abridgments and dictionaries were also numerous, either directly imported from Ming and Qing China or originally written by Japanese scholars. The same information explosion that pushed scholars like Gottfried Leibniz to despise that "horrible mass of books which keeps on growing" or Conrad Gessner to fear that "confusing and harmful abundance of books" compelled Japanese scholars to find strategies to cope with their own overabundance of information.[16]

In the seventeenth century, physicians continued to study *Honzō kōmoku* and, later, *Yamato honzō* for their pharmacological value. Scholars recognized the potential value of *honzōgaku* as an empirical application of Neo-Confucian metaphysics, but they were too preoccupied with surviving as scholars supported by the samurai elite and with ordering the enormous number of texts imported from China and Korea to develop this notion further. Not yet established as an independent discipline, *honzōgaku* at the end of the 1600s remained a hybrid subject of study that engaged scholars for its practical applications, primarily in pharmacology but also in agriculture, philology, poetry, and art.

Practitioners in this early stage of *honzōgaku* research neither were trained in specialized schools nor constituted a community of specialists. Rather, they were often either physicians or scholars well versed in Chinese learning. Their primary concern was to match the Chinese names for minerals, plants, and animals to those found in Japan. Translation was their main preoccupation, and they preferred philological research to direct observation of nature. This preference can, in part, be explained by the reliance of scholars on books imported from China and the Korean peninsula, which were considered to be superior to any research produced in Japan.[17] It was thought that learning from books was not only equivalent to learning from nature—the order of things being identical to the order of words—but even, under certain respects, superior. Words could, in fact, reveal the logic and the structure of reality more clearly than the observation of phenomena.

The study of natural history in seventeenth-century Japan was also affected by the rediscovery of ancient Chinese "encyclopedias." Many scholars that engaged with *honzōgaku* texts believed that solutions to all uncertainties about plants and animals could be found through diligent study of Chinese classics. This attitude resembled that of their European colleagues, especially in the Italian peninsula, who at the turn of the sixteenth century were rediscovering the works of ancient Greek and Roman naturalists.[18] In an effort to restore medicine to what they saw as a purer and uncorrupted state, scholars like Nicolò Leoniceno worked to uncover ancient knowledge and remove from it the influence of Arab and medieval traditions. "Renaissance naturalists," Ogilvie has commented, "were reformers, not revolutionaries."[19] Observation of nature, for these early humanist naturalists, was important only insofar as it could correct or verify the knowledge inherited from the past.

The disciplinary training of seventeenth-century scholars interested in *honzōgaku* was also based on texts, with the authority of one countered by that of others. Observation played a minimal role, Ekiken representing an important exception. Like their Renaissance counterparts, most naturalists were trained as physicians and had studied nature as part of their pharmacological instruction. Many were also Neo-Confucian scholars who saw a thorough understanding of nature as a means to pursue the "investigation of things." Chinese texts providing information about the natural world circulated in large numbers since the seventeenth century. They were both new books imported from the continent and reprints of older ones. Also, they corresponded to different genres, which Japanese scholars often perused in heterodox manners looking for information on plants and animals or any other subject. In the bibliographical notes of many Japanese encyclopedias of the seventeenth and eighteenth centuries, we find category books (*leishu*) like the *Sancai tuhui* (1609), dictionaries (*zidian*) and lexicographic texts of various sorts (*cidian*), encyclopedias (*dadian*), gazetteers (*difangzhi*), *pulu* (monographic texts focused on one or a group of natural or artificial objects), and agronomical treaties (*nongshu*).[20] While none of these can be reduced to our modern, post-Enlightenment idea of the encyclopedia, modeled upon the *Encyclopédie* of Diderot and d'Alembert, they were often encyclopedic in scope in that they offered a comprehensive, all-around treatment of various topics.

Among these, the *Erya* (*Jiga* in Japanese)[21] — the oldest Chinese dictionary or thesaurus compiled in the Warring States period (between the fifth and third centuries BCE) — was "rediscovered" and reedited in the mid-seventeenth century by Japanese publishers.[22] This updated version was used to assist in the correct reading of the canon of Confucian classics.[23] Although it was cataloged as a dictionary or thesaurus (*zidian*), the *Erya* illustrated the meaning of 4,300 words in 2,094 entries arranged in 19 sections that amounted to "a collection of direct glosses to concrete passages in ancient texts."[24] Volumes 13 through 19 explained the names of plants and animals quoted in Chinese poetry, especially the ancient classic, *The Book of Poetry*.[25]

The rediscovery of these dictionaries impelled scholars to update them or to compile new ones. Kaibara Ekiken's nephew Kaibara Yoshifuru published *Wajiga* (A Japanese *Erya*) in 1694, which followed *Erya*'s structure. *Wajiga* is today credited as one of the earliest texts to describe and introduce a number of foreign species, but it is rarely mentioned in secondary sources on the history of *honzōgaku*.[26] Encyclopedic dictionaries like *Erya* and *Wajiga* were a fundamental tool for any serious scholar, much like the *Oxford English Dictionary* or the *Encyclopaedia Britannica* (or *Wikipedia*) in the modern era. Shogunal scholar and official Arai Hakuseki was so fascinated with *Erya* that in 1719 he published *Tōga* (Eastern *Erya*), a dictionary similarly modeled on the Chinese text.[27]

Chinese topographic monographs (*difangzhi*) served as a model for the rich genre of Japanese gazetteers.[28] Beside geographic, climatic, and ethnographical information, these texts detailed the flora and fauna of China, as well as monstrous and imaginary beasts. *Shanhaijing* (The Book of Mountains and Seas, *Sengaikyō* in Japanese) was one such ancient gazetteer, rediscovered and reprinted in the Tokugawa period.[29] *Sengaikyō* enjoyed great popularity during the Tokugawa period, enriching popular imagination with new varieties of monstrous figures.[30] Subsequent natural historical encyclopedias continued the pattern of including monstrous and mythical beasts in their classifications.

Tokugawa "encyclopedias" of Chinese, Korean, or Japanese origins were in great demand throughout the period. They gave structure to nature by dividing its elements into books, chapters, and subchapters that were often taken as mirroring the order of the natural world.[31] They extended knowledge of things (*bowu*), and as such they responded to the need of the "investigation of things" of Confucian origin. Another influential text was Zhang Hua's *Bowuzhi* (Encyclopedic Gazetteer), compiled by imperial commission at the end of the third century.[32] Its ten volumes arranged all available information into different headings: geography; foreign lands and their populations; strange and wondrous beasts, birds, insects, fish, trees, and herbs; medicines; clothes and food; people, books, instruments of various kinds, and music; legends and myths; history; and miscellaneous. Zhang Hua was the first to employ the word *bowu* (*hakubutsu* in Japanese) to define the realm of all that is knowable in the universe.[33] In Meiji Japan, the term came to be used as an equivalent for "natural history," but for Zhang Hua it covered not only the physical aspects of the world but also human civilization, legends, history, and mythology.

Despite the popularity of encyclopedias, however, practical manuals, abridged dictionaries, and textbooks constituted the majority of books imported into Japan, probably because of the costs associated with multivolume encyclopedic works. Most of them were designed to improve agricultural production and were written and illustrated with the needs of cultivators in mind.[34] They were referred to as *zhongshushu* (*shujusho* in Japanese), which can be translated as "books on tree planting." Chinese agricultural manuals of particular value in the Tokugawa period were the *Nongshu* (Manual of Agronomy), written by Yuan scholar Wang Zhen, and the *Nongzheng quanji* (Collected Works of Agronomy), written by Ming scholar Xu Guangqi. Manuals such as these differed from dictionaries and encyclopedias in that they not only classified plants but also provided technical information on how, when, and where to cultivate them. Aside from these two notable manuals, most agronomical texts were monographic treaties that concentrated on one particular species or genus of plant or flower. This focused format was adopted by Japanese scholars both for textbooks and for elegantly illustrated booklets commissioned by wealthy merchants and samurai to celebrate their fine botanical specimens.[35] They were usually slim in size and their titles consisted of the character for the herb, tree, or flower under consideration and the character *pu* (*fu* in Japanese), literally a "chart," or a "list." Kaibara Ekiken, too, indulged in this monographic genre before publishing *Yamato honzō*.

COPING WITH THE INFORMATION OVERLOAD

The stream of information and books coming from China was chaotic, arbitrary, and often contradictory, which drove Japanese scholars to try to order the new knowledge into manageable form by compiling either bibliographies or abridgments. Kaibara Ekiken, as we shall see in the next chapter, attempted to solve some of the confusion in the field of *honzōgaku*, but this was no easy task, on account of the difficulty of matching Chinese names of plants and animals with actual plants and animals of Japan. Also, the sheer number of species treated in Chinese encyclopedias was astonishing.[36] Ekiken cut the number of species in his *Yamato honzō* to 1,362 entries, limiting his analyses to species indigenous of Japan.

As herbal specimens and manuals arriving via Nagasaki from China, via Tsushima from the Chosŏn kingdom, and from Europe via the mediation of both Dutch and Chinese traders grew exponentially, the number of specialists dedicated exclusively to the study of plants also grew. Over time, these early botanists formed a network of scholarly exchange. Prior to Ekiken's work, scholars had attempted various strategies to create order out of the mass of information arriving from the continent. Some tackled systematization in large educational manuals and encyclopedias. Among the most successful attempts was *Kinmō zui* (Illustrated Dictionary for Beginners), the first Japanese illustrated encyclopedia in twenty volumes compiled by the Neo-Confucian scholar Nakamura Tekisai. Printed in Kyoto in 1666, the encyclopedia covered such topics as astronomy, geography, human dwellings, human beings, the human body, clothing, exotic objects, tools, animals, and plants. Originally designed as a reference text for the education of children, *Kinmō zui* soon became one of the most used references for nonspecialists.[37] Each entry consisted of an illustration, the species' name in Chinese characters, its Japanese translation, and a very brief explanatory note.[38] The 668 illustrations alone are reason enough to merit the interest of modern *honzōgaku* scholars, but the text is only cursorily treated in secondary sources.[39]

The structure of Tekisai's illustrated dictionary can be appreciated in figure 4.1 featuring four species of flowers: from right to left, jewelweed (*hōsen*), daffodil (*suisen*), sunset muskmallow (*shūki*), and Garland chrysanthemum (*shungiku*).[40] The jewelweed entry consisted of a picture of the plant, the Chinese name with its Japanese reading, and a brief note that specified four variant names: *kinhōke*, *kyōchikutō*, *shōtōkō*, and *kikuhi*. The daffodils entry specified that there were two species of *suisen*: one univalvular (*tanben*), which was called *kinsangindai*, and one multifoliated (*chiba*), which was called *gyokurōrei*. The sunset muskmallow had a more detailed entry. Tekisai explained that it was colloquially known also as *ōshokuki*, a kind of hibiscus, and that its leaves were always turned in the direction of the sun. He added that a variation of this plant with a single flower

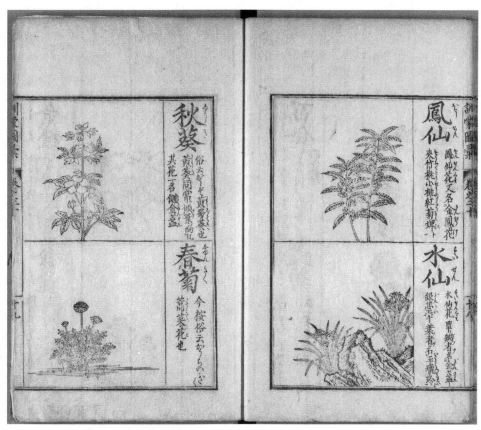

FIGURE 4.1. Nakamura Tekisai, *Kinmō zui* (1666),
volume 20. Tokyo: National Diet Library.

was known under the name of *sokukinsan*. As for the garland chrysanthemum,
Tekisai speculated (*ima anzuru ni*) that it was the same as *kōsaigusa*, also known
as *kōsaika*, a variant of the *yomogi*.[41]

Nakamura Tekisai listed as sources Wang Qi's encyclopedia *Sancai tuhui* (Illustrated Explanation of the Three Realms, *Sansai zue* in Japanese, 1607),[42] Xu
Guangqi's *Nongzheng quanji*, *Tashikihen*, *Honzō kōmoku*, and *Honzō wamyō*,
among others, as references for his interpretations. He also informed the reader
that he had interviewed a number of Chinese immigrants, peasants, fishermen,
artisans, and woodsmen to corroborate the information he gathered from textual sources.[43]

Tekisai explained in the introductory note that he had commissioned many

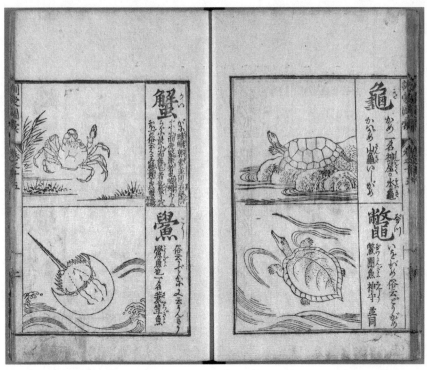

FIGURE 4.2. Nakamura Tekisai, *Kinmō zui* (1666),
volume 15. Tokyo: National Diet Library.

renowned artists to illustrate his encyclopedia.[44] The quality and precision of
these illustrations explained the popularity of the text among *honzōgaku* researchers.

Figure 4.2, for example, is thought to be one of the earliest representations
of a horseshoe crab in the world.[45] The entry consists of the Chinese term *hou*,
read as *kō*, and a brief explanation that the animal was usually called *kabutogani* or *kōgyo*. On the same page, in clockwise order from top right, there is a
tortoise (*kame*), a freshwater turtle (*betsu*), and a crab (*kai*).[46] *Kinmō zui* was a
dictionary or thesaurus in the textual tradition of *Tashikihen* and *Wajiga* in that
it was also a glossary designed to match Chinese and Japanese names. But what
distinguished the *Kinmō zui* from other similar texts was that the illustrations
were the medium through which the Chinese and the Japanese words were correlated. Tekisai attempted to create the perfect dictionary by naming each image
with a Chinese word and also by identifying the image with the corresponding

Japanese names given in previous texts and taken from the testimony of practical experts.

Kinmō zui was one of many texts exemplifying the interest in descriptive images and inspired other similar projects. *Wakan sansai zue* (Illustrated Dictionary of the Three Realms in Japanese and Chinese) by the Osaka physician Terajima Ryōan was published in 1713 in 105 volumes. As in the cases of *Wajiga* and *Tōga*, Ryōan's was a Japanese reimagining of Wang Qi's *Sancai tuhui. Kinmō zui*, quoted among *Wakan sansai zue's* sources, inspired Ryōan to illustrate his category book, but unlike Tekisai, he restricted his focus only to plants and animals of Japan. As he explained in the introduction, the book was the result of thirty years of research. Ryōan frequently made reference to *Honzō kōmoku* and sometimes commented on the pharmacological information of the Chinese pharmacopoeia on the basis of his experience as physician. He also inserted into the encyclopedia sections on Japanese history and religion and was one of the first to add information taken from Western sources. The work continued to be published throughout the Tokugawa period and well into the twentieth century.[47]

The popular success of encyclopedic works like *Kinmō zui* and *Wakan sansai zue* was the direct result of increasing literacy among nonsamurai members of the population.[48] Their success also shows that there was a growing appreciation for descriptive pictures in reference manuals and textbooks. Published in 1666, *Kinmō zui* was reprinted two years later by the same Kyoto bookseller with larger pictures.[49] In 1695, a newly revised and expanded edition was published under the title *Kashiragaki zōho Kinmō zui* (Annotated and Expanded *Kinmō zui*), printed in a smaller format. Yabe reports that these later editions circulated in several different forms: sections were sold independently, and pirated editions (far cheaper but more roughly printed) were also numerous.[50] The *Kashiragaki* edition, in particular, increased the number of entries for animals to 389 and plants from 354 to 395. The newly introduced entries were almost all species from China that had been ignored by Tekisai as irrelevant to the Japanese people.[51] Further editions were used in temple schools (*terakoya*) throughout the Tokugawa period and continued to be printed and sold well into the Meiji period.[52] *Kinmō zui's* images also attracted the Dutch physician Engelbert Kaempfer, who was fascinated with the Japanese encyclopedia. Not only did he purchase a copy of the second edition, but he also inserted more than seventy illustrations taken from it in his *The History of Japan*, published first in 1727 in English translation and then in 1779 in the original Dutch (figs. 4.3 and 4.4).[53]

Despite the enormous success of *Kinmō zui*, little is known about its author, Nakamura Tekisai. Son of a merchant, he became acquainted with Zhu Xi's writ-

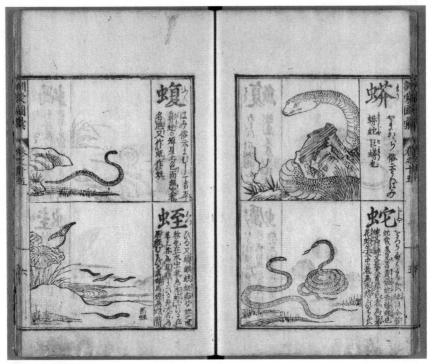

FIGURE 4.3. Nakamura Tekisai, *Kinmō zui* (1666),
volume 13. Tokyo: National Diet Library.

ings as an autodidact. His fame as educator and scholar in Kyoto was on par with
Neo-Confucian scholar Itō Jinsai.[54] The writer Ihara Saikaku quoted sections of
Kinmō zui in his *Saikaku gohyakuin* (Saikaku's Five Hundred Verses), published
in 1679, defining it as a "treasure" (*jūhō*).[55] Later in life, as so often was the case
for scholars in the seventeenth century, Tekisai was hired as an official and educator by the Tokushima Domain, where he entered the service of the Hachisuka
family. A number of works bearing the same title testified to its wide dissemination. *Kokusho sōmokuroku*, a catalog of books written in Japanese and published
before 1867, lists fifteen texts of various genres with the expression *"kinmō zui"*
(illustrated dictionary for beginners) in their titles. Most of them were catalogs
and guidebooks of pleasure quarters or satirical works, the most famous of all
being Shikitei Sanba's *Gejō kinmō zui* (Illustrated Dictionary for Beginners of
the Theatre Quarters, 1803), illustrated by *ukiyoe* artist Utagawa Toyokuni.[56]
In 1690, a catalog of brothels was published under the title of *Jinrin kinmō zui*
(Illustrated Dictionary for Beginners of the Ways of Human Pleasures).

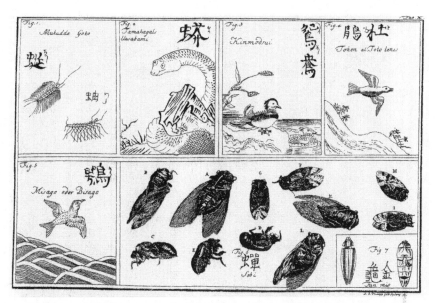

FIGURE 4.4. Engelbert Kaempfer, *The History of Japan* (London, 1727). This illustration was taken from the 1668 edition of *Kinmō zui*, a copy of which Kaempfer had brought back to Europe.

As with Razan's *Tashikihen*, these dictionaries and manuals of materia medica were mostly used by learned individuals such as physicians and Neo-Confucian scholars. Yet they were at the same time well praised and widely quoted for artistic purposes. In particular, they were invaluable sources for painting and poetic composition, especially in the case of Japanese *waka* and *haikai*, where natural objects were a stylistic requirement of the poetic form.

Representations of the natural world were in fact widely used in poetic compilations. Among these, the anthology *Rokuroku kaiawase waka* (Poetic Anthology of Thirty-Six Shells), edited by Kazuki no Amanoko and printed in 1690, associated each poem in the collection with a picture of a specific shell (fig. 4.5).[57] The illustration shows the index of thirty-six poems, divided into two groups, right and left, of eighteen poems each. Each poem, symbolized by a particular shell, contains the name of its representative shell, either directly quoted or in the form of *kakekotoba*, or "pun-words." *Rokuroku kaiawase waka*, however, was something more than an anthology of poetry: it matched poems with fine illustrations of shells, all rigorously classified and named in the fashion of *honzōgaku* encyclopedias.

The use of images of shells for aesthetic purposes or, alternatively, the aes-

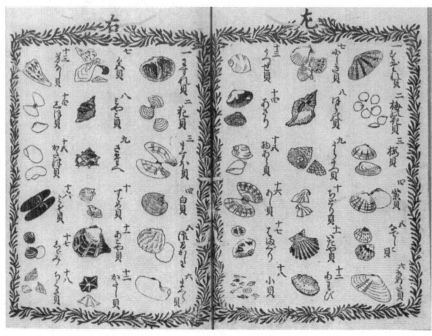

FIGURE 4.5. Kazuki no Amanoko, *Rokuroku kaiawase waka* (1690), the index. Tokyo: National Diet Library.

thetic disposition of images from the natural world in encyclopedias and monographs of natural history was also common in early modern Europe. Emma C. Spary noted that European catalogs of private collections were often artistically illustrated, and taste was the most important criterion to legitimate a collection and the collector's status in polite society. In turn, membership in polite society conferred recognition on the truth-value of one's ideas and research.[58] "In describing the process of making order within the collection," Spary argued, "European naturalists frequently appealed to arguments from design."[59] *Rokuroku kaiawase waka* is similar to, for example, Dezallier d'Argenville's *La Conchyliologie, ou Traité sur la nature des coquillages* (first edition 1757, Conchology, or Essay on the Nature of Shell), in which d'Argenville presented his collection of shells following Linnaeus's taxonomy, but this did not prevent him from artistically displaying them in the fashion of the cabinets of curiosity (fig. 4.6).

The massive production and introduction of new texts in seventeenth-century Japan challenged scholars to organize the influx of knowledge in the form of systematic dictionaries, encyclopedias, bibliographies, and manuals. This explo-

FIGURE 4.6.

Dezallier
d'Argenville, La
Conchyliologie,
ou Traité sur
la nature des
coquillages,
plate XIII,
with original
illustrations by
Jacques Mesnil
(Paris, 1780).
New York:
New York Public
Library.

sion of information occurred in the context of a sharp increase in literacy and in book production in the first half of the Tokugawa period.[60] Physicians were primarily responsible for this multiplication of textual resources and information on the natural world for the practical purpose of developing their pharmacological expertise. Both Tokugawa and domainal authorities often sponsored the publication of great encyclopedic works: as monuments of accumulation of knowledge, encyclopedias were cultural assets that, with or without practical use, enjoyed the patronage of the ruling elites in line with the *bunbu* politics of the bakufu. Also, the proliferation of encyclopedias, educational tools, and fundamental instruments of scholarly research accompanied and favored the process of formation of a professional class of scholars.

The structured order in which encyclopedias presented reality was taken to reflect the ontological order of the universe, in accordance with Zhu Xi's doctrine of the "investigation of things." This principle came to play an important role in the philosophical speculations of the second and third generation of Japanese scholars. A further consequence of this "literati theory of knowledge" based on Neo-Confucian cosmology was that if *ri* was the organizing principle of all things in the universe, then it regulated not only things but also the signs used to express those things. In other words, things and their names were subjected to the same impersonal laws of *ri* and *ki*. It followed that the order of the names of natural objects was identical to the inherent order of those natural objects. Neo-Confucian scholars consumed and produced manuals and encyclopedias of natural knowledge for the purpose of personal cultivation or as teaching tools and practical manuals for the improvement of agricultural and pharmacological techniques. In the next chapter, we will see how Kaibara Ekiken, combining both practical interests and Neo-Confucian ideals, embedded his activities and research in the field of natural history in his role as public educator and domainal official. His *Yamato honzō* was an integral part of his program of moral education, and it stands as the first genuinely Japanese encyclopedia of natural history.

The First Japanese Encyclopedias of Nature
Yamato honzō and *Shobutsu ruisan*

· · · · · · · · · · ·

EKIKEN, THE SCHOLAR-RETAINER

By the end of his long life in 1714, Kaibara Ekiken was probably one
of the most eminent scholars (*jusha*) in Japan. His *Yamato honzō* accompanied
Honzō kōmoku as the most authoritative encyclopedia of material medica and
required textbook in most schools of medicine. The son of a doctor of low samurai rank in attendance at the Fukuoka castle on the island of Kyūshū, Ekiken
faced different obstacles than Razan to establish his reputation of professional
scholar.[1] While Razan needed to invent the very notion of a scholar who was
neither a court noble nor a Buddhist monk, Ekiken's intellectual pursuits were
contingent on the patronage of his domainal lord. To be a scholar, for Ekiken,
was one way to serve the needs of his lord, not qualitatively different from any
other retainers of the Kuroda family.

Although his family claimed Fujiwara ancestry, his father's fortunes were far
from secure. Ekiken's father Kansai spent a number of years as *rōnin*, earning his
living as a village physician.[2] For this reason Ekiken's early life was not dissimilar
to that of a peasant.[3] This background may well have influenced Ekiken's striking concern for the social conditions and education of commoners. Ekiken, in
contrast to the sons of samurai in service of a domainal lord, did not grow up in
a castle compound but had direct contact with the life of commoners. *Kaibara
Ekiken sensei nenpu* reports that Ekiken was a prodigiously precocious child,
a common theme in hagiographic portraits.[4] Ekiken was not trained in a domainal school (*hankō*) but was taught to read Japanese and Chinese by his father
and his older brother Mototada (Sonzai). Mototada, a Kyoto-trained physician,
was particularly influential in his brother's early education. It was Mototada who
introduced Ekiken to the scholarly community in Kyoto.[5] He also encouraged
Ekiken to read Neo-Confucian texts and convinced him to renounce Pure Land
Buddhism, an anti-Buddhist attitude that had characterized Neo-Confucian
scholars since the time of Fujiwara Seika and Hayashi Razan.

In 1647, Kuroda Tadayuki recruited Ekiken to serve as a low-ranking part-
time aide during his required residence in Edo (*sankinkōtai*). Over the next
few years, Ekiken continued to serve Tadayuki as his factotum. He was often

dispatched to Nagasaki, at the time an open city under the joint administration of the Fukuoka and Saga domains on behalf of the shogunate. While in Nagasaki, Ekiken gained exposure to foreign books and ideas: one of his duties was the surveillance of official translators from Chinese and Dutch.[6] After his dismissal from the position in 1650, Ekiken continued to visit Nagasaki to purchase books and informally study medicine with uncertified practitioners. Officially unemployed, he joined his father in Edo and began to study medicine. Through his father, Ekiken was introduced to the inner circle of the Kuroda retainers, and through them he made his debut in the scholarly network of Japan's shogunal capital. *Kaibara Ekiken sensei nenpu* records that Ekiken met with Hayashi Gahō, who was at the time head of the Hayashi school of Neo-Confucian studies.[7]

Back in Fukuoka in 1656, Ekiken entered the administrative service of Kuroda Mitsuyuki. The patronage of Mitsuyuki proved extremely important for Ekiken, as it enabled him to receive a paid leave to study in Kyoto. Mitsuyuki's investment in Ekiken's education is evidenced by the numerous rewards he presented to Ekiken for his scholarly successes.[8] It was then not uncommon for local lords to sponsor talented young men to study in Kyoto (*yūgaku*). The duty of domainal administrations to invest in learning and cultural activities was explicitly sanctioned in the laws for military houses, the *Buke shohatto*.[9] This legislation emphasized the *bunbu* training of samurai, meaning education in the arts of learning (*bun*) and war (*bu*).[10] Since its inception, the Tokugawa regime had openly supported the transformation of military values into civilian ethics as a measure to ensure peace and stability after more than a century of internal warfare. Hayashi Razan and several of his contemporaries were among the first to take advantage of the new possibilities for social advancement afforded by these political measures toward scholarly activities. Ekiken, half a century later, also benefited from these initiatives.[11] Ekiken, who was granted the possibility of a professional career as physician and scholar, spent all his life in the service of the Kuroda family and their Fukuoka Domain, reaching at the height of his career the considerable yearly stipend of three hundred *koku*. During the seven years that he spent in Kyoto, Ekiken enjoyed the culturally vibrant atmosphere of the imperial capital. More important, he built an impressive network of friends and scholarly correspondences with whom he would maintain contact throughout his life. He studied with Matsunaga Sekigo, who was among the most influential Neo-Confucian scholars active in the capital at the time, and Sekigo's pupil Kinoshita Jun'an.[12] Among other scholars Ekiken met in Kyoto, Inoue lists Nakamura Tekisai, the author of *Kinmō zui*; Yamazaki Ansai;[13] Itō Jinsai;[14] Matsushita Kenrin; the physician Kurokawa Dōyū; the *honzōgaku* scholars and physicians Inō Jakusui and Mukai Genshō; and the agronomist Miyazaki Yasusada, the author of the encyclopedia *Nōgyō zensho* (Compendium on Agronomy).

Ekiken's collaboration with the renowned agronomist Miyazaki Yasusada influenced his *Yamato honzō*. Yasusada, a samurai from the Hiroshima Domain, was employed in 1647 by Lord Kuroda Tadayuki as an official in charge of the agricultural politics of his domain. After only five years of service, he retired to the small village of Myōbaru, near Fukuoka, where he dedicated the rest of his life to the study of agricultural techniques. In particular, Miyazaki began experimenting with new techniques to improve soil fertility and develop pest-resistant crops.[15] His lifework, *Nōgyō zensho*, was published in 1697 in ten volumes and remained the most influential manual on agronomy throughout the Tokugawa era and beyond.[16] As Inoue Tadashi has shown, collaboration with Ekiken was vital to the completion of Yasusada's encyclopedic project. Ekiken introduced Yasusada to important Chinese agronomic encyclopedias and helped him with the writing, and, in turn, Miyazaki provided Ekiken with an enormous quantity of data gathered from his empirical research in the field, which Ekiken collected and published in 1704 under the title *Saifu* (On Vegetables). In this text, Ekiken described 316 species of edible plants, detailing the seasonality of each plant, the best way to nourish it, how to prepare it as food, and in certain cases how to eliminate toxic substances from it. The slim text was written for both scholars and educated readers and also intended to provide practical advice to peasants. As Ekiken explained in the introduction, "If we do nothing to improve our life, we would live like birds and animals and just wither like herbs and trees. My intention in writing this little insignificant book is not futile. I did not throw away jewels to collect stones, as the saying goes. I just wanted to add to peasants' knowledge in order to help them improve their lives even if just a little. That is why I am not ashamed nor do I fear the criticism of all other great masters of the way."[17] Throughout his active and productive life, Ekiken never ceased corresponding with his scholarly acquaintances, but he also never put intellectual pursuits before his duty to the domain and to people.

When Ekiken was ordered back to his home as a functionary of the Fukuoka Domain in 1664, he let his hair grow back in the samurai fashion of the time. This meant, symbolically, that he regained formal status in Tokugawa society. As domain official and scholar, Ekiken's days were busy. "He embarked on a rigorous program of lecturing, tutoring, research and writing that was to continue for the next fifty years."[18] He traveled extensively, giving and attending lectures in Fukuoka, Kyoto, and Edo. His audiences included court aristocrats, rural samurai, high-ranking officials of the Fukuoka Domain, fellow scholars, peasants, and commoners.[19] He continued to participate in intense intellectual exchanges with other scholars and repeatedly promoted educational reforms in the Fukuoka Domain.[20]

Ekiken is remembered as the perfect embodiment of the scholar-retainer, a man of learning and public servant. He was a prolific writer whose more than one hundred books can be divided into five categories: Neo-Confucian philosophy and philology, Shintoism, ethics (the largest group), *honzōgaku*, and topographical-historical works, including travelogues such as *Chikuzen no kuni zoku fudoki* (Gazetteer of the Chikuzen Province) and a genealogy. Ekiken prioritized being of practical help to his lord, his fellow countrymen, and, more generally, to all human kind, and this attitude also inspired his work on natural history. He assisted his patrons with administrative matters, wrote on agricultural improvements and on plants and animals, debated Neo-Confucian philosophy, and described the topographical features of the Japanese provinces. As he wrote in the introduction of his *Yamato zokkun* (Precepts for Daily Life in Japan, 1708), "Even though I have written an insignificant work, I want it to serve a purpose in the daily lives of the people and to serve to teach ignorant people and children as well as men and women who are not of high rank."[21] Similar sentiments are scattered throughout his writings, stressing public usefulness and the desire to "not waste the treasures that nature offered us."[22] Ekiken's view of scholars as public servants and educators combined the concepts of feudal loyalty and service to one's lord and Confucian loyalty and service to the people under the lord's rule. For Ekiken, the accumulation of knowledge was not an end in itself but a means to satisfy social needs.

The inseparability of theory and praxis in Ekiken's thought is explained in *Kunshikun* (Instruction for the Prince), a short text in three books he published in 1703 at his own expense, where he clarified his ideas on governance.[23] Grounded in his understanding of Zhu Xi's cosmology, knowledge to Ekiken involved understanding principle (*ri*) in the world as concretely manifest in all things (*ki*), thus rejecting a priori and abstract intellectual pursuit. *Ri* and *ki* could not be separated: a principle without its material realization would not exist; conversely, matter without principle would be a formless mass. Similarly, knowledge without concrete realization in practice would amount to sterile erudition. As *ri* manifested itself in the world through the action of *ki*, Ekiken argued that the mind revealed itself in the actions of the body. In turn, his belief in the inseparability of body and mind provided metaphysical justification for the idea that the cosmological unity of *ri* and *ki* lay in the inseparability of knowledge and action. In logical terms, *ri* related to *ki* as the mind related to the body. Only by acting through their bodies could people manifest their nature. Human nature became explicit only in human labor. On the basis of this philosophical belief, Ekiken rarely mentioned *ri* in his texts or speculated on a priori laws governing reality. He believed that the working of principle (*ri*) could only be discerned by observing its concrete realization in its material form, *ki*, in the real world. Scholars should deduce the principles governing the universe not through abstract reasoning but through painstaking work of observation. This empirical attitude is Kaibara Ekiken's most original contribution to Tokugawa thought, and it constituted the methodological basis of his *Yamato honzō*.

Ekiken believed that individuals, to realize their true nature, had to nurture their minds with virtuous behavior and an unremitting "investigation of things." Scholars had a duty to educate and guide people to that goal. This sense of duty endowed Ekiken's writing with a deep ethical purpose. Ethics was the ultimate aim of all practical learning. In *Yamato honzō*, connection between epistemology and ethics meant that knowledge of the natural world was not an end in itself but a means to achieve a higher morality.

THE STUDY OF NATURE AS PRACTICE AND MORAL ENTERPRISE

Kaibara Ekiken decided to publish his encyclopedic study of Japanese nature, *Yamato honzō*, because "there is much to doubt of the way *Honzō kōmoku* divided species."[24] Li Shizhen's encyclopedia "records the names of about 1800 species that exist in the world." But "*Honzō kōmoku*," Ekiken explained, "ignores many species that have been mentioned in other texts," and "it treats many exotic species that do not live or grow in Japan." For these reasons, "I decided to record in one single text all those species that people can actually see in our country," so that "they can look just at this book for comprehensive information about plants and animals, without wandering about looking for additional information in various books." Ekiken wrote the encyclopedia "with the hope that it can be of concrete help to the people of this country."[25] Thus he excluded all species that did not exist in Japan (this country, *honpō*) or were treated too vaguely in *Honzō kōmoku*. He selected 772 species from Li Shizhen's work, taking only the information that he considered necessary (*setsuyō*). He added 203 species that were treated in other encyclopedias of materia medica but not in *Honzō kōmoku* and identified each with a Japanese name. He also added 358 species of "herbs, trees, insects, fish, birds, and animals" (*sōmoku chūgyo kinjū*) found in Japan but omitted from other manuals and encyclopedias.

Ekiken based his explanations largely on material he had accumulated from his travels and correspondence with other scholars. If a Chinese character was not assigned to a species, he used phonetic characters for the name. In addition, he sought to "give an order to all regional names that refer to the same species."[26] Therefore, he recorded autochthonous species under their names in dialect in the region where they were found. For species that appeared in other texts with different Chinese characters, he chose the oldest written version he could find in Japanese classics (*honchō no koseki*), although he acknowledged that those names might have been written differently in different regions or in later texts.[27] Lastly, he recorded twenty-nine species recently imported into Japan by Westerners (*nanban*).[28]

Yamato honzō was published in 1709 in sixteen volumes by the Kyoto bookseller Nagata Chōbei. Soon afterward, two additional volumes with addenda were published, followed by two more volumes consisting of simple but precise illustrations. In each entry, he gave the standard name and vernacular variations of species; reconstructed the historical treatment of species in previous sources; and described their morphology, physiology, habitat, and seasonality. He also provided additional information on how to grow certain plants and how to prepare food and medicines. Ekiken's encyclopedia covers 1,362 species arranged in a classificatory order that Ekiken had developed, loosely inspired by that in *Honzō kōmoku*. All species were further divided into thirty-six classes, and a separate class was designated for human beings (tab. 5.1).[29]

TABLE 5.1. *The Complete List of* Yamato honzō's *Classes*

1. Waters, *suirui* 水類: 12 species, comprising water, vapor, and hot springs waters but also oil and marine salt

2. Fires, *karui* 火類: 10 species of bonfires, carbon fires, fireworks, and so on

3. Minerals and stones, *kingyoku doseki* 金玉土石: 67 species of various minerals and stones, including coral

4. Cereals, *kokurui* 穀類: 26 species, the most thoroughly treated being rice (*ine* 稲), soybeans (*daizu* 大豆), barley (*mugi* 麦), buckwheat (*soba* 蕎麦), and millet (*kibi* 黍)

5. Yeasts, *zōjōrui* 造醸類: Ekiken copied *Honzō kōmoku*'s entries for yeasts in simplified form

6. Vegetables, *saisorui* 菜蔬類: 67 species

7. Medicinal herbs, *yakurui* 薬類: 79 species, including ginseng (*ninjin* 人参), the Japanese bellflower (*kikyō* 桔梗),[1] licorice (*kanzō* 甘草),[2] perilla (*shiso* 紫蘇),[3] and so on

TABLE 5.1. *(continued)*

8. Useful herbs, *min'yōsōrui* 民用草類: where Ekiken listed 7 species of plants that were utilized for clothes, such as cotton and hemp (*asa* 大麻),[4] and for other human activities, such as tobacco (*tabako* 烟草)

9. Flowers, *kasōrui* 花草類: 73 species of flowers either wild or domesticated

10. Garden herbs, *ensōrui* 園草類: 18 species of flowery herbs grown only in gardens, many of which were introduced into Japan since the second half of the sixteenth century, including cactus (*saboten*) and iris (*shōbu* 菖蒲)[5]

11. "Juicy" vegetables, *urui* 蓏類: 9 species of fruit-bearing plants like strawberries, melon, watermelon, and so on

12. Vines, *mansōrui* 蔓草: 37 species of various vines

13. Perfumed herbs, *hōsō* 芳草: 16 species of scented plants

14. Water grasses, *suisō* 水草: 36 species of plants growing in marshes on along the shoreline

15. Seaweeds, *kaisō* 海草: 28 species of algae but also including some jellyfish

16. Miscellaneous weeds, *zassō* 雑草: 137 species of unclassifiable herbs and lichens

17. Fungi, *kinrui* 菌類: 25 species of mushrooms

18. Bamboos, *takerui* 竹類: 22 species of bamboo

19. "The four trees," *shibokurui* 四木類: 7 species of the most useful trees, including the mulberry (*kuwa* 桑), tea plants, and the Japanese lacquer tree (*urushi* 漆)[6]

20. Fruit-bearing trees, *kabokurui* 果木類: 44 species

21. Medicinal trees, *yakumokurui* 薬木類: 32 species of trees with medicinal properties, like the clove (*chōji* 丁香)[7]—probably imported from the Moluccas by the Portuguese—and nutmeg (*nikuzuku* 桂)[8]

22. Garden trees, *enboku* 園木: 36 species like Japanese cedars, pine trees, camphor trees, and so on

23. Flower trees, *kaboku* 花木: 40 species of trees known for the beauty of their flowers, including cherry

24. Miscellaneous trees, *zatsuboku* 雑木: where he listed 92 remaining species of trees

25. River fish, *kagyo* 河魚: 39 species

26. Ocean fish, *kaigyo* 海魚: 83 species

27. Water insects, *suichū* 水蟲: 21 species of insects living in or near waters; also shrimps, sea cucumbers, and leeches

28. Ground insects, *rikuchū* 陸蟲: 64 species of insects living on the ground; also spiders, frogs, worms, and snakes

29. Shellfish, *kairui* 介類: 54 species of bivalves but also sea urchins, crabs, and tortoises

30. Water birds, *mizutori* 水鳥: 25 species of birds living near waters

31. Mountain birds, *yamatori* 山鳥: 13 species of raptors

TABLE 5.1. *(continued)*

32. Small birds, *kotori* 小鳥: 37 species of small passerines like sparrows, larks, and so on, all small birds indigenous to Japan

33. Domestic fowls, *kakin* 家禽: 4 species of poultry: chickens (*niwatori* 鶏), geese (*gachō* 鵝), ducks (*ahiru* 鶩), and domestic doves (*iebato* 鴿)

34. Other kinds of birds, *zakkin* 雑禽: 10 species of Japanese birds not included in the previous categories

35. Exotic birds, *ihōkin* 異邦禽: 10 species of recently imported birds like the orioles, parrots, cassowaries, peacocks, and so on

36. Animals, *kemono* 獣: 46 species of mammals, either domestic (cows, sheep, horses, dogs, etc.) or wild (deer, wolves, macaques); also *kappa*, water goblins

37. Humans, *jinrui* 人類: here Ekiken treated briefly the different body parts and secretions of the human body, as well as different people; gibbons, too, are treated here

---

[1] *Platycodon grandiflorum.*

[2] *Glycyrrhiza uralensis.*

[3] *Perilla frutescens* var. *acuta.*

[4] *Cannabis sativa.*

[5] *Acorus calamus* var. *angustatus.*

[6] *Toxicodendron vernicifluum.*

[7] *Syzygium aromaticum.*

[8] *Myristica fragrans.*

In terms of classificatory structures, Ekiken rejected Shizhen's abstract systematic order in favor of a more concrete one based on visible characteristics, uses, and habitats that he could directly observe. Realizing his system was artificial, Ekiken did not attempt to reproduce the order of nature. Whereas Li Shizhen struggled to classify trees and herbs on the basis of their morphology and physiology—implicitly revealing the underlying order of *ri*—Ekiken opted instead for an anthropocentric system based on human needs. Herbs were classified first into medicinal (*yakurui*) and useful (*min'yō sōrui*), and then by their aesthetic appeal in the wild (*kasō*) and in gardens (*ensō*). Ekiken based his ordering on the usefulness of species to human beings: food, medicine, and tools took precedence over beauty. Listing first the autochthonous ones and then the exotic, he added wild plants and animals.

In writing his encyclopedia, Ekiken aimed for precision and simplicity. That meant cutting down the number of entries and omitting species that did not live or grow in Japan. It also meant excluding theoretical discussions of species classification unless merited by practical considerations. He omitted supplemental information such as anecdotes, legends, and symbolic meanings. He inserted literary quotes only insofar as they clarified the meaning of the name of a species. Finally, he arranged each entry in an order that took into consideration browsing speed and facility. Classification was intended to facilitate the retrieval of information.

Ekiken's painstaking empirical descriptions and philological clarification of names are evident in the entry on cherry trees (*sakura*), which comes second in book twelve on flowering trees (*kaboku*).[30] Ekiken began by classifying it as an indigenous Japanese species (*wahin*). After quoting early sources mentioning *sakura*, he commented that

the flower that all the abovementioned Chinese sources call *ying* has red petals. The Japanese cherry-tree [*Nihon no sakura*], however, does not exist in China, as Chinese merchants testified when I interviewed them in Nagasaki in the Enpō era [1673–81]. If such a tree did exist, it would have been mentioned in Chinese books or praised in Chinese poems. We should therefore consider their testimony truthful. It is possible that *sakura* grow in the Korean peninsula. Many years ago a Korean boat shipwrecked nearby [near Fukuoka], and people noticed that it was built with *sakura* wood. I interviewed Koreans about this tree, and they told me of a tree called *naimu* [奈木]. It has pale crimson flowers blossoming between the second and third months. They told me that it is much beloved there. In China, books are printed with woodblocks of the Azusa tree.[31] In Japan we use *sakura* because its wood is hard and of good quality. The quality of the *sakura* wood depends, however, on the condition of the soil: this is undoubtedly another effect of the principle of the universe. The Japanese cherry tree lives for more than one hundred years. In some regions, people cut through its bark vertically with a sword to lengthen its life. In ancient Japan, people called the peach flowers "the flowers" [*hana* 花, "flower"], but since the middle ages that appellation has been attributed to the cherry blossoms. This is a first evidence of the love of Japanese people for that flower. There are many different kinds of *sakura* trees. There are some that when they are in full bloom it is almost impossible to count the numbers of petals. Others are simpler. Some have a corolla of eight petals. Often, the variant with eight petals tends to be red: these are called *hizakura*.[32] Others have a bluish coloration: these are called *asagizakura*, and are pungently scented.[33] The *kabasakura* is called by Tesshōki "one-blossom *sakura*."[34] The same tree is also called *kanibasakura* in the *Kokinshū*, where it is described as having a single blossom of pale red. There is a tree called *uzusakura* in the Kurama mountain, where *uzu* is a term used to identify a kind of ornament of saddles. Even though it is call *sakura*, in reality it is not a cherry tree.[35] It is unclear where the meaning of "*uzu*" [雲珠] comes from.

Ekiken listed regional variants of cherry trees. Of the cherry trees growing in the Yoshino region, known also as *yamazakura*, Ekiken wrote,[36] "The *sakura* of Yoshino have been numerous since antiquity. They cover the valleys and the hillsides of the region. They start blossoming at the foot of the hills and then spread to their top. They blossom along the roads and everywhere in the valleys for almost a month. When they are in full bloom they capture the eye." More than a century later, the entry on *yamazakura* would inspire the nativist scholar Motoori Norinaga, who, convinced by Ekiken that *sakura*—and especially the *yamazakura* of Yoshino, near his hometown of Matsuzaka—were exclusively native to Japan, adopted them as a symbol to represent "Japaneseness" in his poetry. Especially famous is a poem in the colophon of a self-portrait Norinaga painted in 1790 to commemorate his sixty-first birthday:

> *Shikishima no Yamatogokoro wo hito towaba asahi ni niou yamazakurabana*
> しき嶋のやまとごゝろを人とハゝ朝日にゝほふ山ざくら花
> "If people ask about the heart of [the people of] Yamato, it is the blossoms of mountain cherry reflecting the rising sun."

The poem persisted in the imaginations of Japanese people well into the twentieth century.[37] Yet it started with a mistake made by Ekiken, for the *Prunus jamasakura*—"Japanese cherry"—can also be found in China and Korea.

Reading through the *sakura* entry, four salient characteristics of *Yamato honzō* emerge. It contains precise verbal descriptions of regional variants (in terms of petal morphology and color in the case of a flowering plant like the cherry tree) that differentiate the various local names or local species or subspecies (in modern terminology) of vegetal and animal genera, descriptions of the utility of plants or animals, and brief references to the cultural meanings and usages associated to a particular plant or animal. Finally, an awareness to the regional distinctions that differentiate Japanese species from those living and growing in China or in the Korean peninsula permeates the work.

Like the majority of Chinese encyclopedists, Ekiken dealt also with creatures that today we would call monsters. In the section on beasts, there is an entry for *kappa*, a species of "water goblin" prominent in Japanese popular imagination. In *Yamato honzō*, Ekiken described the *kappa*, or *kawatarō*, as follows:

Kawatarō live in various rivers and lakes. They are the size of a five- or six-year-old child. It often happens that villagers wandering about alone near a riverside encounter them and lose consciousness as a consequence. These creatures like to wrestle with men to prove their strength. Since their body is particularly slimy, it is difficult to take hold of them and their stench is intolerable. It is not easy to stab them with a knife. If they win the match, they may kill the person by pulling him into the water. If they are not able to overcome the person, they disappear into the water. All those who are able to win the match with the *kappa* fall into an ecstatic torpor. They return home, but their mind remains lethargic. They are sick for about a month, the symptoms being chills, fever, headaches, itching, and excruciating pain throughout the bodies. Not infrequently *kappa* haunt houses and cause a variety of uncanny incidents that annoy their inhabitants. In this, they are like foxes and can cause serious damage. *Honzō kōmoku* mentions the *suiko* (*shuihu*), a species of the amphibian order of the insect class, but the two species have so many differences that they should be classified in different categories.[38]

The disappearance of such monsters in encyclopedic texts is a relatively recent phenomenon, not older than a century and a half. In Japan, as in the West, the history of science includes an impressive array of creatures—unicorns, giants, and vampires—that were described in the same terms as the phenomena of the natural world.[39]

Such a naturalization of supernatural creatures in *Yamato honzō* paralleled Ekiken's exploration of unchartered realms of untamed forests, an early example of breaking through the barrier of human and sacred space. Indeed, Ekiken explicitly claimed that he acquired most of his information from direct observation: "I have visited all villages of the Chikuzen province. I climbed tall mountains. I penetrated into deep valleys. I followed steep paths and walked through dangerous grounds. I have been drenched by rains and lost my way in fog. I endured the coldest winds and the hottest sun. But I was able to observe the natural environment of more than eight hundred villages."[40] Later in the same passage, he wrote, "Sometimes, even after I carefully observed a specific plant or herbs, there remained so many things that I still did not understand! When this happened, I often went back to the original spot and observed it even more carefully and asked around about it. However, there are many things that I still do not understand."[41] Ekiken's empirical attitude resonates throughout *Yamato honzō*: as we have seen in the *sakura* entry, he combined textual references with personal observation and the testimony of people he interviewed. He was also keen to relate species of plants and animals to the territory where they grew and lived, revealing a sense of locality that is rarely found in coeval naturalists in China and Japan.

Ekiken frequently reflected on problems of methodology: "One should not blindly regard all one has watched and heard as true and reject what others say merely because they disagree, nor be stubborn and refuse to admit one's mistakes. To have inadequate information, to be overly credulous about what one has seen and heard, to adhere rigidly to one's own interpretation, or to make a determination in a precipitate manner—all these four modes of thinking are erroneous."[42]

Ekiken believed that one "should value broad learning and wide experience." The two cognitive approaches were for him inseparable since he believed that knowledge accumulated from books should always be accompanied by direct observation and inquiry. "One should remain skeptical in regard to doubtful cases" and "be always fair and objective in one's judgment." The most important task of a scholar was "always to investigate thoroughly and reflect carefully before making any judgment."[43] Ekiken pursued this method to the extreme, in a fashion distinct from most other contemporary scholars: "I followed up on what the townspeople spoke of, salvaged what I could prove out of even the most insane utterances, and made inquiries of people of the most lowly station. I was always willing to inquire into the most mundane and everyday matters and give consideration to all opinions. Forgetting about myself, I listened to others."[44] If the most important mission of the scholar was the full comprehension of the principle (*ri*) at work in the universe (*tenchi*), then there could be no object too low or too humble. "There is no such thing as unworthy and vulgar things that do not deserve to be investigated,"[45] he wrote, since "the universe is full of things that still await to be investigated. The task of scholars is to become knowledgeable about the principle that moves all those things, one by one, until no doubts remain. This is the noblest activity that a human being can undertake. Nothing else gives human beings more pleasure."[46]

Ekiken's theory of knowledge began with a radical doubt: "After one studies one has doubts, after one doubts one has questions, after one questions one can think, after one thinks one can understand."[47] The scholar's mission was to understand the way in which the universe worked. This was a challenging task but was also the most "pleasant" of all human activities. Its reward was an understanding of human nature and the principle inherent in all things. "The way of the universe has always been in motion, it never stops."[48] "Everything moves following a principle (*ri*)."[49] This logic was the "heart (*kokoro*) of all living creatures in the world (*tenchi seibutsu*)," "the heart of all living creatures in the world becomes in human beings their humanness (*jin*)," and "humanness is the principle (*ri*) at work in human beings."[50] This principle requires human beings "to love (*ai*) all things in the universe" and, in turn, "to love all things in the universe means to uncover their way (*michi*)"—that is, the *ri* in action.[51] Just as *ri* (principle) never functions independently and is always contingent on materialization in plants, animals, rocks, and human beings, knowledge too could only be fully achieved if it helps human beings live their life in harmony with the universe. In other words, knowledge in practice was always moral.

Despite the clear reliance of *Yamato honzō* on Chinese pharmacopoeias, and in particular on *Honzō kōmoku*, some peculiar tracts differentiated it from its continental models. The most notable is the emphasis on the particularities of the Japanese ecosystem. There is a constant focus in *Yamato honzō* on indigenous species (their local utilizations and peculiarities) that distinguishes it from other, more universalistic encyclopedic or dictionarial works produced and circulating in Japan at the time. In the following chapters, we will see that this attention to local species, local conditions, and local traditions—by "local," meaning both a domainal and provincial level or a larger "national" level by the second half of the eighteenth century—is characteristic of most original *honzōgaku* production, in contrast to the sinocentrism of other disciplines and schools of thought like those of Itō Jinsai, Ogyū Sorai, and Minakawa Kien. Unlike scholars concerned with more abstract or theoretical issues like ethics or metaphysics, the study of nature forced its specialists to be aware of the materiality of his or her object of study. In the *sakura* entry we analyzed earlier, for example, Ekiken was very keen to distinguish Japanese species of cherry trees from Chinese and Korean ones not only in terms of their different uses but also in their morphological aspect (i.e., the red petals of the tree Chinese sources call *ying* are clearly different from the pale pink of Japanese *sakura*). The same is true for many other entries of *Yamato honzō*.

A second important characteristic of *Yamato honzō* is the relational, rather than essentialist or determinist, explanations given to phenomena. Ekiken did not follow the rigid determinism sometimes associated to Zhu Xi's Neo-Confucianism of the Hayashi school and of those scholars associated to Yamazaki Ansai that saw all phenomena theoretically reducible to a particular permutation of the metaphysical principles of *ri* and *ki*. Rather, he relied on explanations that emphasized processes, flows, and relations rather than essential natures of things. That is evident even in the short quotation of the *sakura* entry, when he mentioned that in Japan cherry trees are preferred to the Azusa trees to make wood blocks for printing because their wood is hard and of good quality, explaining that "the quality of the *sakura* wood" does not depend on its inner, fixed nature but "depends on the condition of the soil": it is the relation of cherry tree seeds and the chemical composition of the earth in which it grows into a plant that develops the potentiality of its timber, because, he commented, "this is undoubtedly another effect of the principle of the universe." The dialectical character of Ekiken's explanations of natural phenomena is distinctly revealed in expressions like "The way of the universe has always been in motion, it never stops."

This second characteristic may resemble the conventional descriptions of Ekiken as a scholar more interested in the concreteness of *ki* than in the abstractness of *ri*. Some historians have even defined his brand of Neo-Confucianism as a "monism of *ki*."[52] In my opinion, however, Ekiken's philosophy is closer to Hegel's notion of "concrete universality" than the caricature of him as a sort of English skeptical empiricist. Ekiken was in fact by no means disinterested in *ri*. He did not refuse the Neo-Confucian idea that all things in the universe are the way they are because of the working of the metaphysical principle *ri*. Ekiken was very precise in arguing that "everything moves following a principle (*ri*)," as he wrote in the introductory remarks of *Yamato honzō* quoted earlier, and that *ri* lies in the "heart (*kokoro*) of all living creatures in the world." *Ri*—a supreme example of dialectical principle putting everything in relation to everything else, in a constant flux—can be discerned only in the way things "moved," in the materiality of natural objects. For Ekiken, scholars have no choice but to infer the logic of *ri* from concrete situations, or particular instances, in the way each thing relates to everything else. It is not that human beings have no direct understanding of *ri* in its abstractness, but *ri* itself can exist only *in action*, conjoined with the material energy of *ki*.[53]

Ekiken's Neo-Confucian principles infused also his view of ethics and politics. He did not advocate a rigid dogmatism or a moral asceticism of the sort associated with contemporaries like Yamazaki Ansai and Satō Naokata. His ethics was situational. Just as the hardness of *sakura* wood depended on the type of soil in which the cherry tree grew, people should live their lives in conformity with their status and perform tasks that allow them to realize their inner potential. A scholar's duty was to clarify this responsibility and help people realize their own nature. The ethical commitment that imbued his political, agricultural, philosophical, and *honzōgaku* writings not only reflected his Neo-Confucian training but was also an expression of his social role as domanial scholars. The indivisibility of knowledge and practical ethics was, in other words, both a consequence of his adherence to Neo-Confucian metaphysics and an endorsement of his role of scholar-official. In fact, Ekiken succeeded in realizing what Razan had aspired to. He became a Neo-Confucian scholar who, as in Ming and Qing China, worked in active service of the government.

That Ekiken envisioned his scholarly production as a service to his domain does not undermine the value of his intellectual enterprise. European scholars to whom later generations would credit the beginning of the scientific revolution often worked in the service of European princes or of the church.[54] Galileo presented his 1610 discovery of Jupiter's satellites as a gift in honor of Cosimo de Medici. Indeed, concepts such as "knowledge for knowledge's sake" or "art for art's sake"—although often perceived as eternal truths—are historical products of modern times.[55]

Both early modern European and Japanese scholars involved in the study of nature saw their intellectual labor in the language of moral improvement and self-cultivation. Ekiken summarized this view of knowledge in his *Yamato zokkun*:

> The way to extend knowledge is first by knowing the way of the five constant virtues and the five moral relations. We should extend this to the principle of regulating the family and governing the people. Next we should seek to know the principles of all things and affairs. Since all things in the world lie within the ken of our mind, we should learn their principles. The way of investigating principles first gives priority to what is primary (the root) and what is close at hand and then follows up with what is secondary (the branches) and things farther away. We should not forget the order of sequence and priority.[56]

Investigating nature was one way to extend knowledge and therefore refine an individual's ethical thoughts and behavior. At the time *Yamato honzō* was published, nature studies were an integral part of Neo-Confucian education, and scholars volunteered to explore either nature or the many books on nature. But their community still remained small and their professional network undeveloped. That would soon change when, seven years after the first edition of Ekiken's masterpiece, Tokugawa Yoshimune became shogun in 1716.

THE CLASSIFICATION OF ALL THINGS: INŌ JAKUSUI AND THE WRITING OF *SHOBUTSU RUISAN*

Inō Jakusui's goal of surveying all species of plants and animals in Japan had important precedents. Kaibara Ekiken's *Yamato honzō* had the ambition to comprehensively list all natural species in Japan. Ekiken had also engaged in a survey of the natural resources in Chikuzen Province commissioned by Lord Kuroda, which he published as *Chikuzen no kuni zoku fudoki* in 1703. The flamboyant Abe Shōō is sometimes regarded as the inspiration for surveying projects.[57] Shōō was born in Morioka, in the northern province of Mutsu, probably around 1650. He claimed he studied medicine and *honzōgaku* somewhere in China for more than ten years after being shipwrecked there during a trip from Morioka to Osaka on a merchant ship. There is no evidence to support these claims, however, and most historians of science believe that Shōō invented this story to justify his authority in the field.[58] In any case, Shōō built up a reputation as a physician and herbalist, whose expertise in collecting medicinal plants in the area of the Inland Sea eventually attracted the attention of the shogunate.

Shōō's interest in plants and animals was limited to their medicinal application. He valued practicality over theoretical speculations and lexicographical research. He encouraged herbalist tours in the countryside rather than the exegetical study of manuals and encyclopedias. As he put it in a letter to a high-ranking official of the shogunate,[59]

> Let me humbly say that I began the study of medicine in my youth, and after I spent many years learning from more than ten encyclopedias of materia medica, there were still many things I did not know. I sought out the help of various people living in the Nagasaki area, and through their help I was able to obtain and study 55 or 56 different species of medicinal herbs from foreign countries. . . . So far I have described a little less than one thousand species. But these were all species of plants that I personally collected from their original habitat, quite different from the idle wordplay of those scholars in the Western provinces.[60]

Shōō's criticism targeted *honzōgaku* scholars from Kyoto, such as Inō Jakusui and his disciple Matsuoka Joan. Jakusui and Joan, today regarded as pioneer naturalists of Tokugawa Japan, never embarked on herbalist tours on the scale of those of Abe Shōō or, later, of Niwa Shōhaku or Uemura Saheiji. They adhered instead to what was the conventional practices of *jusha* and concentrated on philological and semasiological analyses of Li Shizhen's *Honzō kōmoku* and other Chinese encyclopedias, simultaneously limiting their outings to the outskirts of Kyoto. Indeed, Neo-Confucianism in general, and Japanese Neo-Confucianism in particular, never posited an explicit separation between a speculative and an empirical stance, between theoretical and practical knowledge. *Jusha* of different inclinations shared Zhu Xi's metaphysical belief in the comprehensibility of the universe, society, and human mind based on the logical principle, *ri*, and its intelligibility in material reality. That was true for scholars like Satō Naokata, who preferred quiet introspection, and scholars like Kaibara Ekiken, who reformulated Neo-Confucian thought in a dialectic of ethical, hygienic, and educational practices. Knowledge and understanding of the world, whether in a speculative, meditative, encyclopedic, or empiricist mode, always converted into practical action in the form of moral education or political guidance.

Inō Nobuyoshi (Jakusui) was born in a family of physicians and Neo-Confucian scholars in the service of the Nagai family of the Yodo Domain.[61] Jakusui's father was a renowned physician, and, as a result of his connections with the learned society of the imperial capital, his son received intellectual training from the best scholars of the time.[62] He studied Neo-Confucianism first under Kinoshita Jun'an and later under Itō Jinsai, medicine under his father, and *honzōgaku* under Fukuyama Tokujun—a scholar of Chinese materia medica active in Nagasaki in the second half of the seventeenth century. According to the hagiographic biography written after his death by his students, Jakusui precociously manifested a strong interest toward the natural world.[63]

In 1693, the twenty-eight-year-old Jakusui entered the service of Maeda Tsunanori, lord of the Kaga Domain, who provided him with a stipend but allowed him to continue to live, study, and teach in Kyoto. Tsunanori was a vigorous regional ruler who brought great economic and cultural development to his vast territory. He considered himself an amateur scholar and was patron to a large number of scholars and artists. He collected an impressive number of texts for the domainal library and sponsored the publications of manuals and encyclopedias. Tsunanori was well connected to various cultural circles of the capital and was a generous patron of *nō* companies and artists of literati-style painting, a status marker among his peers that gained him the favor of shogun Tsunayoshi. Tsunanori must have been impressed when he received a letter from Jakusui a year after he began his employment, in which the young and passionate scholar pleaded for Lord Maeda to accept his demands for books and financial support. In return, Jakusui promised to put forth every effort to reform the study of nature in Japan. His ambitious plan was to compile a comprehensive encyclopedia of all mineral, vegetable, and animal species of the world, a work so magnificent that it would compel Chinese and Korean scholars to purchase it.[64]

Jakusui died in 1715 before finishing his project, with 362 volumes completed (1,180 species treated) of a massive manuscript he titled *Shobutsu ruisan* (A Classification of All Things), written in Chinese. The encyclopedia was intended to bring glory to Japan as the greatest scholarly achievement of all of Asia. It was certainly the most ambitious encyclopedic project that had ever been attempted in Japan, but Jakusui was never short on ambition or self-esteem. In a passage of the letter previously mentioned, he stated that "there are about 1200 species of [plants and animals] in Japan that are described in Chinese sources, and I cannot think of any other person than myself who has really mastered what we can know about those species."[65]

Jakusui's project aimed to provide precise and final systematization for information about minerals, plants, and animals he collected from 174 Chinese encyclopedias. He intended to correct their mistakes and resolve all ambiguities stemming from synonyms, regional names, and historical and geographical variations. In his original plan, Jakusui wanted to classify all living things in the world into twenty-six groups. He outlined a first set of one thousand volumes with the names of mineral, vegetal, and animal species recorded in Chinese sources and another set of one thousand volumes with the names of minerals, plants, and animals indigenous to Japan. These he thought to record in phonetic order, because it was impossible to match them with their Chinese species counterparts. Jakusui resisted classifying Japanese names of plants and animals in the taxonomy he adopted mainly from *Honzō kōmoku*. This suggests that for him, only the Chinese language could convey the order of nature—or, more pragmatically, that he could authoritatively classify only Chinese names of plants and animals because he had at his disposal sources retraceable to the time of Shennong.

His work was supposed to be "so exhaustive that there will be no need of looking into other texts to find information about anything that exist(s) in the world."[66] For this reason, Jakusui intended to correct another misleading tendency found in all existing encyclopedias: "It is common of most encyclopedias of materia medica not to include all those things which do not have any pharmacological utility. Also, these encyclopedias are written to respond only to the needs of medicinal practice. They do not cover any discussion about the different names that species had in the past nor do they treat newly imported species."[67] *Shobutsu ruisan* was, in other words, purported to become the first comprehensive encyclopedia of natural history in East Asia.

The twenty-six groups (*zoku*) in which he contrived to group all things in the world (*shobutsu*) loosely followed the theoretical arrangement and Neo-Confucian rationale of Li Shizhen's *Bencao gangmu*: (1) herbs (*sō*, 100 volumes), (2) flowers (*ka*, 80 volumes), (3) fish (*rin*, 15 volumes), (4) shellfish (*kai*, 15 volumes), (5) feathery animals (*u*, 20 volumes), (6) furred animals (*mō*, 25 volumes), (7) waters (*sui*, 43 volumes), (8) fires (*ka*, 23 volumes), (9) grounds (*do*, 10 volumes), (10) stones (*seki*, 72 volumes), (11) metals (*kin*, 39 volumes),

(12) jewels (*gyoku*, 25 volumes), (13) bamboo (*chiku*, 36 volumes), (14) crops (*koku*, 39 volumes), (15) beans (*shuku*, 11 volumes), (16) edible grasses (*so*, 166 volumes), (17) seaweed (*kaisai*, 6 volumes), (18) fresh water weeds (*suisai*, 7 volumes), (19) fungi (*kin*, 10 volumes), (20) watermelons (*ko*, 8 volumes), (21) yeasts (*zōjō*, 20 volumes), (22) insects (*chū*, 95 volumes), (23) trees (*moku*, 50 volumes), (24) snakes (*da*), (25) fruits (*ka*, 55 volumes), and (26) spices (*mi*, 2 volumes).

For each entry, Jakusui provided an exhaustive philological analysis of the species name and its uses in classical sources. Morphological descriptions of plants and animals were reduced to a minimum and did not differ much from those provided earlier in *Bencao gangmu* or *Yamato honzō*. Jakusui's concern with "names" rather than "things" closely followed the scholarly pattern of his models. Observation of live plants and animals was not separated from their linguistic treatment in the archives of the Confucian classics. Indeed, observation could at best only complement lexicographical research. Hence Jakusui focused on lexicographical analyses not for their own sake but only because, by systematizing words and retrieving their original, foundational function of ordering reality, he could reconcile a proper relation between the human and the natural without relying on Zhu Xi's abstract metaphysical logic.

The primacy of textual and philological analysis in Jakusui's work was probably influenced by his training in Itō Jinsai's school. Jinsai was interested in a systematic study of words. His scholarly activities focused on recovering the original meanings of words in Confucian classics through accurate philological analysis. A close comparison of Jakusui's *Shobutsu ruisan* and Itō Jinsai's *Gomō jigi* (The Meaning of the Words in the *Analects* and *Mencius*, published in 1705) highlights these methodological similarities.[68] Jinsai struggled to deconstruct the layers of meaning that numerous generations of scholars had piled upon the words of two classical texts of the Confucian canon: the *Analects* and *Mencius*. His operation metaphorically resembles scraping a palimpsest to recover an original underlying text.[69] Time, Jinsai believed, had charged the words of those two great classics with additional meanings and connotations, which risked misleading contemporary scholars. By scraping away the spurious meanings that hundreds of years of commentaries had accumulated, he aimed to restore the original ethical message of Confucius and Mencius.

Similarly, Jakusui's effort focused on resystematizing the minerals, plants, and animals in the Chinese canon of materia medica by deconstructing the superimposed meanings to provide clearer order to the names and, consequently, the things themselves. At first glance both Jinsai and Jakusui works may be classified as examples of lexicographical treaties. Both works, in fact, had as their epistemological foundation a belief in establishing a firm connection between words and things. This belief was also a characteristic of Zhu Xi's Neo-Confucianism, which maintained the pervasiveness of the principle *ri* in all phenomena (natural as well as linguistic). Jakusui, however, proposed something more than mere adherence to the abstract categories of Zhu Xi's doctrine. Like his master Jinsai, he did not want to force words and things into a priori catalogs of correspondences, as he thought the Neo-Confucian scholars were doing. And unlike Ekiken, he did not derive his method from Zhu Xi's metaphysical argument (albeit centered on the concrete universality of *ki*). Jakusui searched instead for a method of interpreting the natural world with words that expressed it without being lifelessly abstract and that could become an active guide to human experience.[70] In other words, Jakusui's intention of understanding the natural order of the world through its concrete phenomena rather than through metaphysical concepts was a perfect elaboration of what historians call the "monism" of some Japanese Neo-Confucian scholars.[71]

According to this interpretation, Japanese scholars, as opposed to their Chinese colleagues, held a preference for *ki* over *ri*, on the basis that principle (*ri*) can be perceived only and insofar as it is materialized in substance.[72] Ekiken was an early representative of this tendency, as we have seen. Jinsai rejected Zhu Xi's "normativism" based on cosmological reductionism and wished to establish an ethical discourse that could recover the "'sense of right and wrong resulting from daytime exertion.'"[73] As he put it in *Gyūsan no ki zenshō kōgi* (Lectures on the Trees of Bull Mountain), "The mind of right and wrong is visible only after there has been contact with things. How can one glimpse it [the mind of right and wrong] before there has been contact with things?"[74] Moral behavior was not a mechanical adherence to instructions but needed to be constructed day by day by solving the concrete dilemmas of life on the basis of the teachings of Confucius and Mencius.[75] Similarly, Jakusui aspired to write a comprehensive guide to the natural world that could enable human beings to be in contact with things and make beneficial and ethical use of them. Contrary to Zhu Xi's a priori belief in the identity of words and things (because the metaphysical action of *ri* constituted both), Jakusui was struggling to give order and precision to a professional language by virtue of which it was possible to construct a methodical and precise understanding of the natural world based on an "original" ordering function of language itself—language being conceived, by both Jinsai and Jakusui, as the interface between human and natural worlds. As he wrote in *Saiyaku dokudan* (Handbook of Herbology),

Whenever I have spare time from my work on the Classics, I like to investigate the nature of birds, beasts, herbs, and trees, why they fly or lurk in the ground, how they move or grow. To that purpose, I climb steep mountains and enter deep valleys and wild fields. I gather [specimens] in woods and valleys and observe them in their minutest detail. If I meet somebody who has passed through remote regions I have not visited, I interview him about what he saw in order to have a sense of the spirits that inhabit the trees and herbs of that place.[76]

"Climbing steep mountains and entering deep valleys" was a metaphorical image, since we know from Jakusui's biographies that he spent his life in cities like Kyoto, Kanazawa, and Nagasaki and never organized herbal expeditions of the scale of Abe Shōō, Niwa Shōhaku, or Uemura Saheiji. Jakusui was interested in systematizing the immense load of new information arriving from China and Korea, yet he did not conceive of his work as a theoretical or speculative enterprise, nor did Jakusui intend to write an encyclopedia that could be used by general populace—as in the case of Ekiken's works and Nakamura Takisai's *Kinmō zui*—or as a practical pharmacopoeia. Jakusui's research was motivated by a different, concrete concern for reality, a new awareness of the world in its phenomenal appearance for which the ordering of names was meaningless unless accompanied by a systematization of the material referents of those names. In Jakusui's work, however, that could happen only through a return to a founding moment, to an original operation of naming of material reality that only a scholar could retrieve through a philological excavation of ancient texts.

From the second half of the seventeenth century through the first half of eighteenth century, a similar critique of Neo-Confucian a priori reductionism of macro- and microcosmos was shared by a number of scholars in different disciplines. In the field of medical studies, it took the form of a critique of the Goseihō methods, deeply influenced by Zhu Xi's system, which ended up dividing practicing physicians into two opposed fronts. This movement was called Koihō (Ancient Medical Method), and it would be influential in the development of *honzōgaku*. The Goseihō was the mainstream school of medicine sponsored by the shogunate, based on Zhu Xi's cosmology.[77] Physicians of this school utilized a series of very detailed charts to record the correspondences of forces dominated by *yin*, *yang*, and the Five Phases with food, organs, and diseases. Furthermore, they prescribed cures on the basis of these charts. By the second half of the seventeenth century, the Goseihō was criticized by a new current of medical thought and practice. Koihō adherents—who adopted different cloth-ing and hairstyles from the Goseihō practitioners (who usually shaved their heads) by dressing like common samurai—called for a return to the spirit and methods of the ancient medical classics.[78] As Sugimoto and Swain explain, "The Ancient Practice [Koihō] advocates were not totally rejecting the broad basic assumptions of correspondence between the human organism and the cosmos, or even the use of *yin-yang* and Five Phases concepts to describe bodily func-tions. . . . What they objected to . . . were schemas constructed with more con-cern for metaphysical symmetry and function than for observable (but seldom observed) physical structures and actions, as well as therapeutic emphases de-rived in one-sided ways from such elaborately abstract theories."[79]

The Koihō school was initiated in Japan by the seventeenth-century physi-cian Nagoya Gen'i. Gen'i emphasized the importance of the actual observation of symptoms in each patient, how each patient reacted to specific treatments, and the circumstances in which diseases were contracted. The treatments he prescribed were intended to reestablish the optimal circulation of *ki* in the hu-man organism—in Japanese, *genki*, literally "restoring the *ki*." Each person, he claimed, contracted a disease in a different and personal manner depending on his or her peculiar *ki*, hence cures needed to be individually tailored and the ef-ficacy needed to be measured separately in each case.

The establishment of Koihō school was a major turning point not only in medical practice but also in scholarly methodology. The influence of the Koihō methods of inquiry on scholars like Itō Jinsai and Ogyū Sorai is hard to dismiss. Ishida Ichirō has argued that scholars like Itō Jinsai and Motoori Norinaga—themselves physicians who approved and adopted Koihō methodologies—should be considered indebted to Nagoya Gen'i's medical treatises, but his claims still await more thorough research.[80] Jinsai and Sorai rejected Zhu Xi's a priori metaphysics and advocated the return to the original meaning of the Confucian classics through painstaking philological research.[81] Norinaga applied a similar philological methodology to the classics of Japanese literature in his efforts to recover an original Japanese ethos.[82] Medical methodology became a metaphori-cal vehicle for the development of this new epistemology, and Jinsai and Nori-naga often adopted it when speaking, for example, of Neo-Confucianism as a disease to be cured.

In the realm of *honzōgaku*, this methodological turn meant that just as every patient had to be cured individually, so each phenomenon and specimen had to be analyzed singularly. Just as every patient got sick in particular ways and responded to treatment differently, so natural phenomena also happened singularly and particularly. Not only was the symbiosis of the micro- and macrocosmos disrupted, but the rules that were supposed to govern that symbiosis were no longer acceptable. Similar to what happened in Europe at the time of Galileo, the book of nature began to be thought of as having its own language, no longer reducible to Latin or Chinese.

Inō Jakusui was probably exposed to Nagoya Gen'i's empirical methodology of medical practice through his master Jinsai or through other physicians and *honzōgaku* scholars active in Kyoto. Among these, Gotō Gonzan, Kagawa Shūan, Yamawaki Tōyō, and Yoshimasu Tōdō were strong and vocal supporters of the Koihō school.[83] Kagawa Shūan recorded in *Ippondō yakusen* (Ippondō's Selection of Medicaments) his observations about the medicinal properties of herbs, and they often contradicted the established assumptions of Chinese encyclopedias of materia medica. Yoshimasu Tōdō was the physician and *honzōgaku* scholar who most emphatically advocated the importance of observation and experiments with herbal and animal substances. "The task of medicine," he maintained, "is to cure people of disease. Spending time in futile and unnecessary discussions is but a hindrance to this simple truth. If one is not able to cure a patient, what kind of physician is he? It is for this reason that those who practice medicine have to practice and polish their practical knowledge."[84] Tōyō's empirical attitude is further evidenced in these statements of methods, taken from his *Yakuchō* (Pharmacological Demonstrations, 1784), which endeavored "to be a medicine for medicine": "I would never utilize a medicine before I first tested it myself," or "It is not sufficient that a cure has been successful once: one needs to see it succeed at least three times before defining it a successful treatment."[85]

Inō Jakusui's monumental endeavor was based on textual authorities and organized in the style of Neo-Confucian lexicography. Nevertheless, it should be understood in the context of the growing awareness of a phenomenal reality no longer reducible to the a priori schematics of Zhu Xi's thought. At the time of his death, Jakusui was able to complete 362 of the planned 1,000 volumes, describing about 1,180 species of fish, shellfish, feathered and furred animals, and fruit-bearing trees. Compared with the Chinese sources he utilized, his major achievements concerned the orders of fish and shellfish, where a large number of indigenous species were added. Ueno commented, "The value of *Shobutsu ruisan* for natural history is not much different from that of *Yamato honzō*. And beside the broad scope of its conception, it is difficult to discern any novelty in the classificatory and analytical treatment of species. Its greatest merit was however that it succeeded in systematizing and ordering the tremendous number of sources into a corpus with an internally consistent organization."[86]

Inō Jakusui and Kaibara Ekiken were both working to develop the discipline of *honzōgaku* in an original way. They each came up with similar solutions to help scholars manage information, but their ideas had some distinctive features. Ekiken developed a manual based on a body of knowledge selected mainly from *Honzō kōmoku* and complemented with personal investigations in the field and with information he collected from fellow scholars, peasants, fishermen, and travelers. His manual was primarily designed to be of practical utility to the population he served, but it was still conceived of in the framework of Neo-Confucian epistemology of *kakubutsu chichi*—an investigation of things aimed at uncovering the laws of the universe, the concrete universality of *ki*, laws that had for Ekiken both a cognitive and a moral value. Jakusui aimed at creating order out of the enormous and sometimes inconsistent mass of information through a detailed philological and lexicographical analysis of the entire canon of Chinese texts. This was done in order to establish a reliable language and classificatory system that would allow future naturalists to identify species with precision without incurring the confusion that names and words might have created. In other words, both Ekiken and Jakusui believed their works would answer practical needs, notwithstanding their differences in approach, style, and purpose.

PART III *Inventorying Resources: 1716–36*

*Every accumulation of knowledge and especially such
as is obtained by social communication with people over
whom we exercise dominion . . . is useful to the state.*
— *Warren Hastings, letter to N. Smith*

In 1696, Hayashi Nobuatsu, also known as Hōkō, the head of the Hayashi school of Zhu Xi's studies, received a grant from the bakufu to move the school to the Yushima ward of Edo. The move to a better and larger location, along a water channel near the modern-day station of Ochanomizu, marked the official enrollment of the school, renamed now Shōheikō, to the rank of official academy for the training of shogunal bureaucrats. It was under Hōkō's leadership that the school's faculty and students were allowed to abandon the custom, imposed by Tokugawa Ieyasu, of wearing Buddhist robes and shaving their heads, a symbolic recognition that scholars (*jusha*) had acquired a distinct socially recognized identity.

If the metropolitan areas of Kyoto and Osaka (Kamigata) were the dominant centers of cultural production throughout the seventeenth and early eighteenth centuries, networks of scholars were also active in other Japanese cities.[1] Each network centered on an inner circle of learned men with similar inclinations and interests. A large portion of these scholars consisted of practicing physicians, as the profession of *machi isha*, established in the early sixteenth century, granted them a constant income. This was a recurring pattern for a large portion of scholars throughout the Tokugawa period, especially for those who did not enter the service of a domainal lord like Kaibara Ekiken. Some of them succeeded in publishing their work, usually paying the printing expenses themselves when they could not count on outside patronage. Their activities included public lectures; visits to other schools; and performances alongside storytellers, fortunetellers, *waka* masters, jongleurs, and street performers (fig. III.1). In this sense, they followed a tradition that had existed since the days of Fujiwara Seika and Hayashi Razan more than a century before.

The most successful scholars were able to collect enough money from one or more patrons—usually wealthy merchants or high-ranking samurai—to open their own private schools (*shijuku*). These institutions instructed pupils in the ideas of the founding master through a system of secret transmission (*hiden*), much like in artisanal and artistic workshops.[2] The professionalization of intellectual expertise transformed in a sense culture into a variety of artisanal craft, often blurring the distinction of intellectual and manual labor.

In the seventeenth century, *honzōgaku* saw steady quantitative and qualitative growth. More than four hundred texts related to *honzōgaku* were produced throughout the century, and even more were introduced from China and Korea.[3] The thorough investigations and the meticulous descriptions of plants and animals in these texts have suggested to modern scholars that *honzōgaku* had by then earned distinction as a discipline in its own right.[4] Yet at the beginning of the eighteenth century, there were still no *honzōgaku* scholars who defined

FIGURE III.I.
A public
lecture on the
Japanese classic
Tsurezuregusa
from Jōkanbō
Kōa's *Imayō
heta dangi,*
published in
Edo in 1752.
Tokyo: Waseda
University.

themselves professionally as such, nor was *honzōgaku* acknowledged as an autonomous field of natural studies. In seventeenth-century Japan, the study of nature was considered part of the pharmacological training of physicians or, to some scholars, an empirical complement to moral self-cultivation through the "investigation of things." Large sections of dictionaries and encyclopedias were devoted to herbs, trees, and animals. Aesthetically compelling depictions of flowers and animals—whether rare specimens from private collections or the magnificent gardens of high-ranking samurai and wealthy commoners—increased in quantity and quality. Still, when Kaibara Ekiken published his *Yamato honzō* in 1709, *honzōgaku* was hardly recognizable as an independent field of study.

This does not mean that nature knowledge enjoyed less consideration than other fields. On the contrary, as shown in the previous chapters, encyclopedias were in great demand. *Honzō kōmoku* and *Yamato honzō* remained the two fundamental textbooks in the training of physicians. Popular encyclopedias like *Kinmō zui* and *Wakan sansai zue* were regularly reprinted and updated. These and other multivolume works, often richly illustrated, were part of a popular genre of illustrated manuals or pictorial dictionaries known in Ming and Qing China under the label of *riyong leishu* (*nichiyō ruisho* in Japanese), or "everyday encyclopedias." While they could not be compared to the more scholarly sophisticated *Honzō kōmoku*, they shared the same concern to order knowledge into a systematic and hierarchical system (the "Chinese classificatory system of knowledge").[5]

In 1716, two years after Ekiken's death, Tokugawa Yoshimune, lord of the Kii Domain, became the eighth Tokugawa shogun. His reign brought fundamental changes to the study of nature in early modern Japan. Under his rule, the bakufu sponsored scholars like Abe Shōō, Noro Genjō, Uemura Saheiji, Aoki Kon'yō, Niwa Shōhaku, and Tamura Ransui as collaborators and executors of a massive reform project aimed at restructuring shogunal control over the land and coping with financial and agricultural difficulties that arose after the economic bubble of the preceding Genroku period (1688–1704). Yoshimune's plan involved these scholars in three main projects: to produce a complete survey of all species of plants and animals that could be found in Japan, to implement agricultural technology in the cultivation of alternative pest-resistant crops and vegetables sustainable in times of famine, and to establish a state-sponsored medicinal garden to supply internal demand for pharmacological substances and enable Japan to end its reliance on Chinese and Korean imports. Moreover, by loosening the shogunal control on the importation of Western books, Yoshimune contributed to expand the reach of *honzōgaku* scholars to goods, information, and texts from Europe.

Through the financial and political support of the shogunate, the discipline of *honzōgaku* acquired enough institutional authority and financial prosperity to sustain a community of specialists. *Honzōgaku* practitioners and the study of plants and animals both acquired cultural value due to natural history's connection to matters of economic livelihood and national prosperity. Governmental support and patronage transformed *honzōgaku* specialists into shogunal officials, employees of the Office of Japanese Pharmacology (Wayaku aratame kaisho) or, later on, of the Institute of Medicine (Igakukan). By the end of the eighteenth century, a division between state-sponsored and amateur naturalists had been established, further contributing to the formation of an autonomous field of natural study.

Shogunal patronage of nature studies in the mid-eighteenth century played a vital role in changing the social composition of *honzōgaku* practitioners, shifting the study of plants and animals from a subsidiary pursuit of generalist scholars and physicians to a specialized discipline of its own. Yoshimune's unusually active rule was largely responsible for this evolution. This had important consequences in the ways scholars studied, understood, manipulated, and explained nature. After a survey of Yoshimune's impact on bakufu's economic politics and scholarly world (chapter 6), chapter 7 chronicles the shogunal sponsorship of colossal surveying projects aimed at collecting, describing, and cataloging all species of plants and animals growing and living in the Japanese archipelago.

6 Tokugawa Yoshimune and the Study of Nature in Eighteenth-Century Japan

∙∙∙∙∙∙∙∙∙∙∙

AN ACTIVE RULER

When the thirty-six-year-old Yoshimune was named shogun in 1716, he found himself at the command of the Tokugawa state without any connection to the clique of senior councilors (*rōjū*) that dominated the government in Edo. Yoshimune, an enterprising and experienced leader who had ruled the Kii Domain as *daimyō* since 1705, was soon able to secure his control over the Edo government by placing trusted men in key offices and avoiding any interference from scheming senior councilors.[1] Throughout his twenty-nine years of rule, Yoshimune adopted a realpolitik that practiced concrete measures over ideological considerations and reigned through a network of trusted men appointed to strategic positions. Conrad Totman, in comparing Yoshimune's policies to those of an earlier shogun, remarked that "whereas Tsunayoshi had, for better or worse, adhered to policies anchored in a sturdy ideology, Yoshimune bobbed and weaved in response to circumstances."[2]

Yoshimune's vigorous regime, at odds with the tendency of previous shoguns to allocate authority to collegial organisms (*rōjū* and *wakadoshiyori*, senior and junior councilors, respectively) and to fragment power among local administrations of liege vassals (*hatamoto*), magistrates (*daikan*), inspectors and superintendents (*bugyō*), and domainal lords (*daimyō*), had a striking impact in the complex structure of the Tokugawa governmental system. While largely upholding feudal structures, he concentrated on various forms of politico-legal coercive control, mainly by extending the bakufu's influence and legal jurisdiction over autonomous domains and intensifying its role in organizing the economy. Under Yoshimune the shogunate tightened its control over the domainal lords. To circumvent the interference of Edo's powerful *rōjū* clique governing on behalf of his underage predecessor Ietsugu, Yoshimune abandoned the collegial system embraced by previous shoguns, which had consisted of collaboration between shogun and councilors in the decision-making process, and relied instead on collaborators he had brought from Kii who served as his special agents (*goyō toritsugi*). He appointed them to the posts of deputies (*bugyō*) and depended on them to implement his policies. He also rewarded direct (*hatamoto*) and collateral (*fudai*) retainers with whom he had personal ties of friendship and often entrusted them with important diplomatic responsibilities.[3] In short, Yoshimune built up a government that had as its pillars capable men who were linked to him by personal ties of loyalty.

Yoshimune's administrative system, completely at odds with Tsunayoshi's reverence for formality and etiquette modeled upon the Chinese imperial state, had to face the problems inherited from his predecessors: currency devaluation, inflation, agricultural shortages, and a negative commercial balance with China. Yoshimune vigorously addressed these difficulties aided by his staff of loyal assistants. He maintained strict control over financial policies and coinage (monetarism) and was active in directing domainal policies, controlling the lords' actions through a complex network of inspectors (*metsuke*) and spies (*niwaban*) under his direct order.[4] He raised the tax burden of the domains to fund agricultural development and public works and extended shogunal control over merchants by codifying wholesale commercial organizations. Called *kabu nakama*, they became licensed oligopolies in the form of chartered trade associations and granted groups of artisans and merchants commercial monopolies for specific goods.[5]

Education received more attention and funding from Yoshimune's government than under any other Tokugawa shogun. He increased the number of village schools (*gōgakkō*) and granted disciplinary monopolies to private teaching schools.[6] He raised state funding to the Hayashi school, the Shōheikō, and extended state sponsorship to Sugeno Kenzan's Kaihodō and Nakai Shūan's Kaitokudō merchant academy.[7] Yoshimune loosened the ban on Western books "to improve access to books containing information on astrology and calendar-making, along with useful learning."[8]

It was, however, in the realm of agricultural technology, administration, and innovation that Yoshimune's reform measures proved the most dynamic. He enacted new regulations for village society and new methods for settling land disputes. He also implemented a series of demographic censuses and product surveys, promoted technological innovations, and encouraged the cultivation of new crops such as sugarcane and sweet potatoes. These policies were all designed to help rural populations prevent or cope with famines caused by crop failures, especially after the rural crisis of 1732–35.[9] It was as part of his agricultural reforms that Yoshimune sought the collaboration of *honzōgaku* specialists.[10]

As Totman commented, "Yoshimune was a 'hands-on' ruler who saw effective administration as the heart of good governance."[11] Historians often describe his rule in terms that evoke political realism, realpolitik, and state control over the economy.[12] Their accounts highlight Yoshimune's ability to design political measures and to adapt them to the contingencies he faced.[13] This political realism is clear in his agricultural and fiscal policies, both of which influenced the history of natural history in early modern Japan.[14]

From the point of view of its economic developments, the contradictions that exploded at the time Yoshimune took power offer an interesting insight to understand the socioeconomic dynamics of early modern Japan. Since its establishment, the Tokugawa state displayed the typical distribution of power and "organic unity of economy and polity" of feudal societies.[15] The main source of wealth for the ruling elites came from the imposition of taxes on agricultural production, enforced by the military, legal, and political coercive power of the Tokugawa state apparatus. To ensure the constant flux of surplus crops, the bakufu had developed a *honbyakushō* system, the division of all cultivable land into small plots for independent peasant proprietors that the samurai authorities could better control and squeeze with taxes on average annual productivity. The organization of peasant population into family units of small landowners, the prelude of which can be traced back to the agricultural reforms of Toyotomi Hideyoshi at the end of the sixteenth century, was further consolidated by a series of edits that prohibited the sale and buying of land (*Denpata eitai baibai kinshi rei* of 1643) and its subdivision (*Bunchi seigenrei* of 1673) and meticulously regulated the lifestyle of peasants (*Keian no furegaki* of 1649).[16] Because each family of peasant landowner (*honbyakushō*) was to be granted a plot of land sufficient to enable their livelihood and reproduction, both central and local administrations organized large-scale campaigns of land reclamation to satisfy the demand of fertile terrains that the steady growth of population required. Agronomical improvements like crop rotation and more efficient fertilizers intensified agricultural production. Local magistrates (*daikan* and *gundai*) organized the extraction of tax surplus in each district, but village headmen (*shōya*) had the responsibility to collect tax revenues annually. This created social divisions and class conflicts inside village population. Moreover, the actual reclamation of land required additional intensive corvée labor from the already squeezed *honbyakushō*. The resulting conflicts punctuating the first half of the Tokugawa period (usually in the form of *daihyō osso*, or "representative suits," and less frequently *sōbyakushō ikki*, or "peasant riots") caused "the amount of tax revenue to decline or at best to fluctuate."[17] It also put in motion a process of social differentiation among the peasant population between peasant landowners (*honbyakushō*) and former landowners that were forced to secretly sell their lands and become tenant farmers (*mizunomi hyakushō*) to fulfill their fiscal and corvée obligations, to pay off debts accumulated with wealthier moneylending peasants (*gōnō*), or to simply survive, transforming them into an expanding population of agricultural tenants and wage laborers.

The Tokugawa authorities also actively promoted the expansion of commercial activities and the adoption of a money economy, favoring a progressive separation of economy and polity that would eventually "erode the foundations of the entire Bakuhan system."[18] The commercialization and monetization of Japanese economic life were a result both of the growth of large urban centers out of medieval castle towns, which were in constant need of supplies from the countryside, and of the expenses that local authorities (*daimyō*) had to face to maintain roads, boats, bridges, and docks in order to facilitate swift communications (required by the 1635 additions to the *Buke shohatto*) and to organize their annual processions to Edo (*sankin kōtai*). Bakufu and domains attempted to maximize their control over commerce by licensing wholesale merchants with monopoly rights to trade in certain cash crops in exchange for *myōgakin*, or "thank-money." This was in part motivated by their desire to extend their control over certain specific products of their own domains—transforming them into monopsonists—and to ensure a constant flux of cash-money to the domainal treasures. However, considering that the conversion of rice and other cash crops into money was controlled by bakufu-authorized wholesale merchants independently from samurai interferences, the ruling elites found themselves increasingly impoverished and forced to squeeze even more taxes from peasants, a dynamic that, in turn, accelerated the stratification of peasant population.

Improvements in fertilizers and agricultural techniques (crop rotation, seed selection, the utilization of more efficient machines) sustained economic growth and limited the effects of samurai impoverishment, at least until the Genroku period (1688–1703). The expansion of big urban centers like Osaka, Kyoto, and Edo, whose population was in constant demand of commodities, caused the domestic trade to swell and reach the size and speed of such proportion that "the samurai control over commerce began to slip away."[19] The establishment in 1697 of the rice market of Dōjima accelerated the autonomy of the market and "economic factors" from political coercive control of the samurai elites.[20] The use of paper money, in circulation since 1710, and the development of new advanced forms of investments in future agricultural yields (*nabemai*) had destabilizing effects on commodity prices (including the exchange value of tax rice), making them fluctuate and creating situation of sudden currency devaluations and inflation of commodity prices. These developments had disastrous effects on the fiscal balances of both bakufu and han, which were forced to raise taxes and thank-money and, after the agricultural crises of the Kyōhō era (1716–35), to increasingly borrow from moneylending merchants and *gōnō* to pay retainer stipends and finance domainal expenses.

Upon installation as shogun, the first major problem Yoshimune had to face was the reform of the currency system he inherited from his predecessors. Tsunayoshi, forty years earlier, had begun a devaluation policy aimed at reaping the profits of a massive minting of gold and silver currencies. In the long run, this measure had disastrous consequences for the economy. Inflation and a sudden collapse of the trading system (an early modern form of stagflation) were the direct results of the depreciation in currency, but the increasing demand for currency resulting from expanding commerce in medicinal and luxury goods with China rewarded, in the short period, the devaluation policy of the shogunate and the irresponsible policies continued by Tsunayoshi's successors.[21] The mounting demand for currency that the commercialization and monetization of Japanese society demanded, together with a sudden decrease in output of silver and gold from Japanese mines and the continuous drainage of currency through trade with China and the Dutch, made it impossible to stabilize the value of gold and silver and control inflation.

The Neo-Confucian scholar and shogunal official Arai Hakuseki firmly opposed Ogiwara Shigehide, the official responsible for the devaluation policy under Tsunayoshi.[22] Hakuseki advocated stabilizing the currency and tightening shogunal control over imports to cut losses. He invoked Neo-Confucian doctrines to emphasize how political measures should not be adopted for the benefit of the shogunate but for the welfare of the entire nation. Through his arguments, he convinced shogun Tokugawa Ienobu to disburse 130,000 *kanme* from the shogunal treasury to restore the value of currency to its level prior to Ogiwara's time. Significantly for the future of *honzōgaku*, Hakuseki suggested that the most successful way to control imports and, consequently, the drainage of currency from Nagasaki was to improve agricultural production of such import goods as raw silk, medicinal products like ginseng, and sugarcane. A restriction on the volume of imports of these goods was initiated concurrently with a proposal to lighten restrictions on the imports of Chinese books, particularly manuals and encyclopedias of practical utility, which were necessary to boost internal production. Hakuseki's reform project, known as *Shōtoku shinrei*—"New Regulations of the Shōtoku Era (1711–1716)"—was met with the strenuous opposition of the majority of senior councilors still committed to Ogiwara's devaluation program. However, his efforts eventually attracted Yoshimune's attention, despite his feelings of displeasure over Hakuseki's role as advisor to his predecessors and his lack of interest in Hakuseki's reputation as a brilliant Neo-Confucian scholar.[23]

BOTANICAL GARDENS, GINSENG, AND SWEET POTATOES: YOSHIMUNE'S SPONSORSHIP OF *HONZŌGAKU*

Yoshimune eventually endorsed Hakuseki's proposal and began a massive program of agricultural reforms.[24] In 1717, only one year after he had assumed the title of shogun, he ordered the expansion of the shogunal medicinal garden of Koishikawa (Koishikawa yakuen) for the purpose of launching domestic production of ginseng. His new initiative did not start until 1719, when Lord Sō, daimyo of Tsushima—the official mediator for shogunal relations with Chosŏn Korea—produced drawings of the ginseng root (*chōsen ninjin—Panax ginseng*) and detailed information about the plant and how to cultivate it. The shogunal supervisor for the ginseng project was Hayashi Ryōki, the official physician of the shogun. According to Imamura Tomo, Ryōki was secretly involved with the Sō lord in attempts to smuggle live ginseng roots into Japan from the Korean peninsula, which was prohibited by Chosŏn authorities under penalty of death.[25]

Yoshimune's promotion of national production of medicinal herbs was focused on two interrelated projects: the cultivation of ginseng and other medicinal herbs in gardens under the shogunal supervision and small- to large-scale surveys of the botanical and animal produce in the Japanese islands. The amateur *honzōgakusha* and professional spy Uemura Saheiji was directly involved in the surveying project, together with a number of other physicians and apothecaries. The project involved long tours collecting the herbs of different regions, the study of the properties and methods of cultivation of different medicinal herbs, and the formalization of the newly acquired knowledge in manuals. All this was sponsored by the shogunate and often closely monitored by Yoshimune himself.

The herbal expeditions and surveys that Yoshimune sponsored led to the establishment of medicinal gardens (*yakuen*). In these gardens, shogunal physicians conducted planting and interbreeding experiments on herbs and trees. Botanical and medicinal gardens were established throughout the archipelago, either under the patronage of the shogunate as proto-official institutions, like the Komaba and Koishikawa gardens, or sponsored by domainal lords eager for Yoshimune's favor. Their managers competed against each other for the most successful cultivation of rare herbs from China or plants from the most remote regions of the Japanese archipelago. These were certainly not "utopia's gardens"—spaces like the Jardin du Roi in Paris defined by rules of conduct and self-presentation and aimed at preserving a community and fraternity of scholars.[26] They did, however, represent a spatial realization of an epistemological ordering of nature modeled upon *Honzō kōmoku*. They were places of practical application of *honzōgaku* theories and descriptions and a concrete reproduction of a loosely Neo-Confucian "order of nature." They also became places where theoretical knowledge was tested in practice and sometimes rejected through struggles to grow sugarcane, sweet potatoes, ginseng roots, or other rare and precious medicinal herbs that would protect the state from famines, crop failures, and the drain of silver coins. In that sense, the administration of the garden could be encoded in a Neo-Confucian language, whereby the government of nature—in the gardens—and the government of society intertwined and the gardens could become part of the central or domainal administrations.[27]

The shogunate had sanctioned two small medicinal gardens on the ground of the Edo castle since the seventeenth century: one on its southern edge, named Azabu goyakuen under the supervision of Ikeda Michitaka, and the other along the northern perimeter, named Ōtsuka goyakuen, under the supervision of Yamashita Sōtaku. The northern garden was closed in 1681, while the southern one was transferred in 1684 to the estate of Lord Koishikawa and renamed Koishikawa goyakuen. Similarly, in Kyoto two medicinal gardens were opened in 1640 in the north and in the south of the Imperial Palace. In 1680, the city magistrate of Nagasaki also ordered the establishment of a municipal garden. Soon, medicinal gardens were constructed in many castle towns of the main islands. Private medicinal gardens were far less numerous and never acquired much notice, with the exception of the Morino yakuen, founded in the village of Matsuyama, near Nara, by the *honzōgaku* scholar Morino Fujisuke. Fujisuke, the son of a wealthy peasant, was hired by the shogunate in 1729 as staff to Uemura Saheiji's herbal expedition in the Yamato region. As compensation for his services, he received seedlings and small plants that became the core of his medicinal garden. Licorice roots, Chinese tallow trees, a Chinese subspecies of Judas trees, Benjamin bush roots, Japanese cornel fruits, and ginseng were the herbal species in which he specialized.[28]

The opening of a medicinal garden was an activity closely related to herbal expeditions, and both often received shogunal sponsorship. Many of the *honzōgaku* experts who collaborated with Yoshimune spent a period of study and research in a medicinal garden. In 1729, Abe Shōō was granted a small plot of land in the Kanda ward of Edo, where he experimented in planting and growing sugarcane, kapok trees, *koganeyanagi*, and ginseng.[29] Niwa Shōhaku was in charge of a medicinal garden on the northeast of Edo, around which a small town soon formed, bearing today the name of Yakuendai. Uemura Saheiji was in charge of the Komaba yakuen.

The most important of all medicinal gardens in Eastern Japan was the Koishikawa, which expanded in the eighteenth century to become the leading horticultural testing and research area for pharmacological plants. Here, various *honzōgaku* specialists on the shogunal payroll conducted experiments with sugarcane (from 1734), sweet potatoes (from 1735), and a large number of other herbs and trees.[30] By the second half of the eighteenth century, Koishikawa became an important step in the formation of many *honzōgaku* trainees, who had the chance to observe the live herbs and trees they had only previously been able to study in encyclopedias.[31] They also had opportunities to experiment with planting and interbreeding techniques.

The history of the Koishikawa garden, from its beginnings as a shogunal *medicinal* garden (*yakuen*) to its transformation into the botanical garden (*shokubutsuen*) of Tokyo University after the Meiji Restoration of 1868—and, later, into the center for botanical research of the University of Tokyo—is interesting and worth researching in its own right.[32] A scholar often associated with the Koishikawa garden during Yoshimune's era was Aoki Kon'yō. As his historical nickname suggests, Kansho *sensei*—or "master sweet potato"—was the scholar responsible for the success of sweet potato's cultivation after 1734. Son of a wealthy fish merchant, Kon'yō studied Neo-Confucianism in Kyoto under Itō Jinsai's disciple Itō Tōgai and eventually opened a private academy in the Hatchōbori ward in Edo. His friendship with Katō Enao, amateur poet and shogunal inspector (*ginmigakari*), enabled Kon'yō to establish fruitful connections with influential personalities of the shogunate and leading scholars of the time. It was through his connections that Kon'yō was asked to write a short essay on sweet potatoes, *kansho* (*ganzhe*, in Chinese)—also known as *Satsuma imo* because it was originally imported from the Ryūkyū Islands through the Satsuma Domain.[33] *Banshokō* (On Sweet Potatoes) was not an original study, but it summarized the information provided in a collection of Chinese encyclopedias of materia medica on sweet potatoes.[34] These were aimed at explaining the nourishing virtues of the tuber during times of famine. Kon'yō's manual was so favorably received that in 1734, Yoshimune ordered him to edit a Japanese version, which was published at shogunal expense in 1735 under the title *Satsuma imo kōnōsho narabi ni tsukuriyō no den* (The Efficacy of Sweet Potatoes and

Their Preparation). In the same year, Kon'yō started his experiments with sweet potatoes at the Koishikawa garden.[35]

The plant that became the focus for the greatest efforts and research was, however, ginseng.[36] Ginseng was the most expensive Japanese import from China and Korea. The thick root of the ginseng plant, whose Chinese characters can be literally translated into "the human root," was steamed and dried. Chinese medicine utilized ginseng in its powdered form as a "heal-all."[37] Its rootstock was believed to increase the body's resistance to stresses such as trauma, anxiety, and bodily fatigue—in other words, to "restore" the proper movements of *ki* in the body—and it was a stimulant and aphrodisiac.[38] Known in Japan since the fifteenth century, ginseng was imported almost exclusively through the mediation of the Sō lords of Tsushima, an island strategically located in the middle of a sea channel between the southern edge of the Korean peninsula and Japan. Until the mid-eighteenth century, Japanese imports of ginseng reached as much as 1,200 kilos per year. To give a sense of its value, the wholesale cost of one dose of ginseng—three to five grams in its powdered form—was about one *ryō*.[39] In 1674, the shogunate established a guild of merchants, the *ninjinza*, headed by Matsuoka Izaemon, as the only merchant group authorized to sell ginseng in Japan.[40]

The Goseihō school of medicine regarded ginseng as one of the most important drugs, because they believed it had the capacity to regulate the flux of *ki* in the body and therefore strengthen its resistance to pathogens. Ginseng was so popular that it became the subject of many *senryū*—a poetic genre that followed *haikai* metrics but focused on ironical and farcical subjects. It was in fact so expensive that the proverb *"ninjin nonde kubi kukuru"*—literally "drink ginseng and then hang yourself"—was very popular in the period. The proverb warned people against acting without taking into consideration the costs.[41]

As the internal demand for ginseng grew and the monetary drainage to Korea and China continued, Yoshimune encouraged and sponsored research to favor its indigenous cultivation. He ordered Lord Sō to import illustrated manuals and encyclopedias on ginseng. In a letter dated the twenty-third day of the tenth month of 1721, Suzuki Sajiemon, ambassador of the Tsushima Domain in Edo, informed the senior councilor Mizuno Tadayuki that he had successfully smuggled three live roots into Edo from the Korean peninsula.[42] The next year, another six were imported, followed by four in 1727 and eight in 1728. All attempts to transplant the roots failed.[43]

The first *honzōgaku* specialist who tried his luck with ginseng cultivation was probably Uemura Saheiji in the Komaba medicinal garden. In 1728, after almost eight years of failures, Saheiji was given sixty ginseng seeds that the Sō of Tsushima again had smuggled into Edo. Saheiji tried to plant them in a different terrain than had been previously tried. Ginseng plants grow better in shady and moist grounds, rich in minerals and well drained. Saheiji chose the Nikkō area, more precisely a field purposely prepared near the village of Imaichi.[44] The seeds eventually grew into plants that bore fruit and, consequently, produced new seeds. In 1733, a second generation of ginseng seeds were planted. *Honzōgaku* scholars were getting close to being able to grow ginseng indigenously, but only a small percentage of seeds took root. In 1737, a new experiment was attempted. Twenty of the newest seeds obtained from the original Imaichi plants were passed to another physician in the bakufu payroll, Tamura Ransui. Ransui succeeded three years later in producing a quantity of new seeds sufficient to distribute to the various medicinal gardens. In 1737, Ransui published a booklet detailing his experiments with the ginseng seeds and providing instructions on how to plant and nourish them. Titled *Ninjin kōsaku ki* (Records of the Ginseng Cultivation), it was widely distributed throughout the country. By the 1780s, Japan was producing so many ginseng roots that it started exporting them to the Qing. The trade was controlled directly by the shogunate through the Tsushima Domain.[45]

STATE NATURALISTS

All the scholars involved in Yoshimune's projects had received medical education and many had practiced privately as town physicians (*machi ishi*). Niwa Shōhaku, before his appointment in 1722 as shogunal physician (*ikan*), practiced medicine in Edo, and the majority of his patients were commoners. Shōhaku, like Noro Genjō, was a physician native to Ise and a disciple of Kyoto *honzōgaku* scholar Inō Jakusui. Niwa, Noro, and Uemura were all from Ise, more specifically from a town near the city of Matsusaka, in Yoshimune's Kii Domain. Their prominent role in shogunal agricultural policy may be explained by the fact that they were all men Yoshimune could trust because of their ties to the Kii Tokugawa family. Together with Abe Shōō and Tamura Ransui, they played a significant role in the state-sponsored project of improving domestic production of medicinal herbs.

Prior to their involvement with the bakufu, these men were not specialists in *honzōgaku*. They acquired their knowledge of medicinal herbs through private practice as physicians or, as in the case of Uemura Saheiji, private study as a personal pastime. It was only through shogunal sponsorship and their involvement in Yoshimune's project that they acquired the institutional title of *ikan* or *saiyakushi* (herbalists). The shogunate's endorsement of the emerging field provided *honzōgaku* specialists with the social and institutional capital necessary to assume autonomy from the fields of medicine and Neo-Confucian studies. Concurrently, shogunal patronage influenced *honzōgaku* methodology by encoding it in terms of usefulness for the entire country (*honpō*). In his *Yamato honzō*, Kaibara Ekiken had already structured the study of nature in terms of public utility, but his tenets had been put into practice only within the Fukuoka Domain.[46] It was Yoshimune who turned the study of nature into a national affair.[47]

The developments in botanical studies under Yoshimune illustrate how he strove to control the administration of the country personally or through a network of trusted and capable men at his direct command. Yoshimune assumed direct control over provinces that belonged to the shogunate (*tenryō*). For the remaining areas, Yoshimune followed a fourfold strategy. First, he started a massive development program of newly cultivable lands (*shinden kaihatsu*), which were all placed under direct shogunal jurisdiction: by 1722 the total percentage of arable land had increased by an impressive 12.2 percent.[48] He reformed taxation in a new system called *jōmen* (fixed-rate), which set the tax rate of each village for a fixed period of three, five, or ten years, after which a new census of land productivity would correct upward the taxation rate. Second, Yoshimune reorganized economic life into self-regulating professional groups and guilds called *kumiai*, which could be more effectively controlled through shogunal rule and imposed limitations on them.[49] Third, Yoshimune increased taxation at the domainal level and forced the *daimyō* to reinvest large portions of their treasuries in public works and agricultural improvements. Finally, he developed a complex system of espionage and information gathering comparable to the one developed by Sir Francis Walsingham for Queen Elizabeth I of England. This network was aimed at controlling possible threats to his authority from local powers as well as the circulation of information around the country.[50] Yoshimune's policy was intended to achieve his goals as well as to maintain control over scholars and the knowledge they produced.

In the early modern world, in Europe as well as in East Asia, states increasingly turned their attention to collecting, controlling, censoring, and manipulating information as a means to strengthen practices of government in internal and foreign affairs and protect the interests of the state from internal and external threats. The history of espionage, or, more precisely, the history of networks devoted to the collection and manipulation of information for government's sake, began with the emergence of early modern states.[51] In Europe, the

Church served as a model of network building and information gathering. The Council of Trent in 1563 determined that every parish should keep registers of births, marriages, and death.[52] In Japan, the Tokugawa shogunate entrusted these matters to Buddhist temples. The emergence of "interrogatories" in Protestant countries during the second half of the sixteenth century and the Inquisition in Spain and Italy may be regarded as first attempts to create archives with information about allegedly subversive figures.[53] The worldwide Jesuit network also functioned as a model for mercantile and diplomatic information-gathering agencies.[54] The Republic of Venice was the first European state to adopt a system of resident ambassadors in the major European political centers, who sent regular dispatches (*relazioni*) to Venice with information they gathered through a network of local informers.[55] Soon, the major European powers imitated the Venetian system and organized similar institutions. The British government established embassies in Milan and other continental capitals. The Spanish government, under its ambassador in Venice Don Diego Hurtado de Mendoza (1503–75), formed one of the most complex spy networks in the Ottoman Empire. During this period, information was collected from inside and outside the state. This system of intelligence gathering contributed to European domination of other parts of the world.[56] "The early modern seaborne empires—Portuguese, Spanish, Dutch, French and British—all depended on the collection of information."[57] "Domestic spies or 'informers' were another tool of government" in most states.[58]

Tokugawa Japan under Yoshimune was not much different. His complex network of spies and informers (*ometsuke* and *oniwaban*) were mostly recruited from among his Kii retainers or among capable men from the same region. They controlled a large volume of intelligence about domainal lords, their administrative decision making, their treasuries, the productivity of their lands, and the ratios of tax entrances and public investments. Yoshimune's control over information affected his policy toward education and learning as well; his shogunate soon became known for its encouragement of scholarly activity and the lifting of the ban on foreign books. While Yoshimune encouraged the expansion of learning, especially in those fields with beneficial and practical applications, he also directed the reorganization and restructuring of the private academies. He laid the foundation for state control over intellectual activities by providing exclusive licenses for the teaching of determined subjects to specific schools. These included sponsorship of the Hayashi, the Sugeno, and Nakai schools. Yoshimune himself set an example when, beginning in 1717, he started attending classes at the Hayashi school.[59]

The first half of the eighteenth century saw a dramatic increase in intellectual production. Neo-Confucian scholars like Muro Kyūsō, Dazai Shundai, Hattori Nankaku, Itō Tōgai, Arai Hakuseki, Hayashi Hōkō, and Sugeno Kenzan, just to name a few of the best known, continued the tradition started a generation earlier by their respective masters. Thanks to Yoshimune's support, they were able to solidify the position of their schools in the field of intellectual production. Kyūsō developed his personal interpretation of Zhu Xi's philosophy from the teaching of his master Kinoshita Jun'an. His reputation expanded after Yoshimune hired him as a Neo-Confucian scholar (*jukan*).[60] Shundai, Nankaku, Tōgai, and Kenzan continued their masters' legacies and schools (and mutual competition), specifically in regard to Ogyū Sorai (Shundai and Nankaku), Itō Jinsai (Tōgai), and Satō Naokata (Kenzan). In the meantime, the scholar and Shinto priest Kada no Azumamaro also enjoyed the support of the shogunate and began what would be later called the School of National Learning (*kokugaku*).[61]

Another repercussion of Yoshimune's politics toward education and learning was the increased number of nonsamurai scholars engaged in the intellectual life. Yoshimune endorsed the opening of the shogunal Hayashi school to commoners, who could enroll in the Shōheikō after paying a fee. Soon, students of mixed social backgrounds alternated days of attendance with elite children in schools officially sponsored by the shogunate (*kangaku*). Beginning in 1723 the shogunate sponsored a large number of *shijuku* that were connected, either directly or indirectly, with the Shōheikō—which thus became protogovernmental institutions, *junkangaku*, as was the case, for example, of the Kaitokudō in Osaka in 1726.[62] Yoshimune also expanded the shogunal library of Momijiyama, ordering the various domainal lords to provide copies of rare books from their own libraries.[63] He ordered the acquisition of large numbers of Chinese and Dutch books and dispatched trusted emissaries and scholars to search for rare books in the provinces.[64] Aoki Kon'yō was sent in 1740 on a three-year mission to collect rare manuscripts from various provinces.[65] Yoshimune then ordered a recataloging of all books of the Momijiyama library into a more logical and efficient order.[66] He also commanded the publication of detailed maps of Japan, most notably Takebe Katahiro's *Nihon sōezu* (A Complete Map of Japan), completed in 1723.[67]

DUTCH STUDIES

Another field of study that advanced because of the support of Yoshimune was *rangaku*, or "Dutch studies," which focused on the study of Western learning.[68] Beginning in the early years of the shogunate, Tokugawa authorities had issued a series of edicts restricting and controlling contacts with Western nations.[69] By the mid-seventeenth century, Dutch traders were the only Westerners authorized to reside in the trading station of Deshima in the Nagasaki bay. In time, a community of more or less competent translators of Dutch formed in Nagasaki, all under direct control of the shogunate. The import of goods and books—mostly Chinese translations of Western texts—although limited, was constant.

After Yoshimune partially lifted the ban on Western books in 1720, the number of texts arriving in Japan skyrocketed. The community of Dutch translators grew, and some of their Japanese translations became bestsellers. It was in this period that the term *rangaku* began to identify a network of scholars who specialized in Western learning, most of whom began as scholars of Neo-Confucian thought. Some of the Dutch learning scholars engaged in an exchange with *honzōgaku* specialists, but it was only in the late eighteenth and early nineteenth centuries that Western natural history really began to influence the ideas and practices of Japanese naturalists.

The first significant encounter between Japanese *honzōgaku* and European natural history dates back to the third month of 1659.[70] The *kapitan* (chief, *opperhoofd*) of the Deshima station of the Dutch East India Company (Vereenigde Oostindische Compagnie, or VOC, literally "United East India Company"), Zacharias Wagenaer, offered shogun Tokugawa Ietsuna a copy of Rembert Dodoens's *Cruijdeboeck* (History of Plants, 1618). Records reveal that the eighteen-year-old shogun was unimpressed by the Dutch manual's small and mediocre illustrations, a sign perhaps of the taste for visual representation that had become so important in later Japanese *honzōgaku*. He nonetheless graciously accepted the gift from his foreign guest. He also requested that Wagenaer bring a more richly illustrated volume on his next visit.[71] Four years later, in 1663, the new Dutch *opperhoofd* Hendrik Indijk brought Ietsuna a copy of Johann Jonston's 1660 edition of *Naukeurige beschryving van de Natuur der vier-voetige dieren, vissen en bloedlooze water-dieren, vogelen, konkel-dieren, slangen en dranken* (Accurate Description of the Nature of Four-Footed Animals, Fish, and Bloodless Water-Animals, Birds, Insects, Snakes and Dragons). This volume was unremarkable as far as content was concerned, but it was decorated with nearly three hundred of the best xylographs of rare and exotic animals produced in Europe. No one in the shogunal entourage could read either of the two texts, and they were eventually cataloged and deposited in the shogunal library of Momijiyama.

Rembert Dodoens (Rembertus Dodonaeus) was a Flemish court physician of the Holy Roman Emperor Rudolph II who then became a medical professor at Leiden in 1582. Dodoens was deeply influenced by his teacher Leonhart Fuchs, who is remembered today as a prominent representative of a generation of phytographers concerned with a comprehensive study "of natural and imported plants and animals of the area where they lived, and identifying them with creatures described by the ancients."[72] His major work, the Flemish *Cruijdeboeck* (History of Plants), was first published in 1554 with 715 illustrations of plants divided into 6 groups.[73] This edition was later expanded in 1644.[74] Dodoens's approach to *res herbaria*, as the study of plants was called at the time, was characteristic of a period of natural studies marked by the rediscovery of ancient texts by such authors as Dioscorides and Theophrastus.[75]

One of the major problems Dodoens and his contemporaries faced was coping with the abundance of new information about plants inundating Europe as a result of the exploration of previously unknown regions of the world.[76] The introduction of Dodoens's *Cruijdeboeck* showed the anxiety that this information overload produced among sixteenth-century European scholars: "The magnitude or difficulty of this science is such that it cannot be comprehended without early and careful examination of all plants and exact reading of many ancient writers—that is, without great labor, long travel, and continuous devotion.... For no one has brought this science to perfection: everyone has omitted many things, thereby leaving to posterity the opportunity to add much to the discoveries and observations of their predecessors and to increase the knowledge of plants."[77]

Johann Jonston, "a well-traveled and prolific writer on natural history who has been described as a weak successor to Aldrovandi," was born in Poland of Scottish origins.[78] A professor of medicine at Leiden, Jonston closely followed in the tradition of Ulisse Aldrovandi.[79] Jonston's *Historiae naturalis de quadrupedibus* (Natural History of Four-Footed Animals), printed in Amsterdam by J. J. Schipperi in 1657, made little contribution to European knowledge of exotic animals. This academically insignificant volume would have been doomed to obscurity had it not possessed some of the best illustrations circulating in Europe since the fifteenth century. Some of these remarkable illustrations have been attributed to such artistic talents as Albrecht Dürer and his atelier.[80]

Dodoens's and Jonston's encyclopedias remained buried in the shogunal library until 1717, when the new shogun, Yoshimune, ordered the two books to be thoroughly investigated.[81] It was Jonston's illustration of an Asiatic elephant that inspired Yoshimune to request the VOC captain Abraham Minnedonk to produce a live white elephant in 1728.[82] Yoshimune's interest in the two texts is corroborated by the diary of Minnedonk's successor, Jan Aouwer, which records how Yoshimune asked the new VOC *opperhoofd* a number of questions about Jonston's book, many of which he was unable to answer.[83] That same year, Yoshimune ordered the scholars Aoki Kon'yō and Noro Genjō to study the Dutch language. Genjō, like most of the scholars and administrators under Yoshimune's orders, was born in a small village in the Ise Province and had studied Neo-Confucianism under Namikawa Tenmin, a disciple of Itō Jinsai, and *honzōgaku* under Inō Jakusui.[84] Genjō's involvement with the shogunate became official in 1724, when he was granted residential status as official physician. He was later appointed personal physician to the shogun (*omemie ikan*) in 1739. In 1741, Genjō began working first on Jonston's *Naukeurige beschryving* and then on Dodoens's *Cruijdeboeck*, appointed with the official title *Oranda honzō goyō* (officer for Dutch materia medica). His research consisted of interview sessions with Jacob van der Waeijen and Philip Pieter Musculus, respectively *opperhoofd* and surgeon of the VOC, in their Edo mansion. Genjō was assisted by the shogunal chief translator (*ōtsūji*) Yoshio Tōzaburō.[85] They met in the third month of 1741 and produced a slim booklet, *Oranda kinjū chūgyo zu wage* (Illustrated Explanations in Japanese of Dutch Birds, Beasts, Insects, and Fish). This publication did not make a significant contribution to Japanese zoological knowledge, since it consisted of little more than a list of names with Japanese translation.[86] Even when a longer "explanation" (*ge*) was added, the information was decidedly rough, as illustrated by the following entry for *ōrihare* (a rendition of the Dutch *olifant*, or elephant), translated as *zō*: "The length of the elephant, measured from the tip of its nose, can reach one *jō* and nine *shaku*;[87] these animals exist in great number in countries visited by the Dutch. However, in none of these countries do they regard their meat as edible. Neither have the Dutch heard that their skins, bones or dung are used as medicine. However the tusks, called *ihōruto* [*ivoor*, ivory], are used for medical purposes."[88] The medicinal specification in this record was related to *Honzō kōmoku*'s entry on elephants, which reports that its intestines were utilized in some regions of Asia for medicinal purposes.[89]

The fact that the text failed to satisfy Genjō can be grasped from his preface of his booklet, where he explained, "This book was not meant as a herbal guide to identify plants with medicinal properties. Therefore it does not give any description about the medicinal effects of plants, but only about their shape or appearance. Even the Dutch said that they could not understand the contents properly because the explanations included so many Latin words."[90] Genjō, it should be remembered, was employed by Yoshimune, along with Niwa Shōhaku, Uemura Saheiji, Aoki Kon'yō, and others, for the specific purpose of developing

Japanese domestic production of medicinal herbs and survival crops. Yoshimune ordered the translation of Jonston's text to collect information about medicinal substances from alternative sources—Dodoens's and Jonston's encyclopedias were probably the only ones available at the time.

Shirahata has suggested that Genjō did not translate Jonston's *Naukeurige beschryving* because "he must have lost interest in it."[91] My guess is that Genjō was not in a position to make such a decision, and it was probably an order from Yoshimune or his entourage. Shirahata correctly recorded that the next step for Genjō was to coauthor *Shin'yū Oranda honzō* (Dutch Materia Medica of 1741) with Yoshio Tōzaburō, the translator Nakayama Zenzaemon, and his assistant Shige Shichirōzaemon. The following year the group also collaborated on *Oranda honzō* (Dutch Materia Medica), a report of their interviews with the Dutch delegation about eleven plants they had selected (roses, grapes, rice, and maize).[92] Shirahata's comment that, after 1741, "nobody showed interest in Johnston's [*sic*] book" is however mistaken.[93] The part that first captured Yoshimune's imagination, Jonston's beautiful illustrations, would in fact inspire later *honzōgaku* scholars and artists. Sō Shiseki, a painter who studied under the Nagasaki painter Shen Nanpin, produced illustrations for his album *Kokinga sōgo hachi shu* (Eight Old and New Species in the Wild, 1771, figs. 6.2 and 6.4) inspired by Jonston's *Naukeurige beschryving* (figs. 6.1 and 6.3).

Fascination with Jonston's illustrations was not limited to Edo and the group of *honzōgaku* scholars connected with the shogunate. Tani Bunchō—a well-known artist of the literati painting style (*bunjinga*, or *nanga*, the "southern school of painting")—found inspiration (fig. 6.7) from one of the most famous of Jonston's images, reproducing the rhinoceros of Albrecht Dürer (fig. 6.6). Dürer had carved the original woodcut in 1515 (fig. 6.5) from a sketch he made of an Indian rhinoceros that the king of Portugal, Manuel I, sent as a gift to Pope Leo X.[94] Dürer never saw the actual animal and based his sketch on a written description.[95] Dürer's rhinoceros was the source of a chain of copies dating from its original publication and lasting until the early nineteenth century.

Dodoens's influence on Japanese natural history was deeper and longer lasting than Jonston's. Noro Genjō and his entourage of translators continued visiting the Nagasakiya, the Edo residence of Dutch envoys, even after their interview sessions on *Naukeurige beschryving*. They met a total of eight times between 1742 and 1750 to translate and discuss Dodoens's *Cruijdeboeck*, the result of which was *Oranda honzō wage* (Explanations in Japanese of Dutch Materia Medica), published in eight booklets, each of which recorded the proceedings of one meeting. *Oranda honzō wage* was not a complete translation of *Cruijdeboeck*

FIGURE 6.1.
Lion, from Johann Jonston,
Naukeurige beschryving
(1660). Edo-Tokyo
Hakubutsukan.

but instead consisted of notes on the morphology, habitat, cultivation, and utilization of 106 species of plants selected from the Dutch encyclopedia. Genjō's overarching practical concern is evidenced by the fact that only the pharmacological use of the herbs was exhaustively translated from Dutch to Japanese.[96]

The private sponsorship of a complete translation of Dodoens's herbal, along with state-sponsored translation of other Western texts in the early nineteenth century, were ineffective attempts to cope with a lack of sophisticated, up-to-date knowledge of European natural history. Or at least this is the general reasoning among historians of science to explain shogunal sponsorship of the translation of two works that at the time had become outdated in Europe.[97] By the end of the seventeenth century, Japanese *honzōgaku* had come into direct contact with Western natural history. European surgeons residing in Deshima established relationships with Japanese scholars, primarily interpreters and *rangaku*

FIGURE 6.2. Jonston's lion from Sō Shiseki,
Kokinga sōgo jū shu (1771). Edo-Tokyo Hakubutsukan.

scholars. Many of these foreign surgeons were also amateur or even professional naturalists. In English and Dutch maritime expeditions, surgeons were often also responsible for the collection of vegetable and animal specimens.[98] It was via these contacts that information about Western natural history was introduced into Japan. Engelbert Kaempfer, surgeon of the VOC between 1690 and 1692, was able to observe Japanese nature while stationed in Deshima and later wrote a best-selling report of his travels. The fifth *fasciculus* of his *Amoenitatum exoticarum* (On the Amenities of Exotic Places), printed by H. W. Meyer in 1712 and titled "*Plantarum Japonicarum, quas Regnum peragranti solum natale conspiciendas objecit, Nomina & Characteres sinicos; intermixtis, pro specimine, quarundam plenis descriptionibus, unà cum Iconibus*" (Of Japanese Plants, which Grow Only in This Country; with Their Names in Chinese Characters, and with Detailed

FIGURE 6.3.
A dromedary and a
camel, from Johann
Jonston, *Naukeurige
beschryving* (Amsterdam,
1660). Edo-Tokyo
Hakubutsukan.

Descriptions and Illustrations), reproduced illustrations from Nakamura Tekisai's *Kinmō zui* and was celebrated by European botanists, as we have previously
seen. It expanded the number of animal and plant species known in Europe and
also confirmed the existence of a species of gingko plant in Japan. This was significant news to Europeans. The Japanese gingko was discovered to be the only
surviving species of plants from the Ginkgophyta family, which once flourished
in the Permian period and was considered extinct by European naturalists. Seedlings of gingko Kaempfer brought back to Europe were planted in the Utrecht
botanical garden and caused a popular craze in eighteenth-century Europe.

Although Kaempfer failed to participate in a natural knowledge exchange
with local scholars during his Japanese residency, this bias was in marked con-

FIGURE 6.4. Jonston's dromedary from Sō Shiseki,
Kokinga sōgo jū shu. Edo-Tokyo Hakubutsukan.

trast to his successor some eighty years later. The Swedish physician and naturalist Carl Peter Thunberg, Linnaeus's protégé on a quest to classify all living species of the world using the binominal system, stayed in Japan for only sixteen months. During this brief span of time, he conducted an intellectual exchange with a group of Japanese scholars interested in Dutch studies and *honzōgaku*. Among them were Nakagawa Jun'an, a physician of the Obama Domain, and Katsuragawa Kuniakira, a shogunal physician, who had also studied Dutch medicine and collaborated with Sugita Genpaku in editing *Kaitai shinsho* (New Book on Anatomy, 1774), a Japanese translation of Johann Adam Kulmus's *Ontleedekundige Tafelen* (Anatomical Tables).[99]

Jun'an and Kuniakira frequented the Dutch residence in Edo (Nagasakiya)

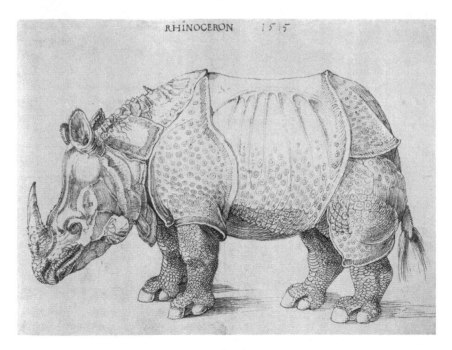

FIGURE 6.5. Albrecht Dürer, *Rhinoceros* (1515).

FIGURE 6.6. A copy Dürer's rhinoceros from Jonston's
Naukeurige beschryving (1660). Edo-Tokyo Hakubutsukan.

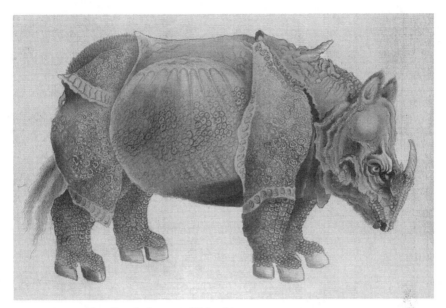

FIGURE 6.7. Tani Bunchō's copy of Dürer's rhinoceros, *Saizu* (1790). Private collection.

on a nearly daily basis beginning in the fourth month of 1776.[100] Jun'an, in particular, sustained a great interest in *honzōgaku*, demonstrated by his friendly exchanges with Gennai during the organization of his fifth pharmacological exhibition in 1762 and his assistance throughout the editing process of *Butsurui hinshitsu* (A Selection of Species, 1763). He collected and grew precious botanical specimens, exhibiting them in Edo and in later *honzōgaku* events in Osaka and Kyoto. The intellectual friendship between Jun'an and Thunberg continued well after the Swedish naturalist returned to Europe, where Thunberg eventually took a professorship in medicine and botany at the University of Uppsala (following in his master Linnaeus's footsteps). His *Flora Japonica*, published in Latin at Lipsia in 1784, was the first classification of Japanese vegetation according to Linnaean taxonomy, recording more than 530 species of plants drawn from those collected during Thunberg's herbalist trips, from specimens sent to him by Jun'an after returning to his home country, and from observations of Kaempfer's collection owned by Sir Joseph Banks.[101]

Thunberg's intellectual exchange with Japanese scholars left hardly any trace in Japan, and it is unknown whether he taught Jun'an Linnaean classificatory methodology. Speculations abound regarding the extent of their intellectual exchange, fueled by the fact that Jun'an and Kuniakira did not leave any notes

FIGURE 6.8.
Gingko's leaves from
Engelbert Kaempfer,
Plantarum Japonicarum
(1712). Tokyo: National
Diet Library.

or recollections about the nature of their discussions with the Swedish naturalist.[102] Thunberg also left no significant records. This deficit may or may not
have been purposeful. Linnaeus's "supposedly objective system" was designed
to be easy enough for anyone to use after only a basic training.[103] However, the
system's practical application would play a very important part in developing
the economic domination of European powers over the world's riches. As Fara
comments, "Some economists argued that God had scattered His riches around
the Earth in order to encourage international trade, but Linnaeus was convinced
that God intended Sweden to prosper by providing all its needs within its own
borders. He sent off his disciples to scour foreign countries for useful materials,
even recommending them to smuggle goods into Sweden if necessary."[104] The
Englishman Joseph Banks exploited Linnaeus's system in his voyages and, "as
head of the Royal Society and confidant of the King, Banks was in a unique
position to show how scientific research could make Britain's growing empire
even more profitable."[105] It is fair to assume that Thunberg, as one of Linnaeus's

direct disciples, conceived of his role of naturalist as a mission for the welfare of his home country.[106] Science, by the late eighteenth century, had been transformed into imperial science, and its "secrets" — especially when they concerned national interests — were not openly disclosed.[107]

Thunberg might have mentioned Linnaeus's classificatory system, even explained it in some detail, only to be met by lack of interest on the part of Japanese naturalists. *Honzōgaku* scholars already had a well-developed classificatory system, derived from the *Honzō kōmoku*, which had withstood the test of time. The Japanese may have seen no reason to adopt a foreign classificatory system. Taxonomies, as Harriet Ritvo has shown, reproduce not only the natural world in a supposedly natural or artificial order but also the cognitive world of classifiers. In Victorian England, it was not rare that naturalists "preferred to inscribe their favorite political categories in the book of nature," and often "nationalistic commitments added further complexity to a technical debate that was already vexed, slippery, and divisive."[108] In other words, *honzōgaku*'s classificatory system had been already shaped by the cultural environment that accompanied and favored its development. Rejecting such a system would not have been a simple matter.[109]

7   *Inventorying Nature*

. . . .

CHANNELING LUCK, CONTROLLING RESOURCES

Ihara Saikaku (1642–93), chronicler of merchant life in early modern Japan, shared his contemporaries' belief that chance was the most important source of business prosperity: "Fortune be to merchants, luck in buying and happiness in selling!" he recited at the beginning of his *Seken munezan'yō* (Worldly Mental Calculations, 1692).[1] On the other hand, Saikaku's urban heroes nicely fit John Stuart Mill's definition of *homo economicus* when they act in his merchant tales as "a being who desires to possess wealth, and who is capable of judging the comparative efficacy of means for obtaining that end."[2] His shopkeepers, artisans, and wholesale traders are well aware that if chance is a fundamental component of success, there are also means at the disposal of ingenious and industrious businessmen to catalyze luck: "Getting rich," Saikaku explained, "is a matter of luck, we say; but this is simply an expression. In point of fact, a man builds his fortune and brings prosperity to his family by means of his own wit and ingenuity."[3]

The preoccupations of Saikaku's townsmen with financial prosperity reveal the degree of how by the end of the seventeenth century, the lives of Japanese of all social classes had increasingly adjusted to the logic of the market and of profit. As we have seen in the previous chapter, the monetization of trade that had begun in the fifteenth century encompassed now almost all economic activities. With the establishment of the Dōjima Rice Exchange (Dōjima kome ichiba) in 1697, the logic of the market came to exercise a strong influence over political affairs simply by controlling the exchange rate of rice (the *formal* unit of measurement of wealth and power among the samurai political elites) into money (the *actual* measurement of wealth in the larger field of economic transactions). In other words, the incremental separation of the political and the economic spheres—a structural creation of the compound regime of the Tokugawa that Yoshimune's reforms accelerated—created a situation in which the wealth of the ruling elites was increasingly at the mercy of the ups and downs of the market.

A modified version of this chapter appears as "Inventorying Nature: Tokugawa Yoshimune and the Sponsorship of *Honzōgaku* in Eighteenth-Century Japan," in *Japan at Nature's Edge: The Environmental Origins of a Global Power*, ed. Brett Walker, Julia Adeney Thomas, and Ian J. Miller (Honolulu: University of Hawaii Press, 2013), 189–206.

Channeling luck into more profit and managing misfortunes with backup plans was certainly a central concern for wealthy peasants, artisans, and merchants, as Saikaku's lively stories show. As the growing volume of trade in international and, especially, domestic markets required the ruling samurai elites to exercise an active role in the economic life of the various domains, the necessity of "managing" luck became a matter that called for political attention.

This chapter reconstructs the organizational effort of a massive survey campaign of all vegetal and animal species living and growing in Japan, carried out under shogunal oversight between 1734 and 1736 and resulting in statistical data unprecedented in world history for its quantitative and qualitative magnitude.[4] Niwa Shōhaku, the mind behind the formidable organization of the resource inventory, acted under the aegis of Tokugawa Yoshimune to mobilize retainers of all domains to survey even the smallest villages. Like one of Saikaku's shopkeepers, Shōhaku seemed to agree that to channel luck and secure profit, one had to take control over one's labor and resources. The first days of the New Year, Saikaku instructed, is "the time for merchants to bind their ledgers, take inventory, and open the vaults to inspect their silver." What Shōhaku achieved was the creation of a "national ledger" where an inventory of the natural riches of the state could be recorded in an orderly fashion.

THE CLASSIFICATION OF ALL THINGS

Shobutsu ruisan, left unfinished by the death of its author Inō Jakusui in 1715, was a first attempt to give a comprehensive classification of all existing species of minerals, plants, and animals. It was, as we have seen, a very erudite enterprise of monumental proportions that attempted to give an ultimate systematization to information about plants and animals contained in the most authoritative encyclopedias of natural knowledge produced in China. It was never published, and after Jakusui's death none of his students continued his work. In 1719, Maeda Tsunanori donated the original manuscript to the shogunal library. At the turn of the nineteenth century, chief librarian Kondō Morishige recorded in his memoirs that on "the 11th day of the 9th month of 1719 Matsudaira Kaga-no-kami [Maeda Tsunanori] offered a copy of *Shobutsu ruisan* as a result of a direct request from the shogun to Lord Kaga-no-kami the 29th day of the 7th month of the same year."[5] The intermediary agent of the transaction was Hayashi Hōkō, the head of the Shōheikō. According to Ueno, Yoshimune appreciated the importance of Jakusui's work and regretted that Tsunanori did not charge anybody with the task of completing his encyclopedia.[6] It was only in 1734 that Niwa Shōhaku, a former student of Jakusui and official physician of the shogunate for twelve years, began the compilation of the remaining 638 volumes of Jakusui's original plan. He named, described, and classified a total of 3,590 species of plants and animals and specimens of various kinds of metals, earths, stones, and jewels. To bring the project to completion, shogunal authorities issued a permit allowing Shōhaku to move freely through the country. He had free access to all territories and was entitled to receive assistance from local authorities whenever needed. Shōhaku's mission was preceded by an ordinance dispatched by the senior councilor Matsudaira Norisato to all domainal lords, temples, and regional magistrates: "In the case that the physician Niwa Shōhaku, in charge of the revision and expansion of *Shobutsu ruisan*, would need it, it is requested that all deputy magistrates [*daikan*] of shogunal territories [*tenryō*], all Lords [*ryōshu*] and estate managers [*jitō*] of private domains [*shiryō*], and all administrative officials of territories controlled by temples and shrines [*jisharyō*], comply to any form of request regarding names, forms, typology of natural products [*sanbutsu*] in all provinces of the realm that the aforementioned Shōhaku may have."[7]

Niwa Shōhaku was born in Matsusaka in 1691, in the Ise Province, at that time a part of the domain controlled by the Kii Tokugawa family of Yoshimune. His father was a physician of samurai origins. Shōhaku was destined to follow in his father's professional footsteps and did so for several years until he decided to study *honzōgaku* under Inō Jakusui. He distinguished himself among Jakusui's students for his interest in fieldwork and herb collecting, activities that differentiated him also from his master's lexicographical interests.[8] He practiced as a town physician (*machi isha*) in Edo between 1720 and 1721, when the shogunate first contacted him to conduct herbal expeditions in the mountainous regions of Nikkō and Hakone. His enlistment as official physician of the shogunate followed in 1722.[9]

Shōhaku is today mostly known for bringing *Shobutsu ruisan* to completion and for organizing the first official survey of Japanese natural resources on a national scale. It is, however, worth mentioning that his first activity as shogunal physician was to establish an Office for Japanese Pharmacology (Wayaku Aratame Kaisho), which was responsible for the importation, cultivation, distribution, and instruction regarding pharmacological herbs. Shōhaku's task was to facilitate an agreement among the major pharmaceutical dealers (*yaku don'ya*) to organize an oligopolistic cartel under the control of the office. In 1729, the Wayaku Aratame Kaisho sponsored the publication of a manual coauthored by Shōhaku that was designed to introduce the general population to the uses of 155 herbs in the preparation of drugs, primarily general analgesics and antipyretics.

In 1734, Yoshimune officially commanded Shōhaku to finish *Shobutsu ruisan* along with a national census of vegetable and animal life in Japan. These surveys were conducted on a regional level in the form of *sanbutsuchō* (product reports) that every domain was required to complete and dispatch back to Edo. Of the original official letters that Shōhaku sent in 1735 to the Edo residences of domainal lords requesting the compilation of a survey of the natural riches of their provinces, many are lost, but we may have a glimpse of their contents from a surviving one sent to chief retainer Hanabusa Iemon of the Fukuoka Domain. The letter reproduced Shōhaku's instructions on how to compile a *sanbutsuchō*:

Produce Report of the ＿＿＿ Province, ＿＿＿ Domain
Compiled by ＿＿＿＿＿＿

—Grains:
    Early rice (*wase* わせ), specify quantity and region of production
    Second rice (*nakate* なかて), "
    Late rice (*okute* おくて), "
    Mochiine (糯稲), "
    Foxtail millet (*awa* 粟), "
    Barnyardgrass (*hie* 稗), "
    Sugarcane (*kibi* 黍), "
    Wheat (*komugi* 小麦), "
    Barley (*ōmugi* 大麦), "
    Buckwheat (*soba* 蕎麦), "
    Soybeans (*daizu* 大豆), "
    . . .

All kinds of crops must be written in this order: additional species should be added at the end in the same fashion.

—Vegetables:
    Leafed vegetable (*sai* 菜), specify quantity and region of production
    Japanese radish (*daikon* 大根), "
    Etc. etc., "

As can be seen from the examples above, the list should include name, form, quantity and region of production not only of herbs and trees, but all edible vegetables. The same should be done for the following:

—mushrooms
—watermelons

—fruits
—trees
—herbs
—bamboo

The lists should be as exhaustive as possible and should follow the order of *Shobutsu ruisan*:

—fish:
    loach (*dojô* 泥鰌)[10]
    etc.
    etc.
—shellfish
—birds:
    crake (*kuina* くいな)
    etc.
—beasts:
    Japanese serow (*kamoshishi* かもしし)
    etc.
—insects:
    snails (*maimaitsuburi* まいまいつぶり)
    etc.
—snakes:
    adder (*mamushi* まむし)
    etc.

Of all previous categories, all species living in the region must be included, without considerations for their edibility.
    . . .[11]

The letter continued with a request to local authorities to investigate among the peasant population all possible usages of "all products generating from the earth."[12] In another passage, Shōhaku reported that he was authorized to inquire about "all species [growing and living] in that region without exception."[13] Local authorities were not to impose geographic limitations on their surveys and were requested to distribute copies of *sanbutsuchō* questionnaires in every village of the domain to be compiled and collected. In addition to the distribution of survey models to all domains, in 1735 Shōhaku requested a meeting with the managers (*rusui*) of the Edo mansions belonging to the various domainal lords to provide further instructions and clarification on the practical execution of the surveys. Yasuda Ken, the professor of agronomy who first discovered and studied the *sanbutsuchō*, reported that the diaries of one such caretaker had been found. Ōkubo Okaemon, an Edo *rusui*, recorded in his diary his meeting with Niwa Shōhaku in the fourth month of 1735.[14] At a certain point in the meeting, Ōkubo asked Shōhaku, "Since plants and animals of the Okayama domain are more or less similar to those of the surrounding countryside of Edo, I was wondering whether we should report only the strangest and rarest of the plants and animals of the Okayama region." Shōhaku replied, "As you rightfully observed, the majority of the species of plants and animals of the different provinces of the realm are the same. Nevertheless, I beg you to ensure that all mineral, vegetable, and animal produce of your region is carefully recorded without exception."[15]

Beginning in the fourth or fifth month of 1735, almost every village in Japan received a copy of the survey questionnaire.[16] As recorded at the end of these surveys, those who actually conducted the censuses were members of the peasant class: *shōya* (village headmen) or members of the *hyakushō sōdai* (village councils). These men were charged with the task of returning them within a few months (two or three months was the average) to a specially established office of the domainal administration, the Sanbutsu Goyōdokoro. This office had the duty of assembling the compiled surveys, copying them into a single booklet, and resolving statistical errors when data were missing or recorded in a confusing manner. The data from the various villages were assembled differently in each domain. Some put them in order of distance from the main castle town of the domain. Others followed the administrative divisions of the province. All surveys had to be dispatched back to Edo by the twelfth month of 1735.[17]

Unfortunately, some of these regional surveys have been lost, either partially or completely. It is known that they were each labeled "Produce Report of __ Domain of __ Province." All entries were sequenced in the order provided by Shōhaku—similar to the order in which entries were arranged in *Shobutsu ruisan*. All names had to follow the standard provided by Shōhaku himself. This meant that every regional Sanbutsu Goyōdokoro had to address inconsistencies and indicate the local forms and names (usually in dialect) of recorded species.[18] We know, for example, that the produce report of the Kaga Domain recorded 386 species of trees, 210 species of herbs, 223 species of birds, and 200 species of fish.[19] These statistics also distinguished regional variations of certain species and noted the distribution of different species by region. For example, the produce report of the Morioka Domain, in the northern part of main island, distinguished 13 species of Scolopacidae, 24 species of Anatidae, while Okayama distinguished 20 different species of crabs, 12 species of bees, 9 species of dragonflies, and 8 species of flies.[20]

Hundreds of new species were added to those previously described in Chinese encyclopedias. The thoroughness of the national survey was, for Shōhaku, more important than the rapidity of their execution, as we gather from a number of letters he wrote granting extensions of deadlines. Delays, imprecision, and negligence were surely only a few of the difficulties that a project of this scale must have suffered. Niwa Shōhaku, once having received a complete report from a domain, checked it for any erroneous, contradictory, or incomprehensible data. He then sent the reports back to the various domains for additional information. He circled any species he wanted to be illustrated and marked with a triangle any entry for which he needed more thorough information. Entries marked with the triangle usually required additional specifications added on the margins. All species marked with the circle required as precise and detailed a drawing as possible.[21]

Figure 7.1 depicts on the left a copy of the *Mutsu no kuni sanbutsuchō* initially sent to Edo and checked by Shōhaku and on the right, the illustration of a giant onion (*hananegi*), sent back in return.[22] The realization of accurate drawings of plants or animals was probably a demanding task for local authorities. The species for which Shōhaku requested an illustration were often rare, difficult to find, or, in case of animals, hard to capture. Each domain had to hire a skilled painter to portray the requested plant or the animal. It was not rare, Yasuda explained, for Shōhaku to request illustrations for one hundred or even two hundred species. For plants or trees, in particular, Shōhaku required the artists to reproduce them in the various phases of their growth or their changes in accordance with seasons. The images were then bound together and titled "Illustrated Albums of Produce of the ____ Domain in the ____ Province." Nothing is known of the authors of the illustration.

Today, these surveys are useful sources of information on species now extinct in Japan such as the *kawauso*, a Japanese subspecies of the Eurasian river otter; the *toki*, a species of crested ibis; and the *ōkami*, the Japanese wolf (fig. 7.2).[23]

These surveys provide important historical documentation of vegetal and animal species but still await full study. They are presently regarded as the largest and most comprehensive surveys of natural resources to have ever been attempted in East Asia and probably in the world. Surprisingly, however, when Niwa Shōhaku presented the 638 volumes that completed Inō Jakusui's *Shobutsu ruisan*, he ostensibly did not include any trace of the immense quantity of data

FIGURE 7.1. *Mutsu no kuni sanbutsuchō* from Yasuda,
Edo shokoku sanbutsuchō, 10. Tokyo: National Diet Library.

obtained from the regional *sanbutsuchō*, apart from a list of regional and dialect
names for various species. The encyclopedia was quickly cataloged in the shogunal library and Shōhaku promoted to head physician of the bakufu.

A further supplement of fifty-four volumes was added to the encyclopedia
in 1745, but the data collected ten years before was still not directly mentioned.
Why did the shogunate devote such a large quantity of resources to complete
this inventory, explicitly designed to support Shōhaku's revision and expansion
of Inō Jakusui's encyclopedia, if Shōhaku did not mention all the information
collected? Jakusui's project was structurally and intentionally lexicographical,
aimed at developing an all-inclusive and rational arrangement of all *honzōgaku*
sources and a definitive classificatory system. Stylistically and substantially, Shōhaku faithfully followed Jakusui's model. What then was the purpose of the surveys? If Shōhaku aimed only to order once and for all the regional and dialect
variations of the names of plants and animals, why bother to request so many
illustrations? Why not insert those images in his *Shobutsu ruisan*?

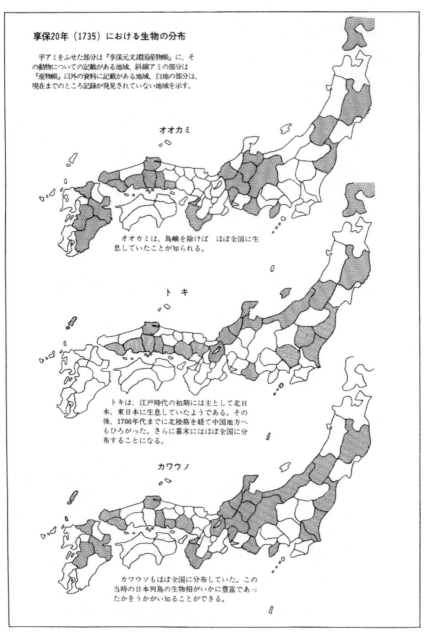

FIGURE 7.2. The distribution of the Japanese wolf, the Japanese crested ibis, and the Japanese river otter (in order) throughout the Japanese main islands as indicated in the 1736 surveys.

Yasuda's view is that Shōhaku exploited shogunal resources and authority to satisfy his own scientific curiosity.[24] Indeed the mystery continues, particularly because there is speculation over the legal status of the surveys, and mere curiosity does not seem a satisfactory hypothesis.[25] Aside from the official 1734 letter assigning Shōhaku the task of organizing the general census of natural resources and the final report summarizing the collected data, there is no trace of the original surveys among the official documents in the shogunal archives.[26] *Shobutsu ruisan* itself, once stored in the shogunal library, was never printed or published, and access to it was limited to those who could obtain authorization from the assembly of senior councilors.

A simple explanation could point to the fact that the survey booklets were a heterodox source without the authoritative status of canonical encyclopedias on the basis of which *Shobutsu ruisan* was compiled. But the secrecy surrounding the encyclopedia, the disappearance of some of the surveys, and the great power and authority at Shōhaku's disposal—unprecedented for a physician—might also suggest a much greater interest in the project on the part of the government and, in particular, of Yoshimune. These surveys may have been one of the strategies Yoshimune adopted to achieve his objective of control of the realm, which might provide an explanation for the secrecy surrounding them. Another motivation for the organization of national surveys of natural resources was probably the rural crisis of 1732–35, also known as "the Kyōhō famine." A general worsening of agricultural output reached its apex in 1732, when bad weather during the winter and spring and a locust infestation in the summer produced a shortage and widespread famine responsible for the deaths of more than one hundred thousand people in western Japan. From a fiscal standpoint, this famine resulted in a cut of tax income for the shogunate and the domains. Yoshimune responded to the famine by ordering a shipment of rice from the eastern reserves, but that measure was not sufficient to contain the effects of malnutrition. Throughout the fall of 1732 and the spring of 1733, incidents and local revolts convinced Yoshimune to increase the amount of rice distributed to the area.

The crisis of 1732–33 induced Yoshimune to adopt an agricultural policy aimed at encouraging the research and production of alternative and pest-resistant crops. The surveys of 1734–36 were arguably a major component of his program of agricultural reform. The famine was a likely motivating force for the expansion of Jakusui's research, as it would be hard to otherwise explain why Yoshimune did not order the expansion and revision of *Shobutsu ruisan* after he acquired it in 1719 or after he hired Shōhaku in 1722 and only did so after the Kyōhō crisis of the 1730s.

Survey campaigns were a common phenomenon in emerging states of the early modern period. In Europe, the young Carl Linnaeus (1707–78) traveled to Lapland with a grant from the Royal Swedish Society of Science to complete a survey of its botanical resources, which he later published in 1737 under the title *Flora lapponica*. This journey, besides having scientific purposes, was motivated by the economic interests of the Swedish government in the area.[27] The value of Linnaeus's classificatory system was exponentially increased by virtue of his students, the so-called apostles of Linnaeus who were sent around the world with the task of recording, describing, and classifying as many new species as possible.[28] Linnaeus's system of classification was simple and malleable enough to accommodate new species of plants. Linnaeus's apostle Pehr Kalm traveled in North America between 1749 and 1751 with a grant from the Royal Swedish Society of Science. Daniel Solander embarked on James Cook's *Endeavor* to gather botanical specimens for the English Royal Society. Fredric Hasselquist gathered specimens in Palestine, while Carl Peter Thunberg traveled in South and East Asia with the Dutch East India Company. It is impossible to separate these scientific enterprises from the economic and political interests of the emerging European colonial empires and to separate pure scientific curiosity from interested commitment. As Peter Raby argued regarding the African expeditions of Joseph Banks, "The 'pure' impulse to add to European knowledge went hand in hand . . . with the utilitarian: the word commerce was not mentioned, but perhaps did not need to be."[29] Scientific knowledge and imperial possession of the most remote parts of the world were the two sides of the same coin.

Similarly, Yoshimune's national inventory of all species of plants and animals can be interpreted as an enterprise to acquire precise data on land productivity and exploitable resources. This is evident from Yoshimune's patronage of *honzōgaku* experts whom he dispatched on herbal tours in the provinces for the establishment of medicinal and botanical gardens under the direct control of the shogunate or of domainal authorities. Niwa Shōhaku is remembered as having acquired the highest official title among them, but other scholars enjoyed the fruits of Yoshimune's patronage as well. Of these it is worth remembering Noro Genjō and Uemura Saheiji, who, like Niwa Shōhaku, both came from the Ise Province.

Uemura Masakatsu (also known as Saheiji) is probably the most interesting of all state-sponsored *honzōgaku* scholars.[30] Born in the small village of Ōzu, in the outskirts of Matsusaka, Saheiji was the son of a family of rural samurai (*gōshi*). In 1710, he entered the service of Yoshimune, then lord of the Kii Domain, as a personal attendant and bodyguard. When Yoshimune became shogun in 1716, Saheiji followed his master in the capacity of bodyguard (*oniwaban*), with access to Yoshimune's private chambers. The title of *oniwaban* was not particularly high in the samurai status ranking but granted him the privilege of direct contact with the shogun. Furthermore, the role of *oniwaban* included duties as varied as espionage and secret diplomatic missions. At the beginning of a 1754 letter he wrote at the time of his retirement, Saheiji explained,

> Regarding my services as a collector of medicinal herbs, I traveled through mountains and valleys of various provinces, beginning from the 15th day of the 5th month of 1720 to the 9th month of 1753. I spent a minimum of 140–150 days and a maximum of 180–190 days per year traveling in various provinces and taking daily notes of my observations. In the same period, I also served as a spy [*onmitsu goyō*], without missing even one year of service. From the spring of 1752, however, I became sick. I pleaded to be discharged from my services, and from the 17th day of the 7th month of the same year I was allowed to retire.[31]

Saheiji never received an education as a physician and we do not know when or how he studied *honzōgaku*. His first mission as *saiyakushi*, or collector of medicinal herbs, was a 1720 herbal expedition to the Nikkō mountains as adjutant of Niwa Shōhaku. There is, however, no extant source on the kind of training he had received prior to that expedition or one that explains why a member of Yoshimune's entourage was chosen to accompany Shōhaku. Niwa Shōhaku's letters or diaries never mention him as a student or trainee. Abe Shōō declared him to be his only disciple, but Saheiji never mentioned the name of Abe or acknowledged him as his master. The only connection between Saheiji and Shōō is the fact that Saheiji moved his residence to Edo on the seventeenth day of the sixth month of 1719 in the Asakusa area while Abe was also residing in Edo.

On the thirteenth day of the ninth month of 1720, Yoshimune entrusted Saheiji with the administration of the Komaba medicinal garden (*goyakuen*). Ueno has argued that the position reflected Saheiji's increasing competence in the field of medicinal herbology.[32] What we know for sure is that the Komaba garden was designed to produce a routine cultivation of the plants Saheiji had collected during his numerous travels throughout the Japanese islands, as well as exotic herbs and trees Yoshimune donated to him. His expertise was also attested to by his title of *yakusō miwake*—a herbologist, or, literally, an expert in distinguishing medicinal herbs. Between 1720 and 1753, he made eighty-six trips from the northern regions of the Honshū island to the Kansai area.[33] In 1722, Saheiji once again joined Niwa Shōhaku on a long herbal tour of the Kinki region. Saheiji's travel diaries and notebooks (*saiyakuki*) reveal how those expeditions were organized. Far from being solitary explorations, they involved large numbers of persons. Every time Saheiji and his staff entered a domain, a number of local guides (*kizukai*) were on hand to facilitate their travel, including reservations at inns, provisions, and the transportation of their materials. Once an area was explored, all specimens were quickly sent back to the Komaba garden in Edo.

As Saheiji disclosed in his retirement letter, he often used his herbal trips to carry on espionage activities. Unfortunately, the majority of his dispatches from these expeditions seem to be lost. One surviving letter reports bad flood management in the southern part of the Kantō plain between the first day and the twentieth of the ninth month of 1742. His report described in detail the damage to property, the inept countermeasures taken by local authorities, and the number of victims. When he died at eighty-three years of age, Saheiji had collected and successfully transplanted a large number of rare and precious plants into his Komaba medicinal garden. These included the *okera* (*pai shu*, *Atractylodes ovata*), a root with diuretic properties; the cutleaf *ōren* (*Coptis japonica*), treating conjunctivitis and stomatitis; the *maō* (*Ephedra sinica*), a stimulant with thermogenic effects; the *uyaku* (*Lindera strychnifolia*), beneficial to the liver and to facilitate digestion; and the *sobana* (*Adenophora remotiflora*), the root of which was used as expectorant and antidote to insect bites.

PART IV *Nature's Spectacles*

The Long Eighteenth Century

(1730s–1840s)

It is as if being was to be observed,
As if, among the possible purposes
Of what one sees, the purpose that comes first,
The surface, is the purpose to be seen.
— *Wallace Stevens, "Note on Moonlight"*

In the second half of the eighteenth century, the theoretical and practical investigations of the natural world that had begun a century and a half earlier fused into a more integrated and at the same time more diversified discipline. *Honzōgaku* maintained its original association with medicinal studies, but it also grew into an increasingly heterogeneous field that relied less on the authority of canonical texts and on the necessities of medical practices. Variously supported by shogunal and domainal institutions; private academies; popular curiosity; and a burgeoning publishing industry of textbooks, encyclopedias, and specialized monographs, the discipline of *honzōgaku* acquired an eclectic character and adopted a wide range of approaches and aims, from lexicographical and encyclopedic to agronomical, gastronomical, aesthetic, entertaining, and natural-historical. Its specialists now had a following large enough to achieve social recognition and relative autonomy from other fields of study. The shogunal support of *honzōgaku* initiated by Yoshimune spread to many domains, where lords, either by sincere vocation or by peer persuasion, began to sponsor specialist scholars or to engage themselves in naturalistic investigations as amateur practitioners. This patronage took various forms, including the establishment of botanical gardens and pharmacological institutes, as well as the direct involvement of the samurai elites in cultural circles devoted to observation, description, collection, and illustrations of plants and animals.

The divergence of Japanese *honzōgaku* from its Chinese model—which under the Qing did not change much from the times of Ming *bencao* professionals like Li Shizhen—did not happen overnight but occurred in relation with changes in scholars' discursive and material practices and their socioprofessional identity, with the involvement of both shogunal and domainal administrations as supporters of product surveys and cultivations of medicinal herbs or alternative crops and with the establishment of new institutions like the Institute of Medicine and bakufu- and *han*-sponsored botanical gardens.

A long period of political stability after the establishment of the Tokugawa shogunate had led to profound social transformations. In particular, the growth of a market-oriented economy deeply affected the Tokugawa social order. Its most noticeable effect was the impoverishment of the ruling samurai elites and a consequential weakening of their actual power. The monetization of the economy, the growth of a domestic and international trade of commodities, the increasing reliance on the market for the acquisition of the means of social reproduction, and the commercialization of agricultural production were all transformations in the socioeconomic life of Tokugawa Japan that favored the emergence of new forms of social domination based on wealth rather than birth.

A steady growth in literacy among the merchant and artisan classes transformed them into enthusiastic consumers of cultural products.[1] The result was a diversification of the market for cultural goods as members of different social classes invested their time and resources in cultural activities. With the emergence of new cultural consumers among the population of large landowning peasants (gōnō), who saw their affluence swelling thanks to money lending and the investment of agricultural surplus in petty industrial manufacturing like textiles, sake brewing, and indigo, culture exited the major urban centers of the Kamigata and Kantō areas and spread to the countryside.

Once regarded by historians as a long period of social immobility, by the second half of the eighteenth century, the Tokugawa era saw radical social changes that fractured the established hierarchy of social relations. On the one hand, samurai elites had been experiencing since the seventeenth century a "civilizing process" that often encoded new rules of peacetime sociability. These rules included what Eiko Ikegami calls internalized "taming," which advocated self-restraint over the medieval ideals of violence and physical force.[2] Moral virtuosity translated into a language of refinement, decorum and artistic expertise. Under this new value system, intellectual erudition became an important marker of status. Distinction in taste and learning conferred status in the eyes of other domainal lords that added to their political privileges. Learned mid- and low-ranking samurai, impoverished by the steady decline in the price of rice, frequently sought to improve their livelihood through employment in intellectual and cultural pursuits. While their success as scholars largely depended on patronage, their employment in domainal administrations contributed to transform the relationship between lord and retainers. Paid consultancy on cultural and intellectual matters had replaced military service in battle, practices consistent with the demilitarization and bureaucratization of the samurai class initiated by the first Tokugawa shoguns.

On the other hand, at the bottom of the formal social ladder, wealthy commoners invested their newly acquired economic prosperity in education and artistic and intellectual training and thereby enhanced their social presence. In so doing, they followed and contributed to transform the now widespread Neo-Confucian moral discourse, which originally praised the activities of moral cultivation even as it condemned the pursuit of money.[3] Involvement in cultural circles and private schools provided commoners indirect and, more rarely, direct contact with members of the samurai elites. The expanding networks of cultural exchange, even though only occasionally bringing samurai and commoners physically together, created intellectual associations by virtue of shared ideas, interests, practices, and often the same scholar-teachers supervising the activities of different groups. The participation of members from all social classes—from samurai to wealthy peasants and merchants—in cultural activities exploded during the second half of the eighteenth century, producing the dynamic popular culture of late Tokugawa Japan.

Honzōgaku was just one of many artistic and intellectual activities that prospered in the eighteenth century. Unlike other pursuits, *honzōgaku* had the advantage of being favored by the Tokugawa authorities for its practical utility for the welfare of the state and its moral benefits due to its connection to the Neo-Confucian idea of the "investigation of things." Being involved in a cultural circle devoted to the study of nature was socially and intellectually beneficial. Much like the wealthy bourgeoisie in nineteenth-century England or the private pursuits of Chinese literati, both samurai elites and wealthy commoners were attracted to such activities because they were interesting, uplifting, and socially sanctioned. The artful illustrations that filled so many *honzōgaku* texts also suited the aesthetic tastes for visual representation that characterized elite and popular culture in the Tokugawa period. In addition, with its public exhibitions of embalmed and live animals and plants imported from remote areas of the world, *honzōgaku* thrived in the lively urban context of public spectacles and entertainment, which enabled its practitioners to both create and exploit a growing popular interest in the natural world.

Even as the discipline of *honzōgaku* maintained its name derived from its pharmacological origins, it was undergoing a profound transformation into something recognizable as "natural history." The main change was that a growing number of scholars often concentrated on the study, observation, and description of plants and animals in their own right. Even though most continued to conceive of their research in terms of practical utility and service, their intellectual practices increasingly strived to produce accurate descriptions of the morphology, ecology, and behavior of vegetal and animal species.

In the process, during the second half of the eighteenth century, *honzōgaku* gradually became a relatively more autonomous field of specialization, taught in state-sponsored schools and research institutions and possessing its own canonical texts. Its practitioners were no longer polymaths collaterally involved in its activities, like Hayashi Razan and Kaibara Ekiken, but specialized scholars who could now be considered naturalists, able to occupy an increasing variety of vocational niches on the basis of that expertise, like Tamura Ransui and Ono Ranzan. Many continued to serve under domainal lords, while others found employment in the shogunal Institute of Medicine. A few thrived as private educators with their own academies, like Matsuoka Gentatsu (Joan). Still others relied on public support through lectures; through cultural circles and clubs; or through employment as gardeners, illustrators, or dealers in exotic species.

Reliance on the public in this broader sense meant that *honzōgaku* specialists found themselves in the unprecedented need to compete with one another to satisfy the tastes of new consumers. They were no longer in the same situation as Hayashi Razan, who, a century and a half earlier, had to struggle for social recognition as scholar outside the traditional institutions of the Buddhist clergy and the imperial court. Now *jusha* were socially accepted professionals. The public, too, had changed. By the eighteenth century Japan had a greater number of large towns and cities than many other urbanized societies, with the population of the shogunal capital of Edo (present-day Tokyo) reaching a million inhabitants. Edo exemplified the new and vibrant urban culture of spectacle, with theaters, pleasure quarters, circuses, exhibitions, and street shows accompanied by a flourishing publishing industry of popular fiction, satire, and guides to tourist attractions.[4] With the emergence of a wealthy population of landowning peasants, the popular culture of cities like Edo, Osaka, Nagoya, and Kyoto began to coexist with a dynamic but distinctive form of cultural networks that appeared in the towns and villages of rural Japan. *Honzōgaku* practices were present in both circuits, although its activities were somewhat different in each context.

Matsuoka Gentatsu, a disciple of Inō Jakusui, gave frequent public lectures in Kyoto, which were attended by large enough crowds that the price of student fees alone was sufficient to support him.[5] Gentatsu's disciple Ono Ranzan boasted of having more than one thousand students from all over Japan enrolled in his Kyoto private academy.[6] In the countryside, circles were smaller and their activities less flamboyant than those in the cities, but they shared similar interests and practices.[7] Indeed, they often shared the same scholar-teachers, who traveled from one group to another, and in so doing linked together unconnected — and in the social structure of the times often unconnectable — members of different social statuses through ideas and cultural practices. Tamura Ransui and his disciple Hiraga Gennai organized the first national exhibitions of plants and animals, in which thousands of rare and precious specimens from private collections were openly displayed to the Edo public and was met with such unprecedented popular enthusiasm that they brought together amateurs and scholars of different social standings from both urban centers and the countryside.[8] Scholars were not the only individuals that benefited from *honzōgaku*'s surge in popularity. A small army of painters and illustrators served at the disposal of domainal lords, cultural circles, publishers, and naturalists to enrich catalogs, essays, and proceedings with their astonishingly sophisticated renderings of plants and animals.[9]

During the second half of the Tokugawa period, natural history enjoyed all the benefits of a popular fashion. A consequence of the increased demand for plant and animal expertise was the swelling of the population of professional *honzōgaku* scholars. The growth in specialist numbers was however counterbalanced by intensified shogunal control over intellectual production, especially after the so-called ban on heterodoxy of 1790.[10] As demand for scholarly works increased, so did their prestige, but earning a living without the patronage and support of the domainal aristocracy was still extremely difficult, as the story of Hiraga Gennai well illustrates.

The popular craze for natural history that reached its peak in the last century of Tokugawa rule was never sufficient to guarantee economic stability to its practitioners. In a sense, the discipline of *honzōgaku* can be said to have fared better than its specialists. On the one hand, natural history became a field with such a sufficiently large following in all social classes that naturalist scholars were able to exert a greater degree of control over their means of intellectual production than their seventeenth-century predecessors. On the other hand, the social homogeneity of the specialists themselves remained largely unchanged until the end of the Tokugawa period. That is to say, most naturalists were still physicians of low or midsamurai rank, often employed in shogunal and domainal administrations. Although the growing popularity of *honzōgaku* did not alter the social composition of its producers, it did mean that their natural historical research was no longer attuned only to the needs of the ruling class. The same was true in nineteenth-century Europe, where being a naturalist was often not a self-sustaining profession, even after the discipline of biology found acceptance as an academic field.[11] The evolutionary biologist Thomas Henry Huxley, nicknamed "Darwin's bulldog," complained that "there is no chance of living by science. I have been loath to believe it but it is so."[12] The yearly income of Richard Owen, probably the most famous comparative anatomist of his time, amounted to £300, "which is less than the salary of many a bank clerk."[13] Natural history in nineteenth-century Europe as in late Tokugawa Japan, despite its success in popularity, remained a gentlemanly activity, counting among its practitioners almost exclusively members of the classes who could afford to engage in it.

The four chapters in this section reconstruct some aspects, events, and consequences of the heyday period of *honzōgaku* in Japan. Chapter 8 describes how nature in its various forms became a *material* commodity for popular consumption, from rare plants and animals to collect and share to various forms of games and pastimes. Chapter 9 focuses on nature as *intellectual* commodity for various cultural clubs and circles, engaging both samurai and commoners or a combination of the two, as was the exceptional case of Kimura Kenkadō's salon. Chapter 10 concentrates on Hiraga Gennai, his humble origins and genius, his successes in organizing the first public exhibitions of pharmacological specimens, and his struggles to survive as a professional scholar. Chapter 11 will finally offer an interpretation of the cornucopia of illustrations of plants and animals that became the main characteristic of *honzōgaku* production in the last century of the Tokugawa period.

8   *Nature's Wonders*
Natural History as Pastime

· · · · · · · · · · ·

On the twenty-sixth day of the fourth month of 1729, a seven-year-old female elephant entered Kyoto. Its arrival was preceded by such excitement that Emperor Nakamikado himself expressed the desire to see the giant animal that shogun Tokugawa Yoshimune had purchased. The encounter between elephant and emperor, celebrated in printed books and gazettes, was a wondrous event.[1]

About a hundred years later, Ogata Tankō retold the story in his *Zō no emakimono* (The Elephant Scroll) and narrated that suddenly and apparently without any direct order from its trainers, the enormous elephant, as it arrived in front of the veranda where the young emperor was sitting, fell into its knees and bowed "in recognition of the awesome presence of the living god the Emperor" (fig. 8.1).[2] The militant fervor of revolutionary emperorism (*sonnō*) that inflamed the southern Chikuzen Province in the first half of the nineteenth century might have inspired the obscure author of *Zō no emakimono*, active at the time in the Kyūshū island. The truth is that after the encounter with the emperor, the elephant continued on its voyage to Edo and its real proprietor, Tokugawa Yoshimune. Contemporary commentators, a century before Tankō, imbued with the patriotic zeal, interpreted the event as the awe-inspiring sign of the emperor's transcendent power, could in fact have understood the encounter of elephant and emperor as the generous concession of the real sovereign of Japan: the shogun. The pachyderm arrived in Edo on the twenty-fifth day of the fifth month. Yoshimune had expressed the desire for an elephant ever since he admired a copperplate reproduction of one in Johann Jonston's *Historia naturalis* (Natural History), a copy of which had been donated to the shogunate by the Dutch East India Company's representatives (fig. 8.2).[3]

Two elephants, a male and a female, had arrived in Nagasaki on a Dutch cargo ship in the sixth month of 1728. Yoshimune had specifically requested a white elephant, which in East Asian traditions is a symbol of just and peaceful ruling. Pictures of white elephants could be frequently found in the mansions and castles of domainal lords (fig. 8.3), but the closest to a white elephant Yoshimune was able to obtain was a young pair of gray *Elephas maximus* from Vietnam.

済し令などの経済政策。「公事方御定
衆の意見の吸い上げ。数々の政策を行
ねとし、木綿物を身にまとい、倹約を
心の持ち主でもあった。

FIGURE 8.1. Ogata Tankō, *Zō no emakimono* (The Elephant Scroll), detail. Osaka: Kansai Daigaku Toshokan.

Of the two elephants sent to Yoshimune in 1728, the male died three months after his arrival in Nagasaki. The female traveled to Edo accompanied by two Vietnamese trainers, their convoy reaching Edo after seventy-four days of travel, which attracted crowds of people eager to see the gigantic animal.[4]

It is reasonable to surmise that the Tokugawa authorities had anticipated the popular sensation of the parade and publicized the passage of Yoshimune's elephant to enhance the shogun's symbolic power. Yoshimune's *kanjō bugyō* Inō Masatake had official ordinances (*ofuregaki*) posted everywhere to schedule the passage of the animal and regulate public behavior during the procession. Onlookers were to look at the animal without making any noise and were prohib-

FIGURE 8.2. Johann Jonston, *Naeukeurige beschryving van de Natuur* (Amsterdam, 1660). Tokyo: National Diet Library.

ited from giving it any food or water.[5] After its arrival in Edo, the elephant was given to a man named Gensuke, together with a yearly grant of two hundred *ryō* from the shogunal treasury for its keeping. The elephant died of malnutrition at twenty-one years of age, thirteen years after its arrival in Japan, probably the victim of a bankrupted bakufu.[6]

CRAZY FOR NATURE

Yoshimune's elephant was but one of many public events involving plants and animals that aroused popular enthusiasm during the last century of Tokugawa rule.[7] Parades of rare and exotic animals imported by Dutch and Chinese merchants were frequent happenings, but popular curiosity (*suki*) for nature's wonders found expression in a vast array of practices and fashions. Collecting shells, leaves, flowers, plants, insects, and exotic birds was a passion shared by individuals of every social standing, with obvious differences depending on the economic and social capital of each aficionado. Gardening, landscaping, and herb-picking picnics in the countryside were also popular pastimes, as were developing new breeds of azaleas and goldfish and enjoying social games involving plants and animals—like *kaiawase*, a game to see who could recognize and cor-

FIGURE 8.3. Itō Jakuchū, *White Elephant* (1768).

rectly name the largest number of shells. A flourishing publishing industry of manuals, catalogs, and specialized monographs evidenced the popular craze for natural history in the eighteenth and nineteenth century.

Fashions for particular vegetal and animal species came and went—often inexplicably—both in Japan and in nineteenth-century Europe. Preferences often shifted seasonally and were also geographically localized. Some were nationwide crazes, while others were based on a sudden burst of enthusiasm for a particular flower, fish, or bird that curiously punctuated the history of local communities. In a social flowchart of such trends, it seems that fashions related to natural history moved from the top down in the social structure. Beginning in the shogun's entourage, they became popular first with domainal lords and their retainers and then among commoners. Traditional sources report that shoguns often displayed a particular indulgence for one species of flower. The *Tokugawa jikki* and other memoirs of high-ranking shogunal officials report how Tokugawa Ieyasu showed special appreciation for cherry trees, which he had planted around his Suruga castle and elsewhere in Edo. His son Hidetada, to whom Ieyasu passed the title of shogun in 1605, is said to have been fond of camellias.[8] Iemitsu, the third shogun, shared his grandfather's tastes and had the Fukiage garden of the Edo castle expanded and enriched mostly by cherry trees and some camellias as a show of filial piety. He even employed a heavily armed night guard in the garden to protect the precious plants. Iemitsu, the resolute ruler who gave the shogunate a more complex and efficient administrative and legal structure, was fond of constraining pine trees into miniature size (*bonsai*). The *Tokugawa jikki* records that one of his most trusted retainers, Ōkubo Tada-kata, put out by his lord's habit of sleeping with a beloved miniature pine tree in a wooden box under his pillow, threw it into the garden and begged his master to either stop that madness or allow him to disembowel himself on the spot.[9] The brief shogunate of Tokugawa Ienobu (1709–12) was marked by a frenzy for multicolored maple trees.[10] The list can continue for all fifteen shogun of the Tokugawa period.

The botanical passions of shoguns would be a mere historical curiosity — and the testimony of uncritical sources like *Tokugawa jikki* would be hard to corroborate — if their personal interests had not been eagerly adopted by others in a process tantamount to a national craze. The transmission of these popular interests down the social ladder and across the country was facilitated by the institution of alternate attendances, which forced domainal lords and their re-tainers to spend half their lives in the shogunal capital of Edo.[11] It was there that people from different regions became acquainted with the latest vogue and brought them back to their homes around the archipelago. A number of sources observed how people of every social standing followed shogunal fashions, be-coming infatuated with camellias during the Kan'ei era (1624–43), with maple trees in the Genroku era (1688–1703), with chrysanthemums in Yoshimune's Kyōhō era (1716–35), and with mandarin orange trees in the Kansei era (1789–1800).[12]

A visible sign of this fascination with natural history was the development of botanical gardens in major cities. Their maintenance involved samurai as well as townsmen and contributed to the growth of a gardening industry and the establishment of professional landscapers. Gardening (*teien*) had always played an important cultural and religious role in Japan, but during the Tokugawa period the number and diffusion of private and public gardens increased exponentially, including the major construction of shogunal and domainal gardens in Edo, Kyoto, and other castle towns.

The main beneficiaries of this widespread enthusiasm were professional gardeners (*uekiya*). They established businesses in specific quarters in Osaka (Shimodera machi, Tenman, and Kōzu), Kyoto (around the Kitano shrine), and Edo (Komagome, Sugamo, and Aoyama) to serve the needs of their customers. *Uekiya* usually sold various kinds of small- to medium-sized plants (fig. 8.4). The larger dealers also traded in small animals such as goldfish, pet mice, and different species of insects for collectors. There were also smaller *uekiya*, who were often street sellers specializing in a particular group of plants, usually from the Edo countryside. These street vendors commuted into the city every day to sell their products (fig. 8.5).

The number of *uekiya* expanded greatly after a fire destroyed a large portion of central Edo in 1657. The devastation required troops of carpenters and gardeners who were called into Edo from the provinces to reconstruct the mansions of the domainal lords. As gardens often served as evacuation areas, their numbers increased after the Meireki fire. Several semiprofessional gardeners, usually low-ranking retainers and Buddhist monks, were also involved in the design and maintenance of private gardens. Eminent professional landscapers were usually connected to the best *uekiya* shops. They published manuals of gardening techniques and botanical catalogs, which helped strengthen their reputations and, consequently, increase the number of clients. Perhaps the most famous manual of the period was *Kinshū makura* (Beautiful Flowers), published in 1692 by Itō Ihyōe Sannojō. This publication provided full descriptions and instructions on how to plant and nourish 174 variants of azaleas (*tsutsuji*) and 163 species of rhododendron (*satsuki*). Second in the list of gardening bestsellers was *Kadan jikinshō* (Beautiful Ornamental Plants), published in 1695, also by Sannojō. It became one of the best known of all general textbooks of gardening techniques and was revised and expanded by Sannojō's descendants in 1710, 1719, and 1733 and reprinted well into the Meiji period.[13]

The popularity of gardening and the development of new horticultural techniques had a profound influence on *honzōgaku*. Many *honzōgaku* specialists were, in fact, involved in the writing of these manuals. Kaibara Ekiken's *Kafu* (On Flowers), completed in manuscript form in 1694 and published in 1698,

FIGURE 8.4. Edo *uekiya* from *Ehon kaga mitogi*
1752). Tokyo: National Diet Library.

provided introductory instruction on how to plant and nourish various species
of garden plants and trees, each species arranged in accordance with its bloom-
ing season.[14] What distinguished Ekiken's manual from Sannojō's was Ekiken's
extensive use of Chinese sources and the Confucian interpretation he gave to
gardening as an activity that could help in exercising one's moral faculties.[15]
Sannojō's manual was technically oriented and was based on his family's gardening expertise and traditions.

Other *honzōgaku* scholars published monographs on specific garden plants.
Inō Jakusui's student Matsuoka Gentatsu published a series of studies on the cul-

FIGURE 8.5. Street *uekiya* from *Shiji no yukikai*
(1798). Tokyo: National Diet Library.

tivation of orchids, cherry trees, bamboos, and plum trees.[16] Gentatsu's disciple
Ono Ranzan coauthored an illustrated gardening encyclopedia with the professional gardener Shimada Mitsufusa titled *Kai* (Catalog of Flowers). Published
in eight volumes between 1759 and 1765, it was based both on Chinese sources
and on the practical expertise of the gardener Mitsufusa.[17] For the authors of
these manuals, whether they were professional gardeners or *honzōgaku* scholars,
publication usually meant an additional source of income. The success of these
texts, especially monographs focusing on a single species of ornamental plants,
depended entirely on the shifting trends of interest in different species.[18]

Visiting temple and public gardens was another popular pastime of the era.
In fact, many of today's popular gardens were established in the Tokugawa period for this purpose, like the Rikugien, near the modern-day Komagome train
station, commissioned by Yanagisawa Yoshiyasu, lord of the Kawagoe Domain
and counselor to shogun Tokugawa Tsunayoshi. Constructed between 1695
and 1702, it consisted of a spatial representation of the poetic anthology *Kokin-
wakashū*.

The ephemeral whims of the urban population of Tokugawa Japan for one
plant or another also meant that public gardens had to constantly renew their

FIGURE 8.6. Ōkubo's azaleas from *Edo meisho zue*
(1834–36). Tokyo: National Diet Library.

collections of plants to keep up with popular tastes. Figure 8.6, from *Edo meisho zue* (Illustrations of Famous Places in Edo, 1834–36), portrays a scene from the garden of the Ōkubo family's mansion, near present-day Waseda University, and shows a group of female visitors admiring azaleas. Circles of horticultural aficionados often organized tours to admire gardens in temples and private mansions. Figure 8.7, for example, taken from *Tōtosaijiki* (Seasonal Events of the Eastern Capital, 1838), portrays a scene from the Somei ward of Edo, headquarters of many *uekiya*. These large shops were open to the public and changed the plants on display in accordance with the current fad.[19]

Camellias, azaleas, maples, peonies, chrysanthemums, cherries, irises, lilies, and hollies were among the most desirable species of ornamental flowers, and florists capitalized on their popularity by publishing illustrated catalogs of their interbreeding experiments. The market for *asagao*, the Japanese morning glory, a flower of great appeal because of its literary associations, was so profitable that it motivated cultivators to experiment in hybrids that would provide the market with many *asagao* varieties (fig. 8.8).[20] Flowers were so important in daily life that Philipp Franz von Siebold commented in a 1829 report he sent to an academic proceeding in his native Germany that Japan was a "garden kingdom."[21]

FIGURE 8.7. Somei's chrysanthemums from
Tōtosaijiki (1838). Tokyo: National Diet Library.

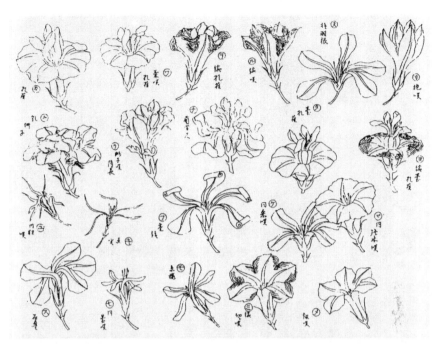

FIGURE 8.8. Different variations of the Japanese morning glories from the catalog *Asagao mizukagami* (1818). Tokyo: National Diet Library.

A variety of leisurely activities revolved around the popular interest in flowers and plants, including botanical games and competitions organized by local circles and groups. In Japanese, these events were usually called *hana awase*, a late Tokugawa modification of a Heian game in which players competed to identify and associate flower species either with a poem or with a specific season. Other popular games were *hanazumō*, *tōkakai*, and *kusa awase*, which were all variation on the theme of *hana awase* (fig. 8.9).[22]

Collecting, exhibiting, buying, selling, exchanging, and playing games relating to natural species was not limited to plants and flowers; it also involved animals. The raising of pets also gained in appeal. Just as trends for flowers moved down the social ladder, passions for pets, especially small birds, developed first among high-ranking samurai and then became popular with townsmen and wealthy peasants.[23]

Fashions for pets also prompted the growth of new markets. Pet shops, pet dealers, illustrated books describing different species of animals, manuals on pet care, and the assorted relevant events hosted by cultural circles and groups all nourished this interest in animals. For example, a prospective bird owner could

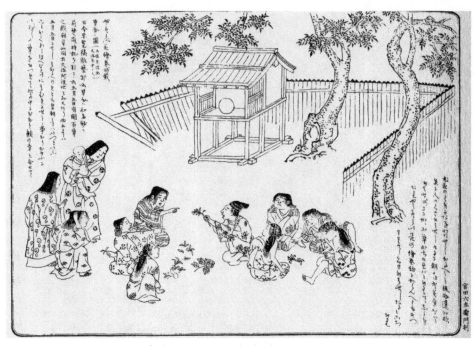

FIGURE 8.9. A scene of a *kusa awase*, in which players compete
to pick up two leaves of the same plant and name it correctly.

visit a *toriya* (bird shop) to choose a desirable bird either as a pet or as dinner.
Figure 8.10, taken from *Settsu meisho zue* (Famous Places of Settsu, 1796–98),
shows one of these shops at Settsu, a town near Osaka. One shop is selling live
birds of various species and sizes as pets, while on its right, another shop, pre-
sumably connected with the first one, is selling dead birds as food.

A flourishing commerce in birdcages developed, with bamboo or wooden
cages sold to townspeople and gold, silver, or ivory cages sold to the military
aristocracy.[24] It was not uncommon to see young men, and occasionally women,
walking in the street carrying a small cage with a favorite pet bird. Many such
enthusiasts attended meetings of bird aficionados, competed for ownership of
the bird with the most beautiful singing voice (*nakiawase*), or organized excursions into the countryside to challenge one another in the ability to recognize
birds by their songs (*kotoriawase*).[25]

Some of the most popular bird species included those traditionally praised in
Japanese and Chinese poetry, such as red crowned cranes, mandarin ducks, hens,
and golden and silver pheasants.[26] With the increase of commercial exchange

FIGURE 8.10. A bird shop from *Settsu meisho zue*
(1796–98). Tokyo: National Diet Library.

with Chinese and Dutch merchants after Yoshimune's reign, rare and exotic birds
began to arrive in Japan through Nagasaki from Southeast Asia, the Moluccas,
New Guinea, Africa, Mexico, and even South America.[27] When a cargo of exotic
birds arrived, Nagasaki's city magistrate hired an artist to draw them and sent
the pictures to the shogun to choose the ones he wanted. The birds rejected by
the shogunal house were distributed to bird wholesalers (*kaidoriya*) and sold to
wealthy amateurs who could afford such a purchase.[28] Among the most desirable
exotic birds were Java sparrows, red avadavat, Western crowned pigeons, canar-
ies, peacocks, barred cuckoo doves, Hill mynas, Temminck's tragopans, guinea
fowls, and a wide variety of parrots and parakeets.[29] Far rarer, but often requested
and very expensive, were cassowaries (fig. 8.11), ostriches, and various species of
hornbills.[30] For those who could not afford to buy exotic birds themselves, some
consolation could be found in the acquisition of a painting or a print portraying
them. Purchases of animal illustrations were still expensive but more affordable
than the live animal itself.

Birds were both the most highly requested and most commonly illustrated
animals of the period, followed by fish and insects. Mammals were rare, with
some extraordinary exceptions like Yoshimune's elephant or other exotic and
feral animals like civet cats, crocodiles, orangutans, dromedaries, porcupines,

FIGURE 8.11. A 1658 illustration commemorating an exemplar of a cassowary donated by Dutch merchants to the shogun in 1635. From Isono, *Egakareta dōbutsu shokubutsu*. Tokyo: National Diet Library.

and kabaragoyas.[31] The pair of dromedaries that Dutch merchants offered to the shogun in 1821 prompted a parade no less magnificent and popular than the Edo voyage of Yoshimune's elephant one century earlier (fig. 8.12).

For those who could not even afford to buy a print, there were teahouses that offered, in addition to a cup of green tea or warm sake, a look at rare and interesting animals on display. *Kujakujaya* (peacock's teahouses, fig. 8.13) gave their customers the chance to observe peacocks, pheasants, mandarin ducks, and other curious animals displayed in cages around the shop. There were also *chinbutsujaya* (teahouses of curiosities), where strange objects were exhibited for the amusement of their clients (fig. 8.14). Public spaces displaying animals and plants were precursors of modern-day zoological and botanical gardens, whose parallels in nineteenth-century England included the club and coffeehouses where amateur naturalists gathered to compare specimens from their collections.[32]

Animals also played a key role in Japanese hobbies and pastimes. Many hobbyists spent hours knee-deep in the surf collecting shells or seaweed, an activity of Tokugawa townsmen as well as of the nineteenth-century English middle classes (fig. 8.15).

Some enthusiasts exchanged shells with one another or played games like *kaiawase* with fellow conchologists. Others spent their time catching butterflies

FIGURE 8.12. The parade of the pair of dromedaries that Dutch merchants offered to the shogun in 1821 from a contemporary broadsheet.

FIGURE 8.13. *Kujakujaya* from *Setsu meisho zue* (1796–98). Edo-Tokyo Hakubutsukan.

FIGURE 8.14. A *chinbutsujaya* from a 1789 painting.

FIGURE 8.15. The humorous 1857 image by John Leech
in *Punch*, titled "Common Objects at the Seaside," clearly
illustrates the widespread popularity of seaside naturalism.

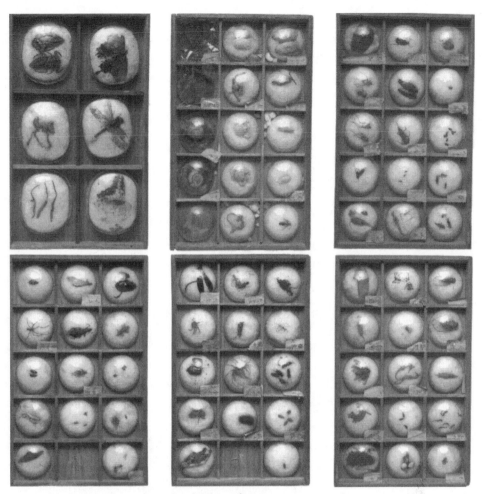

FIGURE 8.16. Musashi Yoshitoki's box of insect specimens. Tokyo: University of Tokyo.

or grasshoppers, which they then carefully classified in insect boxes like the *konchū hyōhon* of the Tokugawa retainer and amateur naturalist Musashi Yoshitoki, shown in figure 8.16.

Although collecting insects or shells was usually a solitary pursuit, it also had important social functions. *Honzōgaku* clubs and circles provided spaces for people to convene, compare specimens, and compete against one another in classificatory knowledge, skillful handling of plants or animals, or descriptive skill. It was in the eighteenth century, at the peak of its popularity, that *honzōgaku*, like other fields of knowledge, could be mobilized as a marker of status. In cultural clubs, distinctions in knowledge were not directly dependent on birth or political and economic capital—the factors that determined position in the social hierarchy. This competition for status was, of course, limited, since cultural clubs and circles were worlds apart from the rest of Japanese society. Nonetheless, these clubs and circles were places where members of different classes were able to meet on the ground of knowledge rather than hereditary status.

The accumulation of more or less systematic knowledge of plants and animals in books, group discussions, and specialized schools for practical goals or theoretical interests developed in the "long eighteenth century" alongside popular divertissements with natural species of various kinds. As we have seen in part III, since the early years of the 1700s, a growing number of scholars had engaged in state-sponsored expeditions in the remotest mountains and valleys of Japan to collect herbs and plants that were then transplanted in medicinal or ornamental gardens. In addition, new exogenous species of roots, tubers, and vegetables were systematically introduced and cultivated. At the same time, popular frenzies for flowers and trees nurtured the development of new businesses that traded in flowers, plants, birds, shells, stones, and small animals to keep as pets. In both cases, natural objects were exchanged, bought and sold, reared, exhibited, collected, represented, rearranged, interbred, bound up to assume aesthetically pleasing forms (*bonsai* trees), organized in gardens instilled with literary associations, caged to decorate the houses of wealthy samurai and merchants, and utilized as symbolic vehicles of satirical, aesthetic, or even philosophical ideas. The commodification of plants and animals as well as of the information about them was often mediated by market dynamics in a consumer society that monetized their value. This happened at the same time that agricultural expansions and reforestation projects of large portions of the archipelago were organized and sustained by bakufu and domainal administrations. All these processes extended human dominion over nature in a quantitatively and qualitatively unprecedented degree. As a result, the metaphysical associations of macro- and microcosmos originally associated to *honzōgaku* writings disappeared almost completely. Also, the sacred spaces of untamed nature ruled by goblins and wild beasts in mountainous recesses began to shrink. In the eighteenth century, these processes happened simultaneously in urban centers and in rural areas, but they were rarely the subject of reflection. We have to wait until the nineteenth century for a thorough conceptualization and philosophical justification of human dominion over nature in the work of Satō Nobuhiro.[33]

9   *Nature in Cultural Circles*

CULTURAL CLUBS AND THE CIRCULATION OF
KNOWLEDGE IN EIGHTEENTH-CENTURY JAPAN

In the first half of the Tokugawa period, the study of nature depended
largely upon the support of the ruling classes, who employed specialized scholars
to oversee agricultural reforms, to supervise the cultivation of medicinal herbs,
or to satisfy a domainal lord's curiosity about natural history. The patronage of
the samurai elites influenced the method and scope of nature studies, a pattern
that continued for the rest of the Tokugawa period.[1] In this respect, intellectual
pursuit in the field of *honzōgaku* could be a form of service that a retainer owed
his lord, as in the case of Kaibara Ekiken. Or it could be an increasingly bureau-
cratized form of labor performed by an employee of bakufu or *han* administrations, where the waning of personal ties of loyalty meant frequent employment
mobility, as with the cases of Niwa Shōhaku and Tamura Ransui. In addition,
honzōgaku specialists continued to depend heavily on the authority and style
of established encyclopedias and canonical texts, above all Li Shizhen's *Honzō
kōmoku*. As exemplified by Inō Jakusui's *Shobutsu ruisan*, *honzōgaku* scholars in
time distanced themselves from the influence of Neo-Confucian metaphysics
and ethics but did not reject the taxonomy of *Honzō kōmoku*. Despite his eclecticism and openness to *rangaku*, the same could be said of Ono Ranzan.[2]

By the second half of the eighteenth century, *honzōgaku*, while maintaining
strong ties with medicine, developed a more pronounced eclecticism. The diversification of its practices, styles, and goals was facilitated by the emergence of
networks of cultural exchange that developed in circles and clubs flourishing in
cities and rural areas among members of different social classes. As Tanaka Yūko
has argued, "Many of the art forms and accomplishments we associate with Japanese culture were forged within a system of small groups and networks, particularly during the Edo period."[3] These clubs and circles were variously called *ren*,
kai, *za*, or, less frequently, *kumi* and *sha*. They were not exclusive to *honzōgaku*
but were dedicated to a wide range of artistic and intellectual activities. There
were groups for composing *haikai* and *kyōka* poetry and for practicing the tea
ceremony (*chadō*). Others focused on *nō* chanting, dancing, ceramics, and the
reading of Confucian classics. Storytelling circles specialized in different genres,
the most popular of which were ghost stories and erotica.[4]

Cultural groups could be found in many rural and urban communities, but because the extemporaneous and amateur nature of their meetings left but scant traces, they constitute a particularly difficult and ambiguous subject for modern historians. These groups provided a venue for the pursuit of many different social and intellectual activities. Members exchanged information, collaborated in artistic performances, and competed in *awase* and *kurabe* contests, which, in the case of natural studies, consisted of games to recognize and name correctly mineral, vegetal, or animal species. Regardless of discipline or location, these groups had probably similar structure. Their members were usually individuals of some distinction in the community, but clubs could be open to people from different social classes. Gender composition differed depending on the cultural pursuit. While it was not rare for *haikai* circles to have women as their members (fig. 9.1), I have found no instances of *honzōgaku* groups with female membership. As far as established social circles and the written record are concerned, it seems that *honzōgaku* remained a "manly" pursuit—as was natural philosophy in early modern Europe, save for some illustrious exceptions. This male centeredness may have been the result of its origins in the samurai class or it may have been linked to a lingering influence of Neo-Confucian styles. Many of these *honzōgaku* adherents were amateur practitioners—individuals "interested in curiosities, wonders, and ancient things," as the amateur naturalist Suzuki Bokushi put it in his *Hokuetsu seppu* (1837–42).[5] Groups recognized merit and talent as well as sincere and serious commitment.

Professional scholars and artists often "visited" different groups as masters. Through the visiting master, circle members gained access to his larger network, even though usually no more than one or two of the members of the circle might have been his direct disciples. This access in turn might have placed a circle in an implicit competition with the others in the network for the distinction of producing the best poetic anthology or pictorial album or botanical monograph, which would be then published in limited copies at the group's expense. Circles in the countryside were generally smaller and composed of notable members from the local community. Village headmen (*shōya*, *nanushi*) and wealthy peasants (*gōnō*) were usually the organizers of cultural activities in a village. Rural *honzōgaku* circles were fewer in number than groups dedicated to religious or artistic activities, probably because natural history required means possessed only by political and economic elites who usually resided in castle towns. Members were mostly low-ranking samurai and wealthy peasant-merchants like Suzuki Bokusui—two social groups that, although distinct by law and status, tended in the second half of the period to associate.[6]

Many of the secondary sources on cultural circles in the Tokugawa period

FIGURE 9.1. A *kyōka* meeting from Yashima Gakutei,
Kyōka Nihon fudoki (1831). London: British Museum.

concentrate on artistic, religious, and literary groups.[7] Eiko Ikegami's study of
what she calls the "aesthetic networks of early modern Japan" focuses on practices that united members of different social classes in "aesthetic socializing."
These networks, she states, resulted in a new form of "civility," which she defines as "the cultural grammar of sociability that governs interactional public
spaces."[8] This cultural grammar of civility "flourishes best in an intermediate
zone of social relationships that lies between the intimate and the hostile" and
"governs social relations across differences in rank and status."[9] She argues that
"aesthetic associations," especially "horizontally structured associations" like *ren*
and *kumi*, played a fundamental role in creating new social identities and, in the
spatial and cultural spheres of group publics, contributed to the formation of
a new political culture that unfolded in its full potential in the second half of
the nineteenth century.[10]

While I am hesitant to share Ikegami's view that "aesthetic networks" symp-
tomize an emerging "public sphere," cultural associations offered a fertile ground
in which cultural capital could germinate and transform into social capital

through acquaintances and collaborations. In these intellectual communities, new forms of sociability loosened the rigidity of social etiquette sustaining the hierarchical order of Tokugawa society.[11] Because of the scarcity of sources, it is very difficult to reconstruct in detail the actual proceedings of circles' meetings, even when the patron of one of these salons, Kimura Kenkadō, kept a detailed diary of his social life.[12] I tend, however, to be skeptical of claims of the alleged liberality and egalitarianism of free exchanges of clubs and circles because the documents of the period do not support this account: the diaries of *daimyō* Matsuura Seizan, guest of Kenkadō's salon, described a rather formal politeness characterizing its convivial events.[13] Moreover, any sign of egalitarianism would have left traces in the subsequent cultural life of Meiji Japan—where the public sphere emerged as a result of the interaction of popular movements and Western notions of "liberalism," "constitutionalism," "individualism," and the like.[14]

It seems to me that clubs and circles had rather a stronger impact on the socioprofessional identity of scholars than on the social order. The multiplication of cultural clubs put into motion three dynamics that affected the social and intellectual trajectories of scholars. First, clubs and networks of amateurs offered to intellectual experts alternative venues for patronage distinct from feudal service, state support, and private academies. In these clubs, the *circulation* of knowledge increasingly replaced its *enclosure*, be it in the name of domanial and shogunal interest or of "secret transmission" (*hiden*) in private academies (*shijuku*). Second, new forms of legitimation of intellectual production began to operate separate from political authority. In turn, scholars were less bounded to the strict adherence to the orthodox styles, ideas, and notions of canonical texts that was conventional of *shijuku* scholars. These new dynamics of legitimation depended instead on the ability of individual scholars to manipulate concepts, information, and—in the case of *honzōgaku*—specimens in a witty, aesthetically appealing, and entertaining fashion. The atmosphere of formal conviviality that seemed to dominate club proceedings facilitated the adoption of an *intellectually* more relaxed attitude. Scholars did not move in an institutional or academic setting and thus enjoyed less control on the kind of ideas and practices they engaged with. Knowledge was pursued for curiosity, for fun, or for aesthetic pleasure: the exchange of specimens and information about newly imported and exotic plants or animals between collectors and amateurs was carried on via letters, or between jokes and sake cups, and often in domestic contexts like Kimura Kenkadō's house. Socioeconomic studies defines this as the strength of "weak ties"—that is, the way in which intellectual "informality" often facilitates the trying out of new forms of thought, new conceptions, and new experiments.[15]

Third, the commodification of knowledge performed by the burgeoning cultural clubs had dramatic consequences in the development of new methods of observation, experimentation, and, especially, accurate visual representation of nature, to the extent that they affected the very notion of species (*shu*). While chapter 11 will focus on the transformations in the perceptions and descriptions of the natural world, Kenkadō's salon and two of the best-known cultural circles of the time—the Shabenkai and the Shōhyakusha—exemplify how the formation of wider networks of circulation of knowledge and new styles of intellectual engagement affected scholars' professional identity and labor.

Ikegami juxtaposes scholars active in cultural circles to scholars with teaching responsibilities in private academies. *Kai* and *ren*, she argues, were horizontal and structurally malleable associations that often did not survive for long periods. *Shijuku*, on the contrary, followed the vertical and rigid structure of the *iemoto* system and could survive for generations.[16] Ikegami argues that, because of their structural differences, cultural associations and private academies entertained a mutually exclusive relationship. This was the result of having "emerged in response to two major socio-political conditions—the political structure segmented by status boundaries and the expanding market economy."[17] Such was probably the case for the *haikai* poetry networks. But in the field of natural history, the situation was far more complex. Being the head of a private academy did not prevent a *honzōgaku* scholar from participating in or even promoting group activity in cultural circles. The outstanding pictorial and encyclopedic works produced in the second half of the Tokugawa period were frequently the result of such communal labor.

Scholars teaching in private academies often protected their teachings with a sort of confidentiality agreement, whereby their students were not at liberty to divulge the secret knowledge they received (*hidenju*). These "secret transmissions" often functioned much like a marketing strategy to attract students to a school, since masters often published many of these alleged "secrets"—for example, Motoori Norinaga's *Kojikiden*. Scholars who owned and managed their own teaching institutions constituted, however, a minority of professional *jusha*. They were entrepreneurs struggling to survive in a varied market of cultural production. Ono Ranzan, for example, frequently condemned the advertising strategies adopted by his former master Matsuoka Gentatsu (Joan), but his critiques of Gentatsu's administrative methods were virulent in letters he wrote to convince potential students to enroll in his school rather than Gentatsu's. His critique arose in the context of competing for students with his former teacher.[18] Gentatsu was one of the few *honzōgaku* scholars able to sustain himself solely from the money earned from his successful public lectures and the enrollment fees to his academy. He established a complex system of gratuities (*sharei*) that his students had to pay in proportion to how much he revealed of his knowledge of plants and animals and of his "secret interpretations" of the major encyclopedias of materia medica. His school was divided into progressively advanced classes, whereby courses could be accessed only after the payment of a fee, which was also progressively determined.[19] Ranzan, on the other hand, set a fixed enrollment fee, and the advancement to a higher class depended on the successful completion of the previous one. As he wrote in a letter to one of his students, Muramatsu Hyōzaemon, from Noto Province,

> In answering your question regarding the organization of my school, it is my rule [*kahō*] to allow [participation to my advanced seminars] to all those persons who had attended an entire circle of lectures on *Honzō kōmoku*. For those who come from distant regions, if they have already studied by themselves for several years, I can accept them as if they had taken an entire circle of lectures. Years of training are the principle around which classes are organized, not the payment of fees. . . . I will not accept any student on the basis of monetary offers [like Master Gentatsu does].[20]

As was customary in most *shijuku* of the period, once a student was accepted into a school, he had to swear not to reveal his master's teaching to anyone. Moreover, in another letter sent to Hyōzaemon, Ranzan specifically warned that "the main activity of *honzōgaku* consists in collecting herbs in the field. It is sometimes worth comparing one's study with the many books that are published, but it must be remembered that basing one's training only on books of other schools is a source of disgrace. There is no need to study anything else [apart from attending my lectures]."[21] Ono Ranzan spent most of his life as researcher and teacher in his own private school in Kyoto, but he also took part in group activities in the botanical circle organized and supported by the Osaka merchant Kimura Kenkadō.[22] His colleagues and many of his students were similarly involved in amateur naturalist circles while managing their own schools.

INVESTING IN CULTURE: KENKADŌ'S HOUSE

Kimura Kōkyō—Kenkadō was the name he adopted for his artistic and intellectual transactions with painters, scholars, and naturalists—was one of the most successful sake brewers in the Osaka area.[23] Under the pseudonym of Tsuboiya Kichiemon, an alias he reserved for his life as a merchant, he speculated in the housing market in the Horie ward and acted as a shrewd moneylender.

In Kenkadō's case, changing names—common practice in Tokugawa Japan—was a useful tactic to distinguish his different social personae. Through money-lending he added a considerable amount of wealth to the one he had inherited from his father, as he bragged to the *daimyō* of Hirado Matsuura Seizan in a letter Seizan recorded in his memoir-like essay collection *Kasshi yawa*.[24]

As was often the case for the sons of well-off merchants, Kenkadō received a first-rate education. He began studying Kanō school painting at the age of six, Chinese poetry under Katayama Hokkai at eighteen, and Chinese literati thought and painting with Yanagisawa Kien and Ike Taiga. In his memoir *Sonsai okina zuihitsu* (Essays of Old Kenkadō), Kenkadō described himself as a sickly kid with a weak constitution. Because of his precarious health, his father allowed him to nourish an interest in the plants and trees growing in the family garden. An elderly Kenkadō recollected that, being often sick, he was visited periodically by the family doctor who introduced him to the works of Inō Jakusui and Matsuoka Gentatsu. He wrote that his passion for gardening led him to move to Kyoto to study under Matsuoka's disciple Tsushima Tsunenoshin, who was also the teacher of Ono Ranzan and an enterprising organizer of small exhibitions (*bussankai*) of rocks, plants, and animals from private collectors of the Osaka-Kyoto area.[25]

Kenkadō's activities in the field of *honzōgaku* can be summarized as follows: he bought and probably read several canonical textbooks; he collected numerous specimens of rare and exotic minerals, plants, and animals—many of them currently displayed at the Osaka Municipal Museum of Natural History; and he was very active in organizing meetings in his house-salon with the finest scholars of the time. If his collection lacked the somewhat flamboyant outlook of many *Wunderkammern*, or cabinets of curiosities, in contemporary Europe, it matched their variety of rare and uncommon species. His connections to maritime traders active in the port of Osaka—directly connected to Nagasaki and, from there, to the world—and the large amount of capital at his disposal allowed Kenkadō to satisfy his curiosity buying from the catalogs of rare birds, precious stones, and other exotic animals compiled under the supervision of the Nagasaki *machi bugyō*.

Although he was a mediocre painter in the literati style; an even a more mediocre composer of poetry in Chinese; a superficial reader of Neo-Confucian texts; and a naturalist with more means than skills, Kenkadō seems ubiquitous in the historiography of Tokugawa culture. His passion for natural history and other cultural pursuits became the catalyst that brought into existence, put into motion, and maintained through much needed financial lubrication a vast network of intellectual interactions and production. In his house, he hosted regular meetings of various experts in different fields. Painters like Yosa Buson, Uragami Gyokudō, and Maruyama Ōkyo; men of letters like Ueda Akinari and Motoori Norinaga; thinkers like Minagawa Kien; *rangaku* scholars like Otsuki Gentaku and Shiba Kōkan; and naturalists like Ono Ranzan and Odaka Motoyasu— the best minds of the period—regularly attended discussions and intellectual convivial events in his salon. Kenkadō was the engine of a small-scale Japanese "Republic of Letters."

The sociological claim that culture gave respectability to wealthy towns-people—dominated classes according to the social hierarchy established by the Tokugawa law, which the commodification and monetization of the economy were rendering increasingly influential—should not induce us to think that Kenkadō and other enthusiastic patrons were shrewdly and consciously exploiting this mechanism.[26] Cultural capital, Pierre Bourdieu and others have argued, exists and acts always as an ensemble of cultivated dispositions that are internalized by the individual through socialization and constitute often unconscious schemes of appreciation and understanding.[27] Furthermore, for Kenkadō and other patrons, cultural capital functioned in embodied objects—such as books, works of art, natural specimens, and scientific instruments—which require specialized abilities to use and appreciate. Kenkadō's obsession with his collections of books, illustrations, and rare specimens is well evidenced in his diaries and memoirs, where he listed, described, and exhibited stones, shells, and stuffed birds, among other specimens.[28]

His conception and pursuit of different cultural activities as entertainment and convivial divertissement enhanced his distinction. Kenkadō was not a scholar: he never pretended to be one, nor did he engage in *honzōgaku* research imitating the doctrinaire or wear the formal fashion of professional scholars. Rocks, plants, and animals tickled his curiosity. He looked at nature in search of wonders and marvels to experience, collect, and possess, not to discern the inner order of the universe. His disposition typified the tastes of the emerging bourgeoisie in the early modern world. His works display playfulness and effortlessness, an almost "postmodern" temperament to let the imagination wander in the labyrinths of *honzōgaku* encyclopedias in search of uncanny homologies and astonishing curiosities. In *Kenkadō zasshi* (Scattered Notes), the playfulness of his engagement with *honzōgaku* and learning in general is particularly evident. Kenkadō sketched objects and events that captured his imagination. He enthusiastically described the new study room he furnished or a bizarre lantern he noticed a few days prior, he sketched a pair of deer horns he saw at a friend's house, or he drew rough but evocative pictures of rare and strange fish he read about in a recently acquired encyclopedia. At one point he recounted a visit to a shop in Hirakawa, in the Kōjimachi ward of Edo, to watch a white cock that was born with two pairs of wings and two pairs of legs.[29] Neither selective nor methodical, Kenkadō described with a fanciful tone curious or gracious objects, both natural and artificial, that were the subject of witty and erudite conversations with his guests. While focused more on natural history and art, Kenkadō's work was not dissimilar from that of Ōta Nanpo or Negishi Yasumori, two chroniclers of rumors and fashions, scandals, haunted houses, and mundane events of Tokugawa Japan's urban life.

Kenkadō's patronage of culture was centered in the activities of his "salon." Kenkadō's meetings became so celebrated that in the fourth month of 1794, the VOC physician Bernhard Keller was the guest of honor. His indefatigable organization of educated meetings served him well in establishing his social position, as demonstrated by the attendance of two *daimyō*—Masuyama Sessai from the Nagashima Domain and Matsuura Seisan of Hirado—in his *honzōgaku* club. That two domainal lords were regulars in a commoner's house was a remarkable novelty for eighteenth-century Japan. Unfortunately, Kenkadō did not describe in any detail the actual proceedings of his *honzōgaku* meetings, but he recorded in his diary, *Kenkadō nikki*, every person of rank entering his house between 1773 and 1801.[30] The domainal lord Matsuura Seizan eulogized Kenkadō's generosity as a host in his *Kasshi yawa*, but he did not elaborate beyond his skills in serving tea and sake and the deliciousness of the cookies.[31]

Kimura Kenkadō sponsored and edited a series of catalogs in his collection, among which the most renown are *Kikai zufu* (Illustrated Catalog of Rare Shells), *Chikufu* (On Bamboos), and *Gyofu* (On Fish). *Kikai zufu* is a large album lavishly illustrated with full-page pictures of colorful shells (fig. 9.2). Most of his other catalogs, however, had small, standardized illustrations, with just the "official" name—the entry name in either *Honzō kōmoku* or *Shobutsu ruisan*—of the species and some scant notes here and there (fig. 9.3). These albums looked like merchants' store inventories and merchandise catalogs, like the ones Katsushika Hokusai illustrated with decorative patterns and textile designs or the ones customers could flip through at Echigoya or other old department stores. The plants and animals Kenkadō collected and described for curiosity or for fun, to expand his knowledge of the natural world or to expand his social reach, were intellectual and material commodities whose circulation in books, exhibitions, and meetings constituted the capital of Kenkadō's social trajectory.

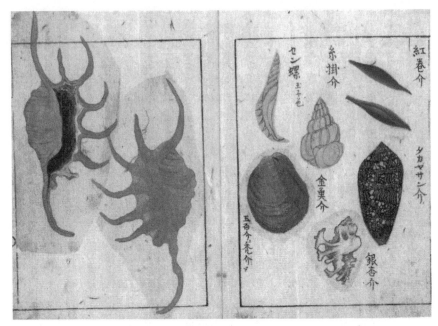

FIGURE 9.2. *Kikai zufu* (1775). Nishinomiya: Tatsuuma Collection of Fine Art.

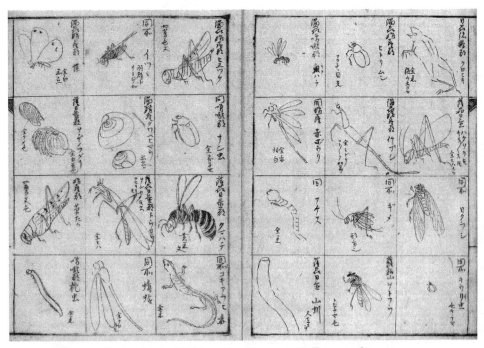

FIGURE 9.3. *Sasshū chūhin*. Nishinomiya: Tatsuuma Collection of Fine Art.

Two prominent *honzōgaku* associations were the Shabenkai, centered in Edo, and the Shōhyakusha, based in Nagoya.[32] Members of the "Association of the Red Rod"—as Shabenkai can be translated in English—were mostly domainal lords, their high-ranking retainers, and shogunal officials. They met regularly to discuss plants and animals and to publish illustrated monographs and albums depicting specimens from their collections. Their name derived from the legend according to which the emperor Shennong used a red (*sha*) cane (*ben*) to whip those herbs that became the basic pharmacological substances in early Chinese pharmacopoeias. The group officially formed in 1838 around Maeda Toshiyasu, lord of the Toyama Domain in Etchū Province, but its members had been informally meeting since the early 1810s.[33] Shabenkai's members included Toshiyasu himself; Kuroda Narikiyo, lord of the Fukuoka Domain in Chikuzen Province; the Tokugawa retainer Musashi Yoshitoki (Sekiju); Baba Daisuke, head of the guards of the Nishinomaru palace; and the shogunal officials Shidara Sadatomo, Iimuro Masanobu, Sabase Yoshiyori, Tamaru Naonobu, and Asaka Naomitsu.[34] The leading *honzōgaku* scholars connected to the group were Kurimoto Tanshū, shogunal physician and son of Tamura Ransui, and his son Kurimoto Joun.[35] Apart from Tanshū and Joun, most of the members came from the upper ranks of the samurai class, which meant that Shabenkai associates had a considerable amount of leisure time and resources.

Some of the Shabenkai meetings were week-long sessions of dedicated research, conducted with established encyclopedias and on specimens from the personal collections of Shabenkai members. But these were not systematic efforts and did not follow any methodical research program. Most of the albums and monographs they produced resulted from extemporaneous interests and whimsical passions, conceived more to tickle members' curiosity for wondrous and bizarre phenomena of nature than to extend knowledge for utilitarian, moral, or encyclopedic ends. Some consisted of detailed and meticulous cataloging of one's private collection, like Musashi Sekijun's albums (fig. 9.4). Others could be more lyrically conceived like Baba Daisuke's charming portraits of fowls, flowers, wild dogs, and badgers caught in their natural environment (fig. 9.5).

In general, the Shabenkai circle produced some of the finest pictorial representations of plants, animals, shells, and rocks. Specimens and materials they did not already possess could be acquired by dispatching their retainers or by utilizing their connections with the shogunate. They hired the best painters specializing in natural subjects and were able to provide them with live or preserved

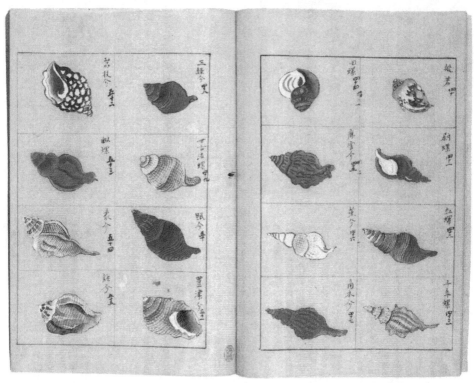

FIGURE 9.4. Musashi Sekijun, *Kōkaigun bunpin*. Tokyo: National Diet Library.

specimens. It is not known how the group's meetings were actually conducted. As in most cultural circles of the period, they probably followed a ritualized agenda in the formal convivial setting of one of the highest-ranking members' mansions in Edo. The group likely met regularly once a month, some times more frequently, each gathering focusing on a predetermined topic. Once they de-cided on a research topic, usually a group of animals or plants, members me-ticulously read sources, commented on them, carefully observed live specimens, and often composed *haikai* poetry that took the chosen animal or plant as its main theme. Once they had exhausted their research, they often sponsored the production of an illustrated and annotated album. These albums were usually donated—rarely sold—to schools and other groups. These "gifts," of course, re-quired recognition and respect in exchange.

Shabenkai hinbutsu ron teisan (Shabenkai Selection of Researched Species) was one such catalog featuring a selection of the species the Shaben group had researched. It consisted of an album of single-leaf pages bound together, each

FIGURE 9.5. Baba Daisuke, *Hakubutsukan jūfu*.
Tokyo: National Museum of Natural History.

page portraying one species of plant or animal from the private collection of a member of the group. Pictures were artistically conceived and composed, but the renderings also displayed an epistemological awareness for the accurate description of specimens.[36] Some were detailed drawings of concrete objects, like the two fish in figure 9.6—a preserved stonefish on the right and a dried red stingray on the left—taken from the private collections of two members of the group, Sabase Yoshiyori and Shidara Sadatomo.[37] This example gives a sense not only of the exotic species taken up by the Shabenkai—chosen specifically for their bizarre aspect—but also of the conscious effort to portray them in an aesthetically refined way to stimulate the curiosity (*suki*) and wonder of fellow amateur naturalists.

Shabenkai's study sessions also produced monographs dedicated to a specific family or group of animals and plants, which usually consisted of a series of large pictures with brief notes. These albums included *Hototogisu zusetsu*, an illustrated essay on cuckoos; *Shōma zusetsu*, on the bugbane rhizome; and *Toneri zusetsu*, on Japanese ash trees.[38] Their most famous works were probably *Chūfu*

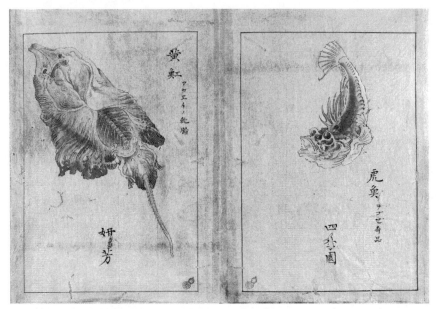

FIGURE 9.6. Red stingray and a stonefish from *Shabenkai hinbutsu ron teisan* (1837). Tokyo: National Diet Library.

zusetsu (Illustrated Manual of Insects), edited by Iimuro Masanobu, and *Mokuhachi fu* (Conchology), edited by Musashi Yoshitoki.

Honzōgaku offered the high-ranking members of the Shabenkai a perfectly commendable pastime: it was highly educated, it had ties with the Neo-Confucian ideal of the investigation of things, it had been receiving approval and support from the shogunate since Yoshimune's rule one century before, it was thought to help moral cultivation, it had connections with the refined literati culture of China (*bunjin*), it assisted in the composition of poetry (at the time more a means of sociability than a purely aesthetic endeavor), it was of practical value to sustain the welfare of the state and its people, and it was a reservoir of never-ending wonders from around the world.

The professional scholars Kurimoto Tanshū and his son Joun profited financially, institutionally, and intellectually from their association with the affluent members of the Shabenkai. Thanks to their financial support and the rare specimens and accomplished artists at their disposal, the two scholars could afford to enrich their own essays with illustrations of live animals and plants. Among Tanshū's best-known publications was *Senchūfu* (A Thousand Insects), a monograph dedicated to 645 species of insects that, on the one hand, followed faith-

FIGURE 9.7. Kurimoto Tanshū, *Senchūfu* (1811). Tokyo: National Diet Library.

fully *Honzō kōmoku*'s classification (figs. 9.7 and 9.8) but, on the other, visually il-lustrated in unprecedented precision the minutest details of insect morphology in the various stages of their life cycle.[39] His participation to Shabenkai meetings gave him privileged access to a great number of rare specimens to observe. Figure 9.7, a page from the monograph, portrays, on the right, a female and male great purple emperor butterflies, minutely detailed and accompanied by instructions on how to distinguish the female from the male and their differing behavior.[40] The figure on the left depicts a group of *Cyana hamata* maenamii (above) and a group of Hesperiinae (below) and describes their life cycle starting from "the first third of the fifth month."

"Insects" (*chū*) was a large class of animals that included, according to Li Shizhen, insects, arachnids, marine and land invertebrates, and different species of amphibians and reptiles. Figure 9.8 depicts a flying fox eating a sweet potato, also classified as an insect. *Yaeyama kōmori*, today known as *kubiwa ōkōmori*, is a *Pteropus dasymallus*, a species of bat common in the subtropical forests of

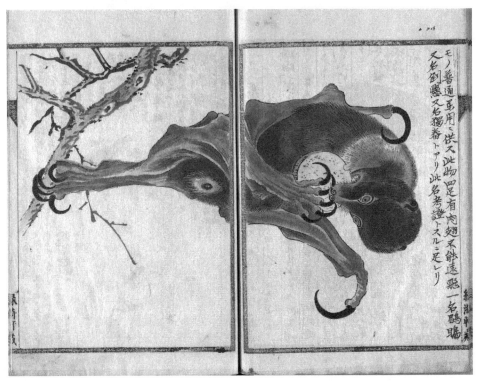

FIGURE 9.8. Kurimoto Tanshū, *Senchūfu* (1811). Tokyo: National Diet Library.

Taiwan, the Philippines, and the Ryūkyū Islands. These flying foxes had been known in Japan since the beginning of the Tokugawa period, when, according to the *Tokugawa jikki*, in 1642 shogun Iemitsu offered the young *daimyō* of the Kii Domain Tokugawa Mitsusada with one exemplar of bat he had received as a gift from the Ryūkyū Islands, which were under the suzerainty of the Shimazu of Satsuma since 1609.[41]

Tanshū and Joun's relationship with the Shabenkai group was by no means parasitic. Thanks to his fluent knowledge of Chinese canonical encyclopedias, Tanshū provided intellectual validation to the group and contributed scholarly excellence to their cultural activities. In turn, Tanshū benefited from the connections and means of its powerful members. He had the opportunity to observe live exotic species that he would not otherwise have had access to, including flying foxes (fig. 9.8), Sunda porcupines, and sunfish.[42]

Kurimoto Tanshū had been involved in other circles before entering the Shabenkai. In 1820, he had joined a team of high-ranking scholars and officials engaged in a comprehensive study of *kappa*, mythological creatures similar to "water goblins" that populated folktales and legends. The team consisted of a teacher of the Shōheikō, Koga Tōan; the governor for the Kantō and Tōkai regions, Hakura Yōkyū; the shogunal official Nakagami Kundo; and Tanshū himself. The research, titled *Suiko kōryaku* (A Study of Water Tigers), was based on hundreds of eyewitness reports and observations of *kappa* corpses and body parts supposedly unearthed in various provinces.[43] The team collected testimonies and carefully analyzed them. On the basis of their study, they were able to map the geographical distribution of *kappa*. Tanshū and his colleagues were persuaded by the veracity of the collected data because the reports displayed a strong consistency with a pattern of geographical variation: people of different social standings in the same region had independently experienced the same kind of encounter with a similarly shaped water goblin. The existence of *kappa*-like creatures was also confirmed by Chinese sources, which mentioned a monstrous creature called *shuihu* with features that resembled the "water-goblins" allegedly inhabiting Japanese rivers and lakes.

Suiko kōryaku was a remarkable achievement. In fact, rather than cataloging legends and literary references on *kappa* like other encyclopedias, the team of researchers treated the water goblins as any other animal, describing anatomical features, size and color, behavior, ecosystem, habits, and the like. These monsters were by them naturalized: they became objects to be studied, understood, classified, and researched in the same manner that exotic plants and animals were. The naturalization of monsters is another common elements that early nineteenth-century *honzōgaku* scholars shared with early modern European naturalists. As Lorraine Daston and Katharine Park have argued, "Originally part of the prodigy canon, with its ominous religious resonances, monsters shifted over the course of the sixteenth century to become natural wonders—source of delight and pleasure—and then to become objects of scientific inquiry."[44] As such, they were subjected to the same analytical procedures as any other natural species. Tanshū, like the botanist Ulisse Aldrovandi, the physician and surgeon Ambroise Paré, and, later, the naturalist *compte* de Buffon in eighteenth-century France, struggled to achieve a precise knowledge of nature's myriads objects and a faithful pictorial reproduction of their appearance without excluding a priori anything.

Some years after the publication of *Suiko kōryaku*, the two naturalists Sakamoto Kōnen and his brother Juntaku contributed data they had personally collected. They produced a broadsheet-sized schema illustrating the morphology and regional variations of *kappa* titled *Suiko jūni hin no zu* (Twelve Types of

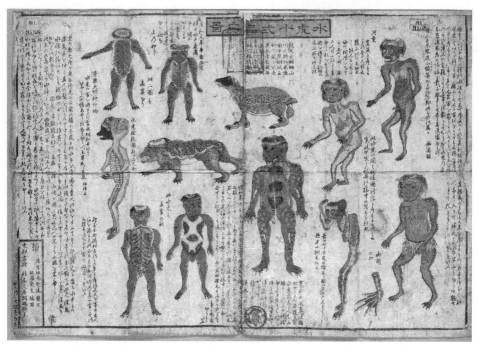

FIGURE 9.9. Sakamoto Kōnen and Juntaku, *Suiko jūni hin no zu.* Tokyo: National Diet Library.

Water Tiger, fig. 9.9). Each illustration, faithfully taken from *Suiko kōryaku*, was accompanied by notes and information about the location of that particular species of *kappa*.

DESCRIBING NATURE: THE "SOCIETY OF THE ONE HUNDRED LICKS"

Kōnen and Juntaku were members of another famous group of *honzō-gaku*, the Shōhyakusha. The "Society of the One Hundred Licks" was formed by mid- and low-ranking samurai from the Owari Domain and physicians from the Nagoya area, and as such it was more a rural than an urban circle of *honzōgaku*.[45] The group leader, a physician named Mizutani Hōbun, had studied medicine and *honzōgaku* under Asano Shundō, a student of Ono Ranzan, and his approach to the discipline was marked by an explicit interest on the medicinal properties of plants.[46] Other members included Owari retainers Ōkōchi Zonshin, his brother Itō Keisuke, Ōkubo Tahei, his son Masaaki, and Yoshida Takanori.

Like the Shabenkai associates, the Shōhyakusha met regularly to discuss and conduct research on plants and animals. In addition to Chinese and Japanese encyclopedias, they also used Western texts as sources. Some of them were familiar with Linnaean classification: Mizutani Hōbun arranged his collection of specimens and sketches of vegetal and animal species in accordance with Linnaeus's taxonomy; his disciple Itō Keisuke, as we shall see in chapter 12, studied under Franz von Siebold and later became the first Japanese to receive a doctorate in science in the newly established Tokyo Imperial University. Shōhyakusha's main activities consisted of recording the results of personal observations of and collegial discussions about specimens that they collected or captured during naturalist expeditions in the Owari region. Their reliance on observation (*jikken*) and accurate, true-to-nature pictorial descriptions of plants and animals (*shashin*) has suggested to historians of science that Shōhyakusha members had developed a scientific approach. Indeed, their epistemological method suggests that their practices were analogous to inductive inferences that today we attribute to the scientific method.

The epistemological issues at the basis of Hōbun's work was still the precise identification of natural species they gathered in order to exploit their healing properties. A deep concern for the practical, therapeutic utility of plants motivated the investigation of Hōbun, a trained physician in both the Koihō (Ancient School) method and *rangaku* surgical techniques.[47] But while they openly question what they called "false learning" (*kyogaku*) and the blind adherence to received knowledge, they never really questioned the *honzōgaku* tradition as a whole. Rather, they supplemented it with information they gathered from their own observations and experiments and from Western books. Shōhyakusha naturalists did not conceive of practices like observation and description as an alternative methodology aimed at substituting the acquired knowledge of Chinese encyclopedias, nor did they conceive of Linnaeus's taxonomy as opposed to and irreconcilable with *Honzō kōmoku*'s.[48] They were not radicals or revolutionaries, nor did they intend to obliterate the received *honzōgaku* knowledge as Keisuke's friend Udagawa Yōan would later do.[49] Hōbun never completely questioned the medicinal efficacy of traditional pharmacopeias, and while he acknowledged the pragmatic advantages of the binominal system, he struggled to make it compatible with the received knowledge of traditional *honzōgaku*. If anything, he and Keisuke utilized it to correct and update the information contained in canonical sources. In other words, established encyclopedias remained for them a valuable source of pharmacological information about the species of plants and animals they observed in nature.[50] Observation alone did not have sufficient epistemological value to replace received information. Rather, their sophisticated observational and descriptive techniques served to more precisely match the plants they studied to the species treated in canonical encyclopedias—supplemented with information they gathered from Western sources. Maki Fukuoka has persuasively defined their epistemological stance as a "triangulation" of collegial observation of actual specimens, their rendition in true-to-nature illustrations, and their reference to multilingual sources (Chinese, Japanese, and Dutch), whereby knowledge resulted from a negotiation between these moments, and received knowledge was, in a sense, "tested" by their experience.[51]

This attitude did not protect them from mistakes. In one such instance, which reveals the reliance that these scholars still had on canonical sources, Hōbun described a medium-sized bird he had captured the fourth month of 1809 off the coast of Atsuta as an *etohiruka*, a tufted puffin, a pelagic seabird living in the northern Pacific Ocean (fig. 9.10).[52] In reality what he had captured was a horned puffin, later known as a *tsunomedori* in Japanese, another species of pelagic seabirds that breed on rocky islands off the coasts of Siberia and Alaska that was yet to be described in East Asian natural historical tradition.[53] The two birds are dissimilar in shape, color, and especially size, but Hōbun forced his description of the bird he had captured (a horned puffin) to fit descriptions from Chinese sources of the tufted puffin. He claimed this despite the fact that paintings of tufted puffins were well known in *honzōgaku* communities, like the one by the talented amateur painter-naturalist Mashiyama Masakata, *daimyō* of the nearby Nagashima Domain (fig. 9.11).[54] *Etohiruka* were even once the subjects of a meeting in Kimura Kenkadō's salon.[55]

In cases of uncertain identification, Western sources—books or conversation with Siebold—were for them a source of verification of their speculations. Information and data gathered from Western texts were not meant to replace the established knowledge of *honzōgaku* encyclopedias and textbooks but to complete, complement, and eventually correct them. This does not mean that Shōhyakusha members' knowledge of Western natural history was shallow or that a "scientific attitude" was yet to emerge in a still feudal Japan. The search for faithful representations of natural species (*shashin*) was as important for Shōhyakusha scholars as it was for contemporary European naturalists: as I have repeatedly shown, *honzōgaku* was as sophisticated a discipline of nature studies as natural history was in early modern Europe. In 1826, VOC physician and naturalist Philipp von Siebold, on his way to Edo, purposely stopped in Miya, a small town near Nagoya, to meet with Hōbun. He was impressed by the

エトヒルカ

エトヒルカ

ヱトミ
ノ圖

文化六巳年四月熱田海中ニテ捕ル
大サハトノ如ク頬ト脇白色甚ハ真
黒色甚美ナリ肯大ニノヨコセヤレ
鳴声ゲーくト云

FIGURE 9.10. A "tufted" puffin (in reality, a horned puffin) from Mizutani Hōbun, *Mizutani shi tori fu* (Mizutani's Bird Handbook). Tokyo: National Diet Library.

FIGURE 9.11. Mashiyama Masakata, *Etohiruka* (Tufted Puffin). Tokyo: National Diet Library.

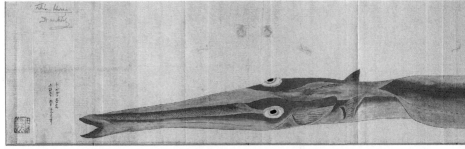

FIGURE 9.12. *Akayagara*. Tokyo: National Diet Library.

precision and accuracy with which these *honzōgaku* scholars had arranged specimens and how they had correctly named them in accordance with Linnaean binominal taxonomy. Hōbun had studied Western medicine (*ranpōi*) under Nomura Ritsuei, who had learned to read Dutch from the Nagasaki interpreter Yoshio Shunzō and was fluent with European textbooks of material medica.[56] In collaboration with Okabayashi Kiyotatsu, Hōbun had edited a glossary of names of vegetal and animal species, *Buppin shikimei* (Clarifications on the Names of Things, 1809), in alphabetical order (*iroha*) and with the Latin names attached. Siebold reported in his diary that the correctness of *Buppin shikimei* surprised him.[57] During the meeting, von Siebold was attracted by a picture of a red cornetfish, a deep-sea fish, more than two meters long, common in the tropical seas of East Asia (fig. 9.12). Drawn by Hōbun's student Itō Keisuke — who later inserted the picture in his *Kinka gyofu* (Itō's Book of Fish) — it was named *akayagara*. Siebold penciled the term *Fistularia tabacaria* in the upper left corner of the paper along with his name, Dr. Siebold.[58] The meeting was an occasion for both Hōbun and Siebold to exchange information about plants and animals as well as on the methods they adopted to study them. As a result of that meeting, Hōbun's disciple Keisuke would enroll in Siebold's school in Nagasaki the following year.

Hōbun, and later Keisuke, developed glossaries that elegantly avoided addressing the issue of classification by listing species in alphabetical order. Hōbun edited a thirty-volume encyclopedia, *Honzō kōmoku kibun* (Notes on *Honzō kōmoku*), which summarized Ono Ranzan's *Honzō kōmoku keimō* with personal annotations and beautiful illustrations in a typical *honzōgaku* style. Although Hōbun knew Linnaeus's binominal system and applied it correctly to name species of plants and animals, in the encyclopedia he simply attached the Latin names to pictures of birds, insects, and flowering plants without deviating much

FIGURE 9.13. Itō Keisuke, *Azarashi*, *Ningyo* (Seal and Siren) from *Kinka jūfu*. Tokyo: National Diet Library.

from the traditional Japanese classificatory system. The Latin name embellished the pictorial representation of a plant or an animal. Providing the Japanese, Chinese, and Latin name of a given portrayed species conferred a certain encyclopedic completeness.

Itō Keisuke, who was destined to be both the last *honzōgaku* scholar and the first Japanese biologist, also reproduced traditional styles and methods in many of his works. His *Kinka jūfu* (Illustrated Guidebook of Beasts) and *Kinka gyofu* (Illustrated Guidebook of Fish), two collection of drawings of mammals and fish, respectively, included sunfish, sharks, and snappers but also mermaids (*ningyo*) and other supernatural creatures (fig. 9.13).

The attitude of Shōhyakusha's members to Western knowledge is evident also in Ōkubo Masaaki's description of a bearded seal he caught in the seventh month of 1833 at Atsuta.[59] His illustration is precise and detailed, but he none-theless continued to classify the seal as a fish. Seals, known in Japanese as *aza-rashi sawagi*, like the one described by Masaaki, were often caught in harbors, attracting visitors from nearby areas.[60] Despite knowledge of alternative taxono-mies being used elsewhere that classified seals as mammals, in Masaaki's descrip-tion, the *azarashi* remained a fish, as dictated by *Honzō kōmoku*.

AN ARISTOCRATIC PASSION

The Shabenkai and Shōhyakusha societies produced probably the more stunning atlases and monographs of the last decades of the Tokugawa period. But other clubs and circles dedicated to the study of nature had long been active since the second half of the eighteenth century, often sponsored by domainal lords and high-ranking shogunal officials in imitation of Yoshimune's patron-age of *honzōgaku*. One such patron was Maeda Toshiyasu, lord of the Toyama Domain and an early member of the Shabenkai. In addition to active participa-tion in his circle, Toshiyasu authored the encyclopedia *Honzō tsūkan* (Compen-dium of Materia Medica), which he left incomplete at the time of his death with ninety-four volumes finished.[61] *Honzō tsūkan*, written in Chinese, was designed to complement Jakusui and Shōhaku's *Shobutsu ruisan*. Toshiyasu also spon-sored the publication of an illustrated abridgement of his encyclopedia, titled *Honzō tsūkan shōzu* (Faithful Illustrations of *Honzō tsūkan*, 1853).

Scholar Okada Atsuyuki wrote in the introduction of *Honzō tsūkan shōzu*,

> My Lord [Maeda Toshiyasu] has just completed his magnificent *Honzō tsū-kan*, where he corrected the mistakes he found in *Shobutsu ruisan*. However, since his work is written in Chinese, it is hard for the people of his domain to understand it. And since there are no illustrations, it is difficult to use. Some of his retainers therefore suggested that he compile another book, easier to understand and richly illustrated. At first his Lordship replied that he could not possibly write a simplified version of the fruits of his extensive experience [in the field of *honzōgaku*], but when we continued to insist, his Lordship changed his mind and decided: "Rather than allowing only a few of my retainers to cultivate in their garden medicinal herbs, I should give these blessings to the larger number of people, and let them know the names and the shapes of medicinal herbs." It was for this reason that he started the compilation of this book.[62]

The text consisted of full-page illustrations of medicinal herbs with their names written above (fig. 9.14). The handbook served as a quick and precise identifica-

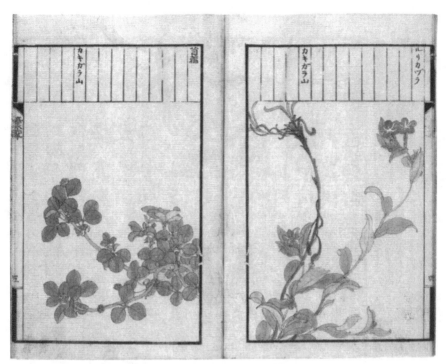

FIGURE 9.14. Maeda Toshiyasu, *Honzō tsūkan
shōzu* (1853). Tokyo: National Archives of Japan.

tion tool for plants. Toshiyasu became domainal lord during the Tenpō famine
(1833–36) and knew firsthand the horror of famine and disease. The experience
of the famine motivated the publication of the slim guidebook, which he distributed to all villages of his domain. Atsuyuki wrote that the decision to add
meticulous illustrations of plants and animals to the manual was made out of
concern for the welfare of the general public in times of hardship and to make it
easier for a less educated readership to recognize specific plants.

The epistemological value of illustrations was apparent also in other atlases
and monographs produced by domainal lords. Matsudaira Yoritaka, lord of the
Takamatsu Domain in Sanuki Province, sponsored some of the most technically
advanced illustrations of nature.[63] Hosokawa Shigekata, lord of the Kumamoto
Domain in Higo Province, was a dynamic and successful ruler who personally
conducted sophisticated research on insect morphology and behavior, which
he minutely illustrated with detailed pictures.[64] Date Munemura, lord of the
Sendai Domain in Mutsu Province, distinguished himself by commissioning

FIGURE 9.15. A *kandai* (*Semicossyphus reticulatus* Valenciennes) from an original sketch of Kurimoto Tanshū in *Shūrinzu*. Tokyo: National Diet Library.

beautiful illustrations of birds.[65] Mashiyama Masakata, lord of the Nagashima Domain in Ise Province, was known for his paintings of birds and butterflies.[66] Shimazu Shigehide, lord of the Satsuma Domain, sponsored a comprehensive survey of the flora and fauna of the Ryūkyū Islands and established an active network of scholars in the castle town of Kagoshima.[67] Satake Yoshiatsu, lord of the Akita Domain, produced outstanding illustrated plates of worms and centipedes.[68] Hotta Masaatsu, a young shogunal councilor (*wakadoshiyori*) and skilled painter, continued his father Date Munemura's project of bird illustrations.[69] Tōdō Takayuki, lord of the Tsu Domain in Ise Province, and the already mentioned Kuroda Narikiyo, lord of the Fukuoka Domain, were both involved in the Shabenkai.[70] These individuals are only a few of the better known domainal lords actively involved in *honzōgaku* study.[71]

The *honzōgaku* research conducted by domainal lords shared a common emphasis on illustration. Their circles competed to produce the finest and most accurate renderings of plants and animals. This was especially noteworthy in the case of Matsudaira Yoritaka, who sponsored the production of arguably the most sophisticated pictorial illustrations of the time. A team of artists under the direction of Yoritaka's retainer Hiraga Gennai produced a series of albums that granted Yoritaka recognition among his peers for excellency in *honzōgaku* achievements—to the detriment of his direct rival amateur naturalists, Lord Hosokawa Shigekata of Kumamoto.[72] Yoritaka's name is associated with the ed-

FIGURE 9.16. A *daruma inko* (*Tanygnathus megalorynchos*), developed from an original sketch of Kurimoto Tanshū in *Shūkinzu*. Tokyo: National Diet Library.

iting and production of richly illustrated albums like *Shūhō gafu* (Illustrated Manual of Ornamental Plants) and *Shasei gachō* (Album of Sketches from Life), which focused on flowering plants. Other albums included *Shūrinzu* (Illustrations of Fish) and *Shūkinzu* (Illustrations of Birds).

The distinctiveness of Yoritaka's monographs, a result of Gennai's technical genius, came not only from the vivid colors of the petals, fish scales, and birds' feathers but also from the sensation they produced when touched. The feathers, scales, leaves, and petals were represented in relief, an effect probably achieved by pasting glue (*nikawa*) or lacquer (*urushi*) over the sketch of the animals (figs. 9.15 and 9.16).[73] The scales of the fish were attached one by one, colored, and then glued with a translucent varnish (fig. 9.15). As the glue hardened, it was incised with the pattern of feathers (fig. 9.16) or with the veining of leaves and then colored.[74] The final result gave the eye a "realistic" visual representation of animals and plants and the hands a haptic representation.[75]

For moral cultivation or just for fun, for utilitarian goals or for the expansion of human knowledge, studying nature in cultural clubs displayed a growing concern for accurate and truthful representation of plants and animals. Both samurai and bourgeois patrons generously invested in scholars and painters to organize public exhibitions of rare and bizarre specimens of plants and animals and to produce the most astonishing and technologically advanced illustrations of natural species in monographs and albums. But what were the functions and goals behind such a concern for faithful, accurate descriptions of plants and animals? How did it affect the study of nature and the way natural species were conceived? Chapter 11 addresses these questions, while next chapter reconstructs the organization of public exhibitions of specimens of plants and animals in Edo and Kyoto.

10 *Nature Exhibited*
Hiraga Gennai

· · · · · · · · · ·

Members of a cultural circle could achieve distinction through the investment of time, resources, and expertise in the activities of the group. For samurai elites, wealthy commoners, and professional or amateur scholars, possessing nature, either in the form of collections of exotic plants and live or embalmed animals or in the form of refined connoisseurship, came to represent a powerful social marker. Dominion over nature symbolically extended into the social. Plants and animals, especially exotic and rare specimens from Southeast Asia and Africa, were objectified as commodities to be enjoyed as spectacles (*misemono*) and entertainments in Japanese towns and cities. There were teahouses that offered displays of colorful tropical birds, parakeets, and peacocks for the amusement of their clients. Public and semiprivate gardens organized tours for club members seeking inspiration for their poems or calligraphy through the admiration of azaleas, morning glories, and camellias. There were street shows of *hakusai mono* (imported things). Parades of unusual animals were a great sensation for curious townsmen and were recorded in many gazetteers, newsletters, and diaries. These included, in the hundred years between 1750 and the 1850, a particularly large sunfish (1762), a small flock of Barbary sheep (1769),[1] a big African porcupine (in Osaka and Kyoto in 1773 and later in Edo in 1775, fig. 10.1), a flock of European sheep (1775), Western weasels (1776), a wolf (1777), a Java badger (1778), two cassowaries (1778), and a civet (1786).[2] An enormous blue whale was transported to Osaka in 1789, and a great white shark was exhibited in Kyoto, Osaka, and Edo in 1792.

Then came two orangutans (1792); wild asses (1792); crowned pigeons (1795); a leopard seal (1795); a white wild boar (1801); a big white constrictor (1802); an octopus so large that it occupied a three-tatami room (1803);[3] seven snakeheads (1804); oriental greenfinches (1805); gibbons (1807); a black sea hare (1808); Eurasian jays and Calandra larks (1808); big American hedgehogs (1809); a two-tailed dog (1811); chestnut manikins (1812); sugarbirds (1812); wildcats (1813); a kangaroo (1814); giant bats (1815); yellow-crested cockatoos (1816); bengaleses (1816); vinous-breasted starlings (1816); warbling finches (1818); Yucatan jays (1819); bramblings (1819); guineafowls (1819); Celebes macaques (1819); fairy pittas (1820); magpie-robins (1820); a couple of dromedaries parading in Osaka

FIGURE 10.1. An African porcupine from *Shijō kawara yūraku zu byōbu.*

(1821), Nagoya (1823), and later in Edo (1825); tragopans (1825); horseshoe crabs (1827); a bicephalous horse (1827); leopards (1827); an albino brown bear (1832); a very long moray (1833); a slow loris (1833); European squirrels (1833); zebra finches (1835); an embalmed narwhal (1836); a giant leatherback sea turtle (1840); guinea pigs (1843); babblers (1844); and a large variety of multicolored parakeets, peacocks, and pheasants.[4] This list does not take into account shows featuring dogs, monkeys, and horses performing tricks or the fantastic public performances launched by the use of microscope in whose sights appeared giant fleas and louses that haunted the dreams of the young man in Santō Kyōden's *Matsu to ume taketori mongatari* (The Pine and Plum Tale of a Bamboo-Cutter, 1809, fig. 10.2).[5]

ONO RANZAN BETWEEN THE MARKET AND THE STATE

Alongside spectacles and entertainments, people from all social classes pursued the study of nature more seriously, with circles counting among their members scholars, experts, and painters capable of producing illustrated manuals of extraordinary sophistication. *Honzōgaku* specialists were in high demand

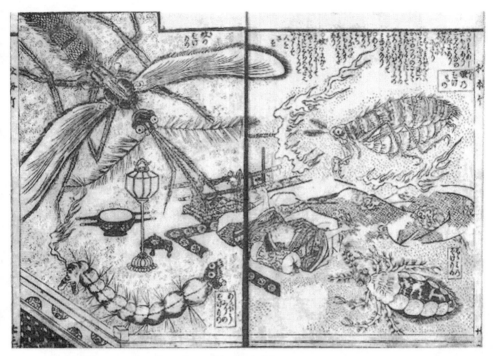

FIGURE 10.2. A scene from Santō Kyōden's *Matsu to ume taketori mongatari* (1809). Tokyo: National Diet Library.

to lend intellectual respectability to such groups. In the last decades of the eighteenth century, Ono Ranzan was probably the best-known *honzōgaku* scholar active in cultural circles, and his erudite lectures were published by his disciples under the title of *Honzō kōmoku keimō* (Clarifications on *Honzō kōmoku*, 1803–6). In this publication, Ranzan followed the classificatory order of Li Shizhen's encyclopedia but provided up-to-date descriptions of a large number of plants and animals, including information taken from Western texts. The medicinal uses of natural knowledge almost took a secondary position in comparison to the pursuit to describe the morphology and life of different species.[6]

The shogunate continued to hire *honzōgaku* specialists as official physicians (*ikan*) and later as members of the shogunal Institute of Medicine (Igakukan), established in 1791. In the third month of 1799, the bakufu ordered a seventy-year-old Ono Ranzan to transfer to Edo and become a teaching member (*kyōkan*) of the institute. Ranzan, who had never left Kyoto, could not refuse the honor. Beginning in 1801, the shogunate ordered him to perform survey expeditions throughout the provinces. Over the eleven years he served at the Institute of

Medicine until his death in 1810, Ranzan accumulated an unprecedented quantity of field notes, which were collected in the form of "herbal diaries" (*saiya-kuki*). He also published a revised edition of Li Shizhen's *Honzō kōmoku* and copied all ninety-six volumes of *Shobutsu ruisan* by hand from the Momiji-yama library, annotating and correcting them as he proceeded. Ranzan's *Honzō kōmoku keimō*, edited by his nephew Ono Motokata, was considered by his contemporaries to be the highest achievement of Japanese *honzōgaku* and was reprinted in 1811, 1844, and again in 1847. His approach was still largely semasiological, but the detailed descriptions of plants and animals were enriched by personal observation of live specimens. Although closely following the taxonomical arrangement of *Honzō kōmoku*, Ranzan's "clarifications" (*keimō*) was not concerned with Neo-Confucian principles but aimed to be a straightforward encyclopedic treatment of plants and animals and their morphology, habitat, life cycle, and practical utilizations (predominantly medicinal) that followed the order of presentation developed by Li Shizhen.

In this context, finding a patron was crucial for the majority of *honzōgaku* scholars. Multivolume encyclopedias and monographs were expensive to purchase and to publish. Hiring professional painters and printers was also very costly. Furthermore, access to rare species of plants and animals was restricted to those who could afford it. In sum, scholarly success in the field of *honzōgaku* was more likely to be attained when backed by an affluent patron of means, usually a high-ranking samurai, more rarely a wealthy merchant like Kimura Kenkadō. Conversely, lack of connections or resources often meant the end of a career. Despite the increasing recognition of the discipline, scholars were subject to the vagaries of fortune. The life of Hiraga Gennai is particularly revealing of the precarious status of scholars in the late Tokugawa period. His is the story of a talented young man of the lowest samurai status who succeeded initially in making a name for himself but eventually failed to survive as an independent scholar once he lost the support of wealthy patrons.[7]

HIRAGA GENNAI AND THE ART OF NETWORKING

Gennai is today appreciated mostly as a writer of humorous fiction (*ge-saku*). In Japan, he often appears as the main character in comic books, television shows, and movies. In these modern works, he is usually represented as a stereotypical scientist, "the Leonardo da Vinci of the Tokugawa period," with strange and straightforward manners—sometimes also as a womanizer.[8] Gennai was a polymath who produced technological artifacts and conducted research on a variety of scholarly subjects from engineering and mining to natural history, from poetry and novels to ceramics and Western-style painting (*yōga*). Sugita Genpaku wrote in his memoir, *Rangaku kotohajime* (The Beginnings of Dutch Studies, 1815), that Gennai "was a naturalist [*honzōka*], very intelligent by birth. He was very talented and always extremely popular."[9] Popular as he might have been, since his early years in the village of Shidoura, in Sanuki Province, Gennai struggled to be accepted as a scholar because, as the son of a foot soldier (*ashigaru*), he occupied the lowest rank of the samurai class. When Gennai assumed his father's duties and stipend in 1749, he lived a life indistinguishable from that of a needy peasant. A yearly income of three *koku* was insufficient to sustain a family, and Gennai recorded in his diaries that all members of his family frequently had to find employment as day laborers in farms of the Shidoura area.[10]

Later hagiographers do not spare adjectives in describing the genius of the young prodigy Gennai, and his rich and adventurous life is difficult to summarize in few lines. Gennai distinguished himself early in his youth by his capacity to produce inventions, such as a pedometer (*manpōkei*), which measured distances, and a thermometer (*ondokei*).[11] Later he would spin the first asbestos cloth in Japan, develop a new technique of glass blowing, and conduct experiments with static electricity (*erekiteru*).[12]

The *erekiteru*, or "electrostatic cabinet," became popular after its introduction as a machine that could produce sparks from a patient's head. It was utilized as a dubious treatment against sultriness and hot flashes (fig. 10.3).[13] As Tachibana Nankei explained in his journal *Seiyūki* (Notes from the West, 1795),

> The device called an *erekiteru* came to Japan twenty-three years ago. It is a machine for drawing fire from a person's body. Wheels are set within a box; it is just under a metre and an iron chain, five or six metres in length and ending in a looped handle, leads off. You have someone grasp the handle and then start the wheels rotating, so that power is transmitted along the chain. This provokes a reaction in the person, and little bits of paper brought up to them will move and dance of their own accord; if someone brings their hand close, you can hear a sound like spitting fat, and see a flame flying out. No-one who has yet to see the amazingness of this device will believe in it.[14]

Gennai was never able to afford any advanced education besides the basic training he received from the Takamatsu domainal school (*hankō*). However, he established a fruitful relationship with the *haikai* poet Watanabe Tōgen thanks to the intercession of Shidoura's *haikai* master Shizukidō Hōzan. This acquaintance developed into a lifelong friendship that would later prove invaluable to Gennai for establishing connections with artists and publishers in Edo. Later

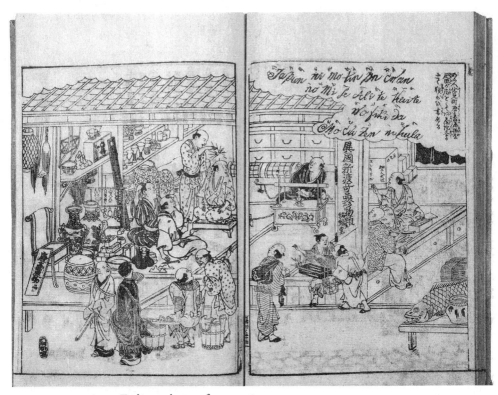

FIGURE 10.3. *Erekiteru* therapy from 1796
Settsu meisho zue. Tokyo: National Diet Library.

on, Tōgen also financially supported Gennai in times of difficulty.[15] Gennai's
rise began around the year 1750, when Matsudaira Yoritaka noticed Gennai's
abilities in cultivating ginseng—an activity in which Lord Yoritaka enthusiastically participated as a way to exhibit his loyalty and commitment to shogunal
politics.[16] Like many other domainal lords of the period, Yoritaka was himself
an amateur naturalist. He had connections with Tamura Ransui in Edo, the physician who first succeeded in growing ginseng plants in Japan. It was through
Yoritaka's mediation that Gennai began to study medicine under Kubo Sōkan,
Takamatsu's domainal physician, and was later accepted as one of Ransui's disciples in Edo (1756).[17]

Gennai was famous for his polymathic brilliance, but as he himself revealed
in a short essay, "Kishū sanbutsu shi" (Agricultural Products of the Kii Province), written around 1761, "Since I was born *honzō* studies has always been my
favorite discipline, that is why I traveled around on herbalist trips and discov-

ered many new species."[18] *Honzōgaku* was his passion. It was also this field that granted Gennai his early successes. While pursuing his training, Gennai established a network of friends accomplished in a variety of fields. He studied Neo-Confucianism under Kikuchi Kōsan and later at the Shōheikō in Edo, *haikai* poetry with Shizukidō Hōsan, and engineering and *honzōgaku* with Watanabe Tōgen. In Edo, his connections expanded to include shogunal officials like Senga Dōryū, Moriyama Takamori, Torimi Genryū, and, eventually, the senior councilor Tanuma Okitsugu, probably at the time the most powerful man in Japan. Other connections were made with Neo-Confucian scholars like Hayashi Nobutoki, head of the Shōheikō; Nakamura Fumisuke; and Shibano Ritsuzan, who in 1790 would be involved in the Kansei reforms. Gennai's acquaintances also included poets and artists like Ōta Nanpō, Suzuki Harunobu, and Shiba Kōkan; nativist scholars like Kamo no Mabuchi and Arakida Hisakata; and *honzōgaku* scholars like Tamura Ransui, Aoki Kon'yō, Kurimoto Tanshū, Sugita Genpaku, and Toda Gyokuzan, as well as amateur naturalists like Hosokawa Shigekata, Satake Yoshiatsu, Hotta Masaatsu, Masuyama Sessai, and Date Shigemura.[19]

PROMOTING NATURAL RESOURCES: THE GREAT EXHIBITION OF 1762

It was under Tamura Ransui that Gennai was able to create a name for himself. Gennai's talent soon made him a favorite among Ransui's students, but his advancement was limited by his obligation to serve as a retainer for Yoritaka. His low status in the Takamatsu Domain meant that he could not establish a career as domainal scholar. Yoritaka recognized the obstacles Gennai was facing and tried to promote him but was faced with such opposition from other higher-ranking Takamatsu domainal scholar-retainers that he abandoned his efforts.[20]

The best strategy for Gennai's social advancement as a scholar would have been active engagement as an independent man of learning in a *honzōgaku* cultural circle, but this time it was Lord Yoritaka who worked against him. In a letter dated the twenty-first day of the ninth month of 1761, Gennai pleaded with Yoritaka to accept his resignations as retainer of the Takamatsu Domain. Yoritaka accepted but with the binding conditions that Gennai never enter the service of any other lord and never engage in cultural circles frequented by other lords.[21] These mandates prevented Gennai from receiving the patronage of other samurai households. In effect, Yoritaka's conditions signified the end of Gennai's official career. He tried to sustain himself through writing popular fiction, which he was able to publish thanks to the personal ties he had established with Edo publishers.[22] But the concept of royalties had yet to be established and his earnings did not last long.[23]

The peak of Gennai's fame was reached during the years of his collaboration with his master Tamura Ransui, especially after the organization of nationwide exhibitions of medicinal substances (*yakuhin*). Literally the "meetings of medicinal substances," *yakuhinkai* were actually exhibitions of rare and precious plants and animals—they were also known as *bussankai* (produce meetings), or *honzōkai*. The first of these exhibitions was organized by Ransui in 1757 for the purpose of "establishing an association of [collectors of] medicinal substances [*yakubutsu*], beginning with a large meeting to be held in Yushima, in Edo."[24] The role Gennai played in the organization of the first exhibition is unclear, since by that point Gennai had only been a student of Ransui for a few months, but Ueno has argued that the talented Gennai may even have been the source of Ransui's inspiration.[25]

During a trip to Kyoto in 1754, Gennai had the opportunity to meet Toda Gyokuzan, under whom—as Gennai reported—he studied medicine for a brief period and probably learned about meetings of local *honzōgaku* scholars that were regularly held at Gyokuzan's house. Gyokuzan, a former retainer of the Okayama Domain, had retired to practice as town physician (*machi isha*), first in Kyoto and then in Osaka. He began his *honzōgaku* studies as an autodidact and later used his own garden, named Hyakukien, to plant and study different herbs. Rumors about the excellence of Gyokuzan's garden had started to circulate in the Kansai region, eventually reaching the first disciple and successor of Matsuoka Gentatsu, Tsushima Tsunenoshin in Kyoto, at the time master also of Kimura Kenkadō and Ono Ranzan.[26] Tsunenoshin decided to pay a visit to Gyokuzan and his garden in the nearby Osaka in 1751. Soon thereafter a friendship developed between Gyokuzan and Tsunenoshin, who, in 1752, began organizing yearly informal gatherings of local amateur and professional *honzōgaku* scholars called *yakuhinkai*.

When Gennai visited Osaka in the eighth month of 1754, he went to the Hyakukien and met with Gyokuzan and Tsunenoshin. He was likely told then about their meetings.[27] These *yakuhinkai* consisted of comparing and exhibiting medicinal substances along with various specimens of curious and rare plants, animals, and minerals. Meetings such as these were not rare in Tokugawa Japan and involved persons with an interest in nature from various social classes. But unlike later circles such as the Shabenkai and the Shōhyakusha, most of these early clubs were locally based and minimally connected to a wider network of scholarly exchange. What Ransui—with the collaboration of Gennai—

achieved in 1757 was the creation of the first national gathering of collectors, scholars, and amateurs.

The first meeting of Ransui's Tōto Yakuhinkai (Eastern Capital Meeting of Medicinal Substances) in 1757 led to four more exhibitions in 1758, 1759, 1760, and 1762, which was the largest. They all enjoyed wide participation and were significant events in the intellectual life of late eighteenth-century Japan. People from throughout the provinces traveled to Edo with the specific purpose of observing the rare and exotic plants and animals exhibited at these meetings. While there, visitors could also appreciate Gennai's organizational displays, with specimens arranged in accordance to the *Honzō kōmoku* classification system and labeled with a tag providing the species name and habitat and the name of the owner. These exhibitions were widely acclaimed in *kawaraban* (broadsheets), gazetteers, newsletters, and popular writings and served to further fuel the popular craze for natural history.[28] Ōta Nanpō described the meeting in his *Yakko dako* (Yakko Kites), a record of events and daily life in Edo, published in 1818:

> Hiraga Gennai was a man from the Sanuki Province. Since he loved natural products [*bussan*], he studied with Tamura Gen'yū [Ransui] and organized an association. The first meeting of this association was held by Tamura Gen'yū in 1757 in Yushima. The following year they met at Kanda. The year after that, Mr. Hiraga organized a third meeting again in Yushima. In 1760 Mr. Matsuda organized the fourth meeting at Ichigaya. In 1762, Mr. Hiraga organized another meeting once again in Yushima. [For this meeting] the specimens exhibited were sent by collectors and members of Tamura's association. Of the more than two thousand specimens they gathered from thirty provinces, they selected the best for the exposition, and a catalogue of a further selection was published in 1763 with the title of *Butsurui hinshitsu*.[29]

It may be questionable whether Gennai actually inspired the meetings, but he was definitely responsible for their success. That being said, the ultimate beneficiary of the popular sensation of the *yakuhinkai* was Tamura Ransui. In the 1750s, Ransui was a town physician with no particular connection to the world of *honzōgaku* studies. He studied briefly with Abe Shōō, but he never enrolled in any specialized school.[30] His fame was linked to his success in cultivating ginseng plants in Japan, but he had no connections with the major schools of *honzōgaku* or with other important naturalists. Ransui's experiments with ginseng seedlings produced the first fertile plants able to grow in Japan in 1740. In 1748, he published the results of his studies on *Panax ginseng*, a text that became a bestseller in its genre.[31]

Historians of science usually regard Ransui's expertise with ginseng cultivation as the reason the shogunate hired him as an official physician, which came with a substantial yearly income of three hundred *koku*.[32] However, his enlistment as bakufu physician (*bakufu ikan*) did not come until 1763, at forty-six years of age.[33] This was a full twenty-three years after his first successes with ginseng and fifteen years after the publication of his monograph. His enormous celebrity among *honzōgaku* scholars and the sensation caused by his *yakuhin* exhibitions probably played a decisive role in convincing the bakufu to hire him at the institute. His special nomination (*senbatsu*) in 1758 followed in fact the 1757 opening of the first exhibition, and his position as shogunal physician in 1763 followed the success of the fifth exhibition of 1762. The timing of these events suggests that the sequence was more than coincidence.

The fifth Tōto Yakuhinkai was held in 1762, two years after the fourth and least successful of the meetings.[34] In 1760, Toda Gyokuzan had organized a rival national exhibition in Osaka. This event was far larger than those Gennai and Ransui (and Chōgen) had thus far organized in Edo, with a display of 241 plants and animals from private collections all over the country. An illustrated catalog was also published the same year under the title of *Bunkairoku* (Proceedings of the Cultural Meeting).[35] The exhibition outshone the Edo meetings. Gyokuzan boasted that the major *honzōgaku* scholars from Kyoto, in particular Ono Ranzan, took part in the event. Kimura Kenkadō also participated with specimens from his collection.

Gennai's response to Gyokuzan's challenge was two years in preparation, and it was unprecedented in scale, grandeur, and the sensation it caused. Collectors from more than thirty provinces brought around 1,300 species of rare and exotic minerals, rocks, plants, and animals to the Yushima ground through a complex system of transportation developed by Gennai. It was one of the largest naturalist exhibitions ever organized thus far in the world.[36]

Gennai advertised the event in a series of manifestos he posted throughout Edo and sent to provincial castle towns with the help of his network of friends and associates (fig. 10.4). The manifesto did far more than publicize the event and call for collectors to send their best specimens. It was in fact a powerful statement in defense of *honzōgaku*, stressing the vital importance of studying nature for the prosperity of the Japanese people. It provided a genealogy of nature studies in Japan that linked Inō Jakusui and Kaibara Ekiken to Gennai's master Ransui. Furthermore, it stressed that *honzōgaku* not only should extend the knowledge of plants and animals and their potential pharmacological utility or merely constitute a pastime for idle people of means but should be a practical field engaged in developing new farming and breeding techniques that could enhance the welfare of the Japanese nation:

EASTERN CAPITAL MEETING OF MEDICINAL SUBSTANCES
Opening Day: the 10th day of the 4th intercalary month of 12th year of Hōreki[37]

Our Great Japan is a land blessed by the gods with fertile and gorgeous mountains and river valleys. Its people are sensitive and their customs are beautiful. All things living and growing in this fertile land, herbs and trees, birds and beasts, fishes and shells, insects, jewels and stones, and grounds, sustain a healthy and strong population. If one compares Japan with other nations, it is the most excellent place to live. In particular, since antiquity herbalists [saiyakushi] have been traveling around this land, [UNREADABLE] and collected a bountiful amount of substances. As a result, people of Japan have known various products [sanbutsu] since ancient times.

However, in regard to [developing] medicinal substances [yakubutsu] [from these products], Japanese people are far behind. Even when they collect precious substances, they do not know how to prepare medicines with them. They continue to believe only in old traditions, foolishly neglecting to investigate things further. Instead, they rely only on imports from abroad and believe only in things written in foreign books. They are comforted and put at ease by merely mimicking foreign customs. Only luck has prevented diseases from worsening under these circumstances. What would happen if a cargo full of medicine was caught in a storm and could not enter our ports? Would the medicines we have be enough to cure all illnesses? If you think that some herbs may have a pharmacological use just because of their resemblance with known medicine, well, that is a big mistake, a very big mistake indeed. We need to learn honzōgaku, because if we do not, we will deceive other people and get into trouble. But if there are pharmacological substances and we do not know [how to recognize and treat] them, then we would deceive others and ourselves at the same time. This is a matter of vital importance. The discipline of honzōgaku was first studied in Kyoto by Master Inō during the Genroku period. After him, Master Kaibara and Master Matsuoka continued his studies. Their studies have proved useful and have produced a large number of texts that we are now benefiting from. It is thanks to their activities that the number of persons relying on foreign medicines has decreased considerably. Moreover, our knowledge of indigenous species of medicinal plants and animals has greatly improved, thanks especially to the great achievements of these three masters. Their legacy has now been received in Edo by master Tamura Ransui, of whom I am a humble student. Master Ransui once said, "it is not so difficult to master the secrets of various species of rare, strange and exotic plants and animals from texts or even investigate them in steep mountains and deep valleys. The hardest part is actually cultivating them." How true this is of plants and animals living in Japan! Even though there are many persons that practice *honzōgaku* today, how few of them really love it passionately! This is the reason we have not yet exhaustively surveyed all the products that our provinces generate. If we had really exhaustively investigated all plants and animals in Japan, we would not need anything from Chinese and Dutch ships; we would be self-sufficient. It was for this reason that in 1757 [Master Ransui] organized a meeting here in Edo inviting friends from different provinces to exhibit precious specimens from their collections. [UNREADABLE] that in the past were not known but now are, as, for example, the ten species of teas from Hizen or the alabasters of Dewa. Now we are still far from what we originally intended to achieve with our friends, but I hope that the situation will change with the meeting I am planning that will take place at the beginning of summer next year. I therefore beg all fellow scholars in Japan, for the sake of enlarging our knowledge, to send their most precious specimens to the nearest delivery station [*toritsugidokoro*]. I do not have any other desire than to give the opportunity to all those scholars of similar inclination to exhibit their best specimens. As soon as the specimens are properly named and classified, they will be returned to their owners.

In the Winter of the 10th month of 1761

Hiraga Kunitomo [Gennai]

In the next section, Gennai informed the reader that his master Ransui would display fifty new species of plants and animals, and another fifty would be exhibited by Gennai himself. None of these one hundred species had ever been displayed before. Then in the following section, written in Japanese, Gennai provided further instructions:

- As explained in the Introduction, the reason behind the organization of a new Eastern Capital Meeting of Medicinal Substances is the conviction that, once the remotest areas [*shinzan yūkoku*] of our country have been exhaustively explored, there will not be substances among those we currently import from China [*karawatari*] that are not found here. It is

FIGURE 10.4. Gennai's 1761 manifesto of the Eastern Capital
Meeting of Medicinal Substances. Tokyo: Waseda University.

true that traveling around the remotest roads and provinces [of Japan] is a very difficult endeavor and many plants and animals are indeed hard to find. But if we ask persons [with similar inclination] who live in those provinces to investigate plants and animals living in their regions, I am confident that we will discover that we would not need to import half of the substances analyzed in all Chinese and Japanese encyclopedias of materia medica and in Dodoens's *Cruydeboeck* [*Dodoniyōsu koroitobokku*] because they are found here in Japan.[38] These discoveries would be of great help in the practice of medicine. Thus far we have organized four editions of the Eastern Capital Meetings of Medicinal Substances, in which more than 700 species of medicinal substances were put on display. Other such exhibitions have also been held in other cities. For the next Edo exhibition I beg the assistance of all persons with the same passion living in all regions [of Japan].

- Any kind of plant, mineral, bird, beast, fish, shell, or insect is welcomed, even those specimens whose name is not known. For those persons who live in distant provinces but wish nonetheless to send their specimens, we have established delivery stations, a list of which is given below, that will transport the specimens to Edo without difficulty.

- In the previous four exhibitions, more than 700 species have been displayed. [UNREADABLE] If a specimen is sent from distant regions, it will be displayed even if a similarly named plant or animal had been previously exhibited, since there are interesting regional variations to observe. When dispatching a specimen, [UNREADABLE] please write its standard name as well as the regional names under which it is known in the village and district or in the mountain or marsh from which it has been collected. Please, write also the reason why it is so particularly easy to find the specimen in that place.

- Regarding specimens sent from the vicinity of Edo, please send them by the first day of the fourth intercalary month directly to my home address. If I receive a specimen the last minute without any previous notification, that specimen will not be put on display.

- All of those persons who wish to participate in the exhibition are kindly requested to arrive at the exhibition area early in the morning, even if it rains. Anyone wishing to take part to the exhibition is strongly advised to send a letter with his last name. All those who do not reserve a place will not be allowed entrance.

- As in the previous meetings, we will not provide beverages or food. All those participants who come from distant regions are urged to bring their own food. It is not allowed to party in the exhibition area.

The following sections listed the names of the clerks in charge in Yushima and a list of thirty delivery agents in various provinces responsible for collecting the specimens from their provinces and ensuring their safe delivery to Edo.[39] As Gennai explains at the bottom of the list,

All those persons interested in displaying specimens from their collections may send them to the nearest delivery agent who will provide to their delivery to Edo. Once the Meeting is closed, all specimens will be returned as soon as possible at our expense. When you send your specimens, please attach to it a letter or a piece of colored paper with "pay on delivery" written on it. All persons sending specimens from the Chūgoku area [Western part of Honshū island] and Kyūshū are urged to use the Osaka delivery agent, who will dispatch all specimens to Edo via express [*choku hikyaku*], which will take only 12 or 13 days. Please, make the necessary arrangements for live plants not to wither.

Gennai's manifesto provides an understanding of the complex organization involved in planning this event, which exploited the advanced system of highways and mail services present in Tokugawa Japan.[40] Furthermore, in view of the fact that the organizers were willing to pay for delivery and the accommodation expenses of the exhibitors, it is fair to assume that Gennai and Ransui were backed by a powerful sponsor, probably the shogunate itself, considering that Yushima was a ward close to the bakufu headquarters where various shogunal institutes were located (the most famous probably being the Shōheikō), and that the shogunate adopted Gennai's organization in exhibitions held later in the period by the Seijukan, a school of medicinal studies opened in 1765 by Taki Genkō and later transformed into the shogunal Institute of Medicine.[41]

The most revealing part of the manifesto, however, is the introduction in Chinese, where Gennai provides a justification for the organization of the national meeting. Of the many topics that Gennai touched upon, three shed light on the importance of the discipline of *honzōgaku* for the benefit of the state and on the strategy adopted by Gennai and Ransui to establish themselves in the field. The first is Gennai's insistence that *honzōgaku*'s ultimate goal was the development of domestic production of pharmacological substances. He stated that Japan was a land blessed with bountiful natural riches, and it had sustained throughout its history a "healthy and strong population." He lamented that many plants and animals with important medicinal potential still remained undiscovered because of the underdevelopment of nature studies. Second, Gennai denounced what he perceived to be a misguided attitude of Japanese naturalists, who "believe only in things written in foreign books" and were "comforted and put at ease by merely mimicking foreign customs." A precise knowledge of plants and animals would allow the Japanese to exploit the natural resources of their land. It was because of the activities of pioneer scholars like Inō Jakusui, Kaibara Ekiken, and Matsuoka Gentatsu that knowledge of indigenous species of medicinal plants and animals had greatly improved. According to Gennai, that was not enough. Contemporary scholars were still limiting their intellectual activities to philological analyses of Chinese texts—an obvious reference to those *honzōgaku* scholars, starting from Ono Ranzan, who were still heavily influenced by *meibutsugaku* methods and scopes as well those who engaged in *honzōgaku* practices for fun—instead of being actively involved in herbalist explorations and cultivation projects, like his master Ransui and, obviously, himself. Gennai's critique resonates with the importance that, decades later, Shōhya-kusha naturalists would place direct observation to complement textual sources. Gennai took issue with traditional *honzōgaku* scholars who "continue to believe only in old traditions, foolishly neglecting to investigate things further." Because "they rely only on imports from abroad and believe only in things written in foreign books," they did not contribute to advance knowledge of natural species, nor did they exploit the pharmacological properties of herbs, plants, and animals.

This led Gennai to a third point: the legacy that those three pioneering scholars had left was inherited and continued by Tamura Ransui, Gennai's master, who not only was able to master the traditional knowledge of Chinese encyclopedias of materia medica but also added new species through expeditions in the provinces. Ransui also had, Gennai continued, gone a step further: "It is not so difficult to master the secrets of various species of rare, strange and exotic plants and animals from texts or even investigate them in steep mountains and deep valleys. The hardest part is to actually cultivate them." Ransui, Gennai argued, was able to put successfully into practice what his predecessors conceived of only as a scholarly activity. It was because of Ransui's talent and genius that Japan would eventually be able to fully exploit its riches, and that was the reason behind the organization of the Edo exhibition. Gennai twice highlighted the necessity for Japan—and he specifically wrote "Japan" (*Nihonkoku*), not "the shogunate" or "the realm" (*tenka*) as it was often the case at the time—to achieve self-sufficiency in agricultural and medicinal production that would cut its dependence from continental imports. We will see in part V how Gennai's instrumentalist conception of knowledge as a practical means to better exploit the natural environment for human needs would develop in the nineteenth century. But already in this text the injunction to extend human reach to the natural environment and exploit it as a resource is clearly advanced: "The reason behind the organization of a new Eastern Capital Meeting of Medicinal Substances," Gennai explained, "is the conviction that, once the remotest areas [*shinzan yūkoku*] of our country have been exhaustively explored, there will not be substances among those we currently import from China [*karawatari*] that are not found here." Gennai's explicit advocacy of an extension of human domination over nature to "the remotest areas (*shinzan yūkoku*) of the country" is one of the earliest in the Tokugawa period. It prefigured notions of systematic exploitation of natural resources for economic needs that would fully develop half a century later in the writings of Satō Nobuhiro but that already bore witness of the relentless erosion of that separation of human and natural spheres characteristic of earlier periods.[42]

On a personal side, Gennai cleverly used the occasion of the meetings and the enthusiasm they generated to promote Ransui's position in the field of *honzōgaku* studies and, by extension, his own. Ransui, it should be remembered, was not connected to any major network or circle of natural studies. He did not study in Kyoto under the well-known scholars of the time. He was not a proper *honzōgakusha*, just a town physician who had succeeded in the crucial task of producing fertile offspring of ginseng plants. These were the years in which Ono Ranzan and his network of *honzōgaku* scholars dominated the intellectual field of natural studies. His acclaimed private school boasted more than one thousand students, and Ranzan was considered the only person in Japan able to name, recognize, and argue about all species of plants and animals quoted in the classic texts of materia medica.[43] Ransui had nowhere near the knowledge necessary to compete with the members of Kyoto's *honzōgaku* circles. Furthermore, Ransui lacked connections, offered no competitive curriculum, and had his only success linked not to philological or semasiological expertise, nor to the publication of a major encyclopedia, but to the manual task of cultivating ginseng. In other words, Ransui lacked all the basic conditions of entry into the field of *honzōgaku*. Competition for recognition and distinction in the field required, above all, textual competence in Chinese manuals. Herbalist experience in the field came only second.

The heterodox strategy that Gennai used to boost his master Ransui's credibility in the field of natural studies was subversively effective in manipulating the established social order. The magnificence of his exhibitions, and the fifth in particular, became associated with the name of Ransui. His expertise was vouched for by the wide participation of amateur and professional *honzōgakusha* of Japan: Toda Gyokuzan and his circle sent their specimens to be exhibited, as did many of Ono Ranzan's disciples and members of various *honzōgaku* circles, including Kenkadō's. Although Ransui lacked the support of cultural groups, Gennai created a network of support through participation in the exhibition. This network ended up becoming far more than a regional *honzōgaku* circle. He created a *nationwide* network, an expanded network connecting all the local groups.

Once he succeeded in linking *honzōgaku* scholars to the Edo meeting by exploiting their desire for distinction through participation, Gennai pushed his strategy even further. He established a genealogy of *honzōgaku* studies with Inō Jakusui listed as its founder, followed by Kaibara Ekiken and Matsuoka Gentatsu. This genealogy was clearly intended to allow Gennai to claim Ransui as the rightful successor of that lineage of study. Ransui was depicted as the scholar who not only excelled in the intellectual pursuit of natural history but had successfully put it into practice by growing the most resistant and precious of plants, ginseng.

The Edo exhibition was not only a public display of rare and exotic plants and animals but also a public competition for intellectual distinction among *honzōgaku* scholars and collectors. Gennai and Ransui had created a public arena for scholars to present shocking new species of plants and animals and to compete in their knowledge of them. Unusual specimens brought fame to their collectors. This is why the catalog of specimens on display not only included the expected "medicinal substances" (*yakuhin*) but also listed curious species (fig. 10.5).[44] According to a catalog Hiraga Gennai published in 1763, *Butsurui hinshitsu* (A Selection of Species), the specimens on display were not only medicinal herbs (licorice, harebell, ginseng) but also especially stunning and uncommon plants and animals without medicinal use: lizards (fig. 10.6), crocodiles, various species of mice, parrots, frogs, and a multitude of insects.[45] Among the species displayed in later exhibitions of medicinal substances, rare and bizarre plants and animals were increasingly the norm, especially after the intensification of commercial exchange with Western countries. These exotic specimens included giant crabs, sea otters, two-headed tortoises, carnivorous plants, tiger

FIGURE 10.5. A *yakuhinkai* organized in 1844 by the Institute
of Medicine of the Owari Domain. From *Owari meisho zue*
(Famous Places of Owari, 1844). Tokyo: Waseda University.

skins, giant butterflies, stuffed "sirens" (*ningyo*, fig. 10.7), water goblins (*kappa*),
and a *tengu's* clawed hands.

Decades after the 1762 exhibition, numerous other meetings were organized
in various provinces by different groups of *honzōgaku* scholars, physicians, and
amateurs. From the first *yakuhinkai* in Edo held by Ransui to the end of the
Tokugawa period, more than 250 were organized in different cities.[46] Figure
10.5 portrays one such meetings, organized in 1844 by the Institute of Medicine
(Igakuin) of the Owari domains, the same institute where Shōhyakusha naturalists were affiliated. Smaller gatherings were also regularly held there the seventh
day of every month.[47] "These display practices," Maki Fukuoka has argued, "became significant for the group as a way to address collectively the issues of the
naming, availability, and efficacy of plant specimens."[48] *Yakuhinkai*, *honzōkai*,
or *bussankai*—as they were variously called—were at the same time convivial
events; entertaining spectacles; chances for scholars to meet and compare the
results of their investigations as well as to exchange specimens and drawings;
discursive spaces to discuss issues of identification, naming, and descriptions;

FIGURE 10.6. A lizard in alcohol from Gennai's
Butsurui hinshitsu (1763). Edo-Tokyo Hakubutsukan.

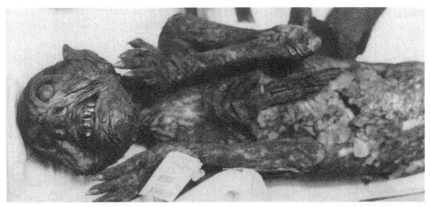

FIGURE 10.7. A stuffed siren. This was a specimen—fabricated, of course—that
was sold to a Dutch merchant in the late Tokugawa period and is now the property
of Leiden's National Museum of Ethnology (Rijksmuseum voor Volkenkunde).

venues to improve one's position in the field by establishing associations with other scholars and visibility of one's expertise; and symbolic means to affirm human dominion over the natural world. "Observation of the specimens permitted collective calibration of the names and efficacies attached to the item exhibited, while the unknown objects would remind the participants of the perpetual existence of the novel and unique brought from their regions and beyond."[49] Gennai's 1762 exhibition anticipated these developments by mixing together cognitive, pharmaceutical, agricultural, economic, symbolic, and recreational issues.

Despite Gennai's efforts to develop his reputation via that of Ransui, success did not arrive for himself. After the triumph of the fifth Edo meeting, Gennai became involved with Senga Dōryū, a shogunal physician, and developed an interest in mining. Ransui, fulfilling a new official role of bakufu physician, could not continue to help Gennai in light of his agreement with Lord Yoritaka. It was through Dōryū that Gennai established contacts with the powerful Tanuma Okitsugu.[50] Okitsugu sponsored Gennai as an outside consultant for two iron-mining projects in Chichibu and Akita in 1773, which both ended in failure.

In the last years of his life, Gennai often visited Okitsugu, who, rumor had it, was particularly fond of *erekiteru* treatments. However, because of the binding conditions he had accepted in order to be discharged as a retainer of Matsudaira Yoritaka, Gennai was unable to enjoy any direct patronage of other high-ranking samurai and could not afford to open his own school. He died of tetanus in prison after slashing to death two carpenters who, in his drunken state, he mistakenly thought had stolen some construction plans.

Despite Gennai's troubled life, the tremendous success of the 1762 exhibition bears witness of the widespread interest for natural history in Tokugawa Japan. Most importantly, as Gennai's manifesto clearly stated, it emphasized the instrumentality of *honzōgaku* for the sake of economic prosperity—in line with a contemporary genre of manuscripts that focused on *kokueki*, or national prosperity.[51] The conclusive part V will show how in the nineteenth century, instrumental reason and concerns of practical utility began to dominate the field of *honzōgaku*, on the one hand replacing, other aesthetical, epistemological, ethical, or entertaining ends for studying and collecting plants and animals and, on the other, anticipating the role science had in the modernization and industrialization of Meiji Japan.

Representing Nature
From "Truth" to "Accuracy"

. . . .

A fine line splits *honzōgaku* textual production in two distinct phases, the turning point roughly coinciding with Yoshimune's regime and the survey he sponsored in 1736. What distinguished the vast assortment of catalogs and monographs of the late eighteenth and early nineteenth centuries from the encyclopedic works of Kaibara Ekiken and Inō Jakusui was a burst of pictorial representations of rocks, plants, herbs, animals, fish, and insects. Canonical lexicographical encyclopedias such as Kuroda Suizan's *Komeiroku* (Records of Old Names, published posthumously in 1885) continued to be produced, but by the last third of the Tokugawa period, the primacy of illustrations became a defining characteristic of *honzōgaku* scholarship.

This chapter argues that accurate and detailed illustrations of plants and animals developed as a new cognitive apparatus to identify species and solve the old problem of matching Chinese names with actual plants and animals. If on the one hand they indeed put an unprecedented emphasis on morphological descriptions, on the other they could also reinforce, rather than dismantle, the taxonomical edifice of *Honzō kōmoku*. In the process, however, the pervasiveness of detailed illustrations in *honzōgaku* texts strengthened the inductive labor of scholars in observational, descriptive, and representational practices that contributed to abstract plants and animals from their ecosystem and turned them into idealized species ready for commodification and manipulation by human for economic, political, and cognitive ends. While these epistemological transformations undoubtedly favored the metamorphosis of many *honzōgaku* scholars into modern scientists in the early decades of the Meiji period, their development resulted from internal dynamics of the field of nature studies itself rather than from earnest adoption of Western scientific theories and practices.

ILLUSTRATING THE BOOK OF NATURE

Richness in visual production was a common feature of the entire culture of the Tokugawa period, from the flamboyancy of its popular culture to the most aesthetically refined works of art. In the field of *honzōgaku*, the transition from an eminently lexicographical discipline into a largely descriptive one was the effect of three distinct but interconnected dynamics. First, as seen in part III, the strategic importance for the welfare of the state of natural resources such as alternative subsistence and commodity crops, textiles, and medicinal herbs favored the development of more refined techniques of visual representation necessary to precisely identify species of plants and animals. Second, popular curiosity in the natural world as a form of entertainment, spectacle, and cultured pastime and the ensuing competition for distinction in erudition and connoisseurship among amateur collectors transformed natural species into cultural commodities to collect, possess, exhibit, and exchange, a widespread craze that favored the production of lavishly illustrated catalogs and atlases. Third, as the recruitment of *honzōgaku* scholars in shogunal and domainal administrations became more substantial and as the expansion of a market for cultural goods offered new opportunities to profit from professional expertise, the population of naturalists swelled to an unprecedented size; this trend favored specialization and the development of more sophisticated analytical skills, in particular the ability to identify new and rare species and to produce accurate visual and verbal descriptions.

The pressure to develop exact and truthful descriptions was primarily related to the practical need to identify a species with certainty, especially as the global trade in plants and animals expanded the number of known species in both Europe and Japan.[1] Philological and semasiological research, the method generally preferred in the seventeenth and early eighteenth centuries, could help *honzōgaku* scholars orient themselves in a thick forest of synonyms and homonyms of *known* species but was utterly useless at ordering the growing number of species imported by Dutch and Chinese merchants. In chapter 7, we saw how Niwa Shōhaku, in organizing the 1734–36 national survey of natural products, relied on illustrations of plants and animals to cope with the confusing diversity of regional names. This practice contributed to the development of pictorial techniques capable of producing more faithful representations of specimens supposedly taken from life (*ikiutsushi*). As a result, schools of painters specializing in accurate reproductions of plants and animals grew in size and influence.

Such a shift toward visual accuracy also occurred in early modern Europe, where "the sixteenth century saw a pronounced tendency in botanical description toward an almost exclusive concentration on morphology—that is, on visual elements, a tendency reinforced by illustrations."[2] As the naturalist Fabio Colonna remarked in 1616, "In addition to learning a variety of subjects and languages, the investigator of nature, in order to become better, should acquire skill in painting and drawing, or at least a knowledge of it. Someone who is entirely ignorant of the art of painting cannot make true images of things whose descriptions and *differentiae* are clear in his mind."[3] In Europe, too, workshops of specialized painters found a profitable niche assisting natural historians and apothecaries in illustrating manuals and encyclopedias, including, for instance, the ateliers of Albrecht Dürer, Edward Topsell, or Georg Hoefnagel.[4] Though the development of Linnaeus's binominal system in the eighteenth century certainly facilitated the tasks of European naturalists, the demand for skillful painters continued to grow steadily in response to both new aesthetic ideals and the needs of the emerging field of comparative anatomy.[5] The development of "realistic" paintings of plants and animals cannot be reduced to utilitarian necessities: on the one hand, Dürer's illustrations of rhinoceros, hares, or simple herbs responded to the aesthetic taste of his new bourgeois clients;[6] on the other, the mechanistic conception of nature that emerged in the sixteenth and seventeenth centuries as the new paradigm of explanation among natural philosophers promoted quantitative and descriptive methods of analysis over traditional investigations of ultimate causes.[7]

In Japan, where Linnaeus's binominal system was probably known already by the late eighteenth century but was hardly applied, the development of techniques of "realistic" illustration found fertile ground in the context of the "quiet revolution in knowledge" of the eighteenth century.[8] The great expansion of travel—either for political purposes (*sankin kōtai*) or for commerce, religious pilgrimage, sexual tourism, aesthetic tracings of earlier poetic travelogues, or peripatetic practices of scholars moving from one circle to another—supported the development of illustrated guides, maps, gazetteers, travel diaries, picaresque novels, souvenirs, and "postcards" of popular places.[9] Truthful representations of celebrated views in the form of printed albums called *meisho zue* (illustrated guides to famous places) paralleled the fame of "true views" painting (*shinkeizu*), an aesthetic movement associated with artists like Ike Taiga and Yosa Buson that revolutionized the more expressionist style of Chinese literati paintings (*bunjinga* or *nanga*).[10] A proliferation of maps of various kinds was another characteristic of the period.[11] Accurate representations of birds and flowers were by no means limited to *honzōgaku* circles: artists as diverse as Maruyama Ōkyo, Matsumura Goshun, Itō Jakuchū, and Katsushika Hokusai produced depictions of plants and animals with an entirely novel concern for truthful representations.[12]

In this rich and complex cultural milieu, aesthetic, cognitive, moral, and leisure components were frequently intertwined and therefore difficult to separate in retrospect. Members of cultural circles who observed and described rare plants and animals were both assembling natural facts about their subjects and contributing to the network of meanings associated with them. In this sense, a natural fact was at the same time a cognitive, ethical, and aesthetic datum. Matsudaira Yoritaka, in sponsoring the production of the lavish illustrations of Hiraga Gennai and Kurimoto Tanshū, was interested not only in their cognitive function but also in their aesthetic significance. Motivated by a complex mix of cognitive, aesthetic, utilitarian, and moral goals, these new forms and styles of observation and representation came to dominate *honzōgaku* methodologies and had tremendous consequences for the way nature and natural species were conceived.

THE EVOLUTION OF NATURE'S ILLUSTRATIONS

Despite these methodological changes, the authority of canonical texts like *Honzō kōmoku* was never in peril, nor was the primacy of the pharmacological utility of *honzōgaku* ever questioned. Illustrations often complemented very conventional treatments of plants and animals. In 1810, for instance, Iwasaki Tsunemasa published an encyclopedia titled *Honzō zusetsu* (Illustrated Materia Medica), which consisted of large one-page illustrations of all plants and animals listed in the *Honzō kōmoku* in roughly the same order. As late as 1853, Lord Maeda Toshiyasu sponsored the publication of *Honzō tsūkan shōzu*, an illustrated guidebook of fundamental medicinal herbs to be distributed to all village headmen in his domain.[13] Still, though detailed and accurate illustrations attached to eighteenth-century encyclopedias and monographs may have only accompanied conventional content, as in the case of Gessner's *Historia animalium*, their accuracy conveyed a self-sufficient epistemological meaning.[14] Representational fidelity to an actual object—expressed in terms like *shasei, shajitsu*, or *shashin*—had a precise cognitive function that complemented the verbal content of each entry and affected new modes of conceptualizing species (*shu*).

Even in a cultural context rich in visual representations like that of Tokugawa Japan, a comparison of pictures taken from encyclopedias and manuals from the seventeenth through the early nineteenth century reveals a major qualitative change that cannot be simply explained by advancements in pictorial techniques. Most manuals and encyclopedias compiled or printed through the mid-eighteenth century, like Ekiken's *Yamato honzō* and Terajima Ryōan's *Wakan sansai zue*, were illustrated, but their pictures tended to be simple, small, and marginal to verbal explanations. They accompanied and enriched the text but did not add anything to its verbal treatment of the various species of plants and animals, nor were they meant to replace verbal descriptions with pictorial ones. Signs of an urgent need to illustrate the book of nature abounded as early as seventeenth-century Japan. For example, Japanese reprints of *Honzō kōmoku* added original illustrations to the text. Li Shizhen's *Bencao gangmu* was not originally intended to be illustrated, and pictures had, in fact, played a marginal role in earlier Chinese encyclopedias of materia medica.[15] However, when the first edition of the *Bengcao gangmu* was published in Nanjing in 1596, Li Jianyuan, Li Shizhen's son and editor of the encyclopedia, suggested that pictures would augment the value of the text.[16] Two volumes of pictures were subsequently produced, although their authorship is unknown. Many of these illustrations were schematic at best, and the contrast between the 1596 Chinese illustrations and those produced in the 1672 Japanese edition of *Honzō kōmoku* is remarkable.

Both editions displayed the same animals, as seen in figures 11.1 and 11.2: clockwise from top left, a leopard (*bao*, *hyō* in Japanese), a lion (*shi*, *shi*), a tiger (*hu*, *tora*), and an elephant (*xiang*, *zō*). Produced only seventy-five years earlier, the Chinese images pale in comparison with the more elaborately and artistically conceived Japanese copies. In the 1596 drawing, the elephant is recognizable only for its trunk, but the 1672 illustration includes details of tusks, trunk, legs, and ears. Even so, these pictures were a mere embellishment of the texts and were not conceived as bearing cognitive or explanatory function: they were not even mentioned in the introductory remarks of any Japanese edition but were attached as separate addenda volumes to the main body of the encyclopedia.

The same argument holds for Tekisai's *Kinmō zui*: here the illustrations were certainly larger and played a more important role than in other encyclopedias.[17] They had an evident cognitive value, but it was connected to the *educational* purposes of the encyclopedia and did not deliberately convey any supplemental morphological information on the species portrayed. In the same vein, *Yamato honzō*'s drawings could hardly stand by themselves and would be useless for a birdwatcher to identify birds (see fig. 11.3). In stark contrast are the pictures attached to the 1736 survey sent to Niwa Shōhaku, like the one shown in figure 11.4, from the *Bizen no kuni Bitchū no kuni no nairyō sanbutsu ezuchō*; these illustrations could be used to identify birds even today.

In other words, the illustration in manuals and encyclopedias that followed the style and convention of Chinese textbooks did not add anything to the verbal descriptions. They were not supposed to convey further information about the morphology and anatomy of plants and animals, nor were they attached to help the reader identify the species portrayed. Their function was to accompany, complement, and embellish a self-sufficient text, not to complete it.

Illustrations started having a precise cognitive function only after the sur-

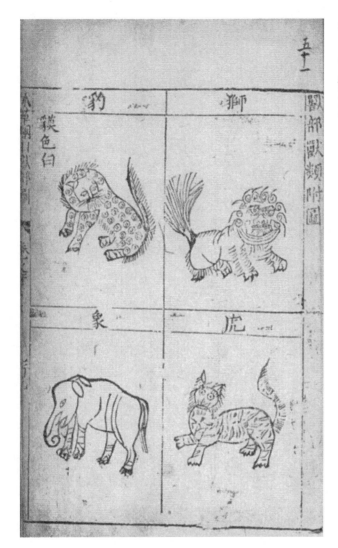

FIGURE 11.1.
1596 Nanjing
edition of *Bencao
gangmu*. Tokyo:
National Diet
Library.

veys of 1734–36, when Niwa Shōhaku expressly requested adding pictures to the lists of plants and animals to help him precisely identify species indigenous to each district and to clear possible confusions originating from homonyms. The drawings also gave him further details about regional variations of the same species.[18] The emergence of accurate visual representations of plants and animals was therefore connected to the inventory of natural resource.

Illustrations in shop catalogs of rare birds and plants for sale and in catalogs of private collections had a similar function of conveying precise information

FIGURE 11.2.
1672 Japanese
edition of *Honzō
kōmoku.* Tokyo:
National Diet
Library.

about specific plants and animals, either to help identify species or to simply
exhibit the visual appearance of a plant or an animal to a less knowledgeable cus-
tomer. It was also not uncommon for collectors to commission renowned painters to draw specimens from their collections of flowers, plants, birds, or shells.
One of the earliest works is *Sōmoku shasei* (Sketches of Herbs and Trees), which
consists of four scrolls by an artist who signed his work with the name Kanō
Shigekata. Each scroll presented a collection of flowers representing one season.

FIGURE 11.3. A *kashidori* (*Garrulus glandarius*) from *Yamato honzō* (1709).

All plants in full blossom were arranged one after the other with their name and a brief explanation attached. *Sōmoku shasei* was produced between 1657 and 1699 and portrayed 284 flowers, most of them grouped together in the two scrolls of spring and autumn. The original owner of the scrolls, the purpose of their commission, and the kind of collection it represented remains unknown.[19] Many of the species represented on the scrolls were exotic plants introduced into Japan between the late Muromachi and the early Tokugawa periods. Because they were first cultivated in the Kanō region of the Mino Province (now Gifu Prefecture), Isono Naohide has argued that the scroll may represent a collection of the Kanō Domain.[20] Figure 11.5 shows a segment of the "spring" scroll. On the left side, there is a sketch of a *shungiku*, the Garland chrysanthemum; on the right, there is a sketch of a single, slender plant of *araseitō*, a stock or gillyflower, a plant that originally grew only in the Mediterranean basin.[21] It is unclear when the plant was first introduced into Japan. The note on the side reports the third month of 1660 as the date when it was first planted in the Kanō region.[22]

As we have seen in the previous two chapters, amateurs and scholars active in *honzōgaku* clubs devoted their energies to producing richly illustrated albums that cataloged specimens from private collections or reported observation of

FIGURE 11.4. A *daikashira* (*Numenius arquata*) from *Bizen no kuni Bitchū no kuni no nairyō sanbutsu ezuchō* (ca. 1735). Tokyo: Tokyo National Museum.

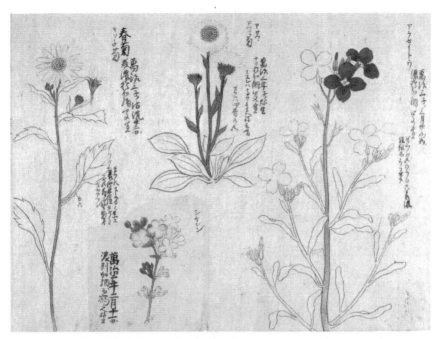

FIGURE 11.5. Kanō Shigekata, *Sōmoku shasei*,
"Haru no maki." Tokyo: National Diet Library.

vegetal and animal species in the wild. In all these cases, the visual representations of plants and animals had the specific function of describing the observed
species in the minutest detail. They not only complemented or enriched a self-
sufficient text but, on the contrary, conveyed a precise cognitive value in themselves, and often the text played a complementary role: written descriptions appeared now less frequently and focused on clarifying the various names of the
species, on reporting the circumstances in which the specimens were acquired,
and on detailing who observed them and how. For example, the accompanying
text to a beautiful sunfish portrayed by Tanshū (fig. 11.6) reported when and
how the fish was captured and where it was put on display and gave precise measurements of its size and weight but left the description of the fish—the proportions of the different body parts, the shades of colors, the shape of the fins, and
so on—to the picture, which acquired now the double function of reporting
the results of observational practices and of extending that experience to the
watcher, making him or her visually participate to the cognitive construction
of the species "*manbō*." The various techniques of pictorial representations developed by Tanshū and Hiraga Gennai extended the experience of the observer

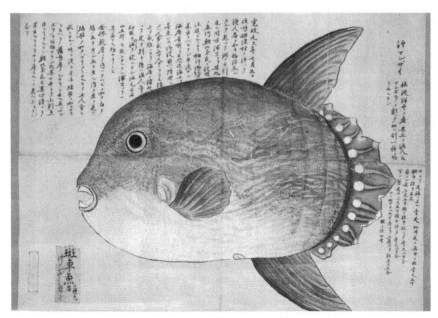

FIGURE 11.6. Kurimoto Tanshū, *Manbō*. Tokyo: National Diet Library.

to other senses, using glues to augment the translucency and tactile sensation of the scales of the fish.[23]

In the mid-eighteenth century, Hosokawa Shigekata, lord of the Kumamoto Domain, and his group of amateur *honzōgakusha* also resorted to illustrations to report the results of their long observational sessions of butterfly and moth life cycles (fig. 11.7). In these studies, Hosokawa *shows* rather than *describes* the metamorphoses of a caterpillar (larva) into a cocoon (pupa) and then into a winged moth or butterfly. The text is limited to recording the time of the observation.

A further example is the work of Iwasaki Tsunemasa, who brought the technique of botanical illustration to perfection.[24] Tsunemasa, a low-ranking shogunal official (*kachi*), began studying *honzōgaku* in his early teens under Yamaoka Shuzen, an Owari retainer who was trained by Tamura Ransui. When he was later hired by the shogunal Institute of Medicine, he had the opportunity to study under Ono Ranzan during the last years of his life. *Honzō zusetsu* (Illustrated Materia Medica, published in 1810) was Tsunemasa's major work, consisting of sixty illustrated volumes based on Li Shizhen's *Honzō kōmoku* (fig. 11.8). After the success of his gardening manual *Sōmoku sodategusa* (The Cultivation of Herbs and Trees), Tsunemasa went on to distinguish himself as a talented gardener and ended his career as associate curator of the Koishikawa garden.[25]

As these few examples show, a growing concern for descriptive accuracy in illustrations of plants and animals was not merely an epiphenomenon of the visually rich Tokugawa culture or the result of an improvement in pictorial techniques but was directly connected to new cognitive values attributed to visual representations over verbal descriptions. Without dehistoricizing "science" into a universal category, it is fair, I think, to say that the effort to produce faithful pictorial renditions of observed specimens characterizing *honzōgaku* texts in late Tokugawa period indicates a surprising convergence of epistemological approaches in natural historical research in early modern Europe and Japan. This convergence was the product of distinct and autonomous transformations rather than direct influence. In most cases, in fact, the usage of visual representations to convey information of plants and animals in *honzōgaku* texts still aimed to tackle semasiological concerns of traditional works like those of Ekiken, Jakusui, Ranzan, and others. In other words, illustrations were a new technological device to solve the traditional predicament of identifying a species with precision and matching Chinese names of plants and animals in canonical sources with actual plants and animals living and growing in Japan or brought there by foreign merchants. As we have seen, scholars like Mizutani Hōbun, Itō Keisuke, and other members of the Shōhyakusha group made extensive use of refined techniques of visual representations to complement information they gathered from multilingual sources and from collegial discussions and experiments in order to precisely identify species of herbs and plants for medicinal use.

Whatever the reasons that led to true-to-nature illustrations, the question of "realism" in visual representations of nature mobilizes deep ontological and epistemological issues. On the one hand, it manifested a new commitment of scholars to record their observations of actual specimens, stressing the *procedures* of cognition and the inferences that they collegially developed on the basis of those procedures rather than a reduction of the observed specimens to a priori categories.[26] In other words, the properties of a species of plants or herbs were not simply deduced from general principles but induced from the analysis of the body of the specimen itself through shared protocols of analysis. On the other hand, as Michel Foucault argued in the case of European natural history, the adoption of new systems of description and classification extended far beyond the pragmatic reorganization of empirical data on plants and animals; rather, it activated a discourse that presupposed the training of the expert's observing gaze to the necessities of the system itself. Images conformed to the language of the new natural historical paradigm and as such were the result of new techniques

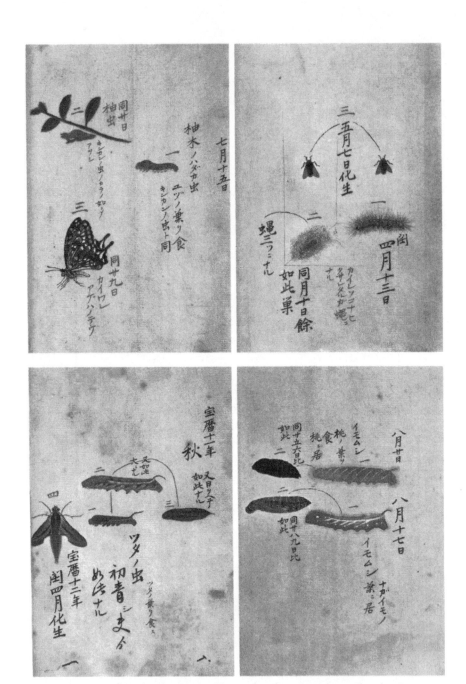

FIGURE 11.7. Life cycle of moths and butterflies from Hosokawa Shigekata's *Chūrui ikiutsushi* (Sketches of Insects from Life, 1766). Tokyo: Eisei Bunko.

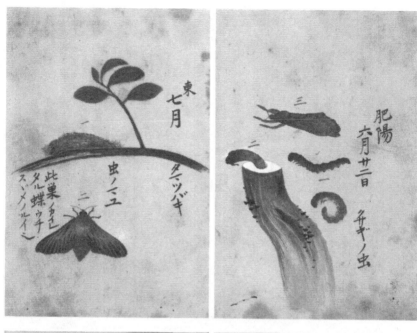

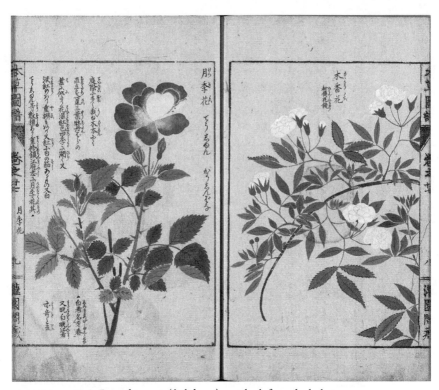

FIGURE 11.8. A *Rosa chinensis* (*kōkibana*), on the left, and a baksia rose (*mokkōbana*, *Rasa baksiae*), on the right. From Iwasaki Tsunemasa, *Honzō zusetsu* (1810). Tokyo: National Diet Library.

of "seeing systematically": "With seeing what, in the rather confused wealth of representation, can be analysed, recognized by all, and thus given a name that everyone will be able to understand: 'All obscure similitudes,' said Linnaeus, 'are introduced only to the shame of art.' Displayed in themselves, emptied of all resemblances, cleansed even of their colours, visual representations will now at last be able to provide natural history with what constitutes its proper object, with precisely what will convey in the well-made language it intends to construct."[27]

To make this point clearer, it is useful to look more closely at the conceptual network associated with these illustrations, in particular at the notions of *sha-sei*, *shajitsu*, and *shashin* and the "training of the gaze" that they produced. *Sha* (*utsushi*) means literally "to copy," "to transcribe," like in the traditional ritual of copying Buddhist sutra (*shakyō*). *Sei* means "life," "to live," or "to be alive"; *jitsu* stands for "truth," "factuality," "actuality"; and *shin* refers to the abstract categories of "truth," "originality," "authenticity," and so on. *Shasei*, literally translatable as "to transcribe life" or "to copy from life," is an important notion in the history of Tokugawa art that substantially diverged from its original Chinese *xiesheng*. Art historian Kōnō Motoaki has argued that *shasei* was used more ambiguously than *xiesheng* in reference to a wider range of paintings, and it generically emphasized accuracy and faithfulness in the representation of objects, whether or not the artist was actually sketching the object while looking at it.[28] In the field of *honzōgaku*, *shasei* usually appeared in association with self-standing illustrations or in the title of illustrated albums like *Sōmoku shasei*. Much more relevant is perhaps *shashin*, the modern term for "photography." The concept of *shin* was often mobilized in early *honzōgaku* texts of lexicographical leaning to refer to the essential properties of a species or the correctness of the attribution of a specific name to an actual plant. In the early nineteenth century, Shōhyakusha's textual production made wide use of the term *shashin* in connection with the faithful pictorial representation of a vegetal or animal species after it had been carefully observed and precisely identified. In effect, it seems that the notion of *shin* referred to the ontological status—the "essence" or "inner nature"—of a species' exemplar, while *shashin* referred to the epistemological practice of correctly and faithfully reproducing that essence. Satō Dōshin has argued that while the concepts of *shashin*, *shasei*, and *shajitsu* were more or less used in the Tokugawa period as synonyms and all gravitated around a notion of "perceptual realism," the three expressions had distinct connotations insofar as *sei* emphasized "the livelihood of the living entity," *jitsu* emphasized its specific objectivity (i.e., its location in time and place), and *shin* emphasized the universality of its ontological status.[29]

Generally speaking, both *shin* and *jitsu* characters appeared in various compound forms in the jargon of various scholarly fields. Both referred to the notion of "real," or "truth," but while *shin* denoted the unchanging, ontological essence of things, *jitsu* emphasized their objectivity, their factuality, especially in connection with cognitive practices. *Shin*, in other words, was the truth of a thing or phenomenon in itself, while *jitsu* referred to the truth of an object or phenomenon correlated to human cognitive practices. Both *shin* and *jitsu* could be used in opposition to the concepts of *kyo* (counterfeit), *gi* (mistake), and even *kū* (void) or *mu* (nothingness), but what distinguished *jitsu* from *shin* was its explicit stress on the concrete operations that a scholar performed to discern the nature of an object or a phenomenon: like a seed—*mi*, another meaning of the character *jitsu* (実)—that reveals its inner nature only by transforming itself into a plant, the "truth" that the word *jitsu* denoted had to be allowed to "germinate," to be expressed, realized, acted upon through concrete objects and practices. Used in compound form, *jitsu* in fact acquired the meaning of "realizing truth," or, better put, a truth that is objectified, realized in concrete objects or actions. Examples of this semantic function of *jitsu* included terms like *jittoku* ("true virtue" in the sense of virtuosity in action, realized), *jitsuri* (the "true *ri*," the Neo-Confucian principle seen in action in concrete phenomena), *jisshin* (the "true heart," truthful intentions manifested in concrete social practices), *jitsugaku* and *jitchi* ("practical learning" and "true knowledge," respectfully, both referring to applied learning, the practical, concrete effects of knowledge in human life and the environment), and *jikken* ("actual observation," or, more literally, "seeing truth," a term that often appeared in *honzōgaku* texts).[30] The semantic universe of *jitsu*—of a "truth" that manifests itself as a result of certain operations performed upon material reality and in correlation with human observers—included also the important concept of *jisshō*, or "verification," a procedure that actively made truth appear.

Through their careful philological analyses of Chinese and Japanese texts, thinkers as different as Itō Jinsai, Ogyū Sorai, and Motoori Norinaga founded the entire edifice of their thought on the concept *jisshō*. Words, they claimed, change over time and their transformation implied also a transformation of the world that human beings inhabited. For different purposes, they all insisted on the importance of excavating the original, primeval meanings of words. In *Gomō jigi*, for instance, Jinsai argued that "meanings derive from philosophical lineages [*ketsumyaku*]. Scholars should therefore first identify the lineage of any new ideas they encounter. Without an appreciation of their lineage, attempts at fathoming their semantics will remain haphazard, like a rudderless boat drifting in the dark with no certainty about where it goes."[31] Similarly, *honzōgaku* scholars, many of whom were familiar with the methods of verification (*jisshō*) practiced in the Confucian schools of Jinsai and Sorai, could succeed in correctly identifying a plant (i.e., finding out its true essence, or *shin*) through an active research that implied observation (*jikken*) of a vegetal or animal specimen, verification of its name in the canonical sources (*jisshō*), and eventually a truthful representation (*shajitsu*) capable of revealing its true nature (*shashin*).

Inō Jakusui, a student of Jinsai, thought that the best way to achieve correct knowledge (*jitchi*) of plants and animals was to recover the original meaning of their true names (*shin*) in the canonical texts. In writing his *Yamato honzō*, Ekiken aimed at authenticating a plant or animal by correcting the many discrepancies between *Honzō kōmoku*'s entries with the profusion of different names by which plants and animals were known in the different provinces of Japan.[32] In turn, Ono Ranzan applied new observational skills and his knowledge of Dutch texts to correct Ekiken's mistakes and retrieve the *shin* of natural species of Japan.[33]

In the work of Shōhyakusha members, we find the most original reinterpretation of *shin*. Mizutani Hōbun, Itō Keisuke, and others produced atlases and catalogs that offered a faithful representation (*shashin*) of the plants and animals they studied. As Maki Fukuoka rightly argues, "The concept of *shashin* emerges at the very intersection between visuality and knowledge. For Hōbun and other members, the role of direct observation began to gain more currency and value in this process of synthesizing their text-based understanding and their observed physical conditions. In the various types of pictorial representations they produced and studied, the group questioned not only what is represented but also *how* it is represented and *where*."[34] For this reason, she renders the concept of *shashin* as the "transposition (*sha*) of the real (*shin*)" in order to emphasize the central role that the observed specimen played in the cognitive practices of the Shōhyakusha naturalists. Their efforts to render as faithfully as possible that object in illustrations—utilizing a variety of techniques, including plastering dried specimens on the paper, ink rubbing (*in-yō-zuhō*), and shadowing methods (*shin'ei*)[35]—spelled out "the efficiency of *shashin* in attesting to the existence of a particular specimen and the direct observational experience of the specimen."[36] To stress the importance of actual objects as material verification of their cognitive claims, Fukuoka renders *shin* as "the real," thus reproducing in her own analysis the epistemic labor of Shōhyakusha naturalists.[37] The fact is that previous scholars like Ekiken or Jakusui never questioned the actual, material existence of the objects they inquired about. But instead of deducing the properties of objects from received texts, as early *honzōgakusha* did, Hōbun and Keisuke induced them from the objects themselves. Thus what happened to the concept of *shin* in the works of the Owari naturalists is tantamount of a semantic shift from the ontological (the elucidation of the essential properties of things, their *true* nature) to the ontic (the evidence of their actual, real *existence*). The object, for them, came first and played an integral role, with texts and observations, in constructing truthful knowledge of its properties. In their works, *shin* thus experienced a semantic shift from the realm of "truth" (i.e., *shin* as the essential properties of things) to the realm of "certainty" (i.e., *shin* as the result of discernment through measurements, observation, experimentations, etc.). The truth-value of the knowledge of plants and animals rested, for them, in the certitude of their analytical *procedures* rather that in a more or less explicit system of metaphysical relations.

Generally speaking, then, the cognitive value of illustrations in *honzōgaku* texts was the product of a dialectics of two conceptions of "truth," *jitsu* and *shin*, the first epistemological and procedural and the second ontological and universalistic. A picture of a plant or an animal was effectively truthful (*shajitsu*) only insofar as it was capable of rendering (*sha*) the inner essence, the "truth" (*shin*) of the species (*shu*) it portrayed. This was true also for Shōhyakusha members, for whom the true nature of a specimens was embodied in its concrete materiality rather than in a metaphysical sense. Furthermore, the growing relevance of illustrations as vehicles of knowledge in *honzōgaku* texts was a clear sign that a precise correspondence between words and things could no longer be established with certainty: it challenged, in other words, the centrality of lexicography (*meibutsugaku*) in *honzōgaku* research. Inō Jakusui, Kaibara Ekiken, and in certain measure even Ono Ranzan were convinced that by ordering words, one could hope to achieve a true and correct understanding (*jitchi*) of the inner essence (*shin*) of species, since it was thought that the name of a thing internalized and reflected all those inherent characteristics (shape, color, size, odor, behavior, properties, etc.) that singular plants and animals shared and made them all individual members of a distinct species.[38] Starting with the 1734–36 survey and up to the activities of circles like the Shōhyakusha and the Shabenkai in the first half of the nineteenth century, illustrations increasingly complemented and sometimes even substituted for language as the privileged vehicle to express the *shin* of a species.

In a strictly epistemological sense, however, there remained a firm belief in the correspondence between thought and things, since "truth" was still conceived as a form of *adaequatio* of thought—in the form of either language or pictorial representation—to things. Thus *shashin-shasei* illustrations can be understood as "realistic" only in the dogmatic sense of being representations resembling or, better put, *adequate* to—in its etymological sense of being "equal to"—an object subsisting "in itself" and not in the modern sense of being representations of individual specimens of plants or animals, with the particular, individual shades of color and shapes that they could only have in that particular time and place as a result of their interaction with the observer. The function of these pictures was, in other words, to express the general characteristics of a species—for example, what made all individual *kōkibana*, as seen in the *Rosa chinensis* reproduced on the left side of figure 11.8, identical members of the species *kōkibana*. The *shashin* illustration had to be true not only to the individual specimen observed but to those characteristics—its *shin*—that made it representative of a particular species. These pictures reproduced *species* rather than individual *specimens* of plants and animals. In other words, the idiosyncratic appearance of an individual specimen was thus sacrificed to transform it into a representative bearer of *species*-specific characteristics, much like the early modern European natural philosophers' epistemological and moral imperative to be "true to nature" led them to produce standardized representations of natural species and phenomena.[39] As Lorraine Daston and Peter Galison put it, "These images were made to serve the ideal of truth—and often beauty along with truth."[40] John James Audubon's birds and Ernst Haeckel's pseudopods and mollusks, just like Tanshū's bats and Gennai's fish, were not merely representations of specific specimens but a distillate of what a particular species should look like.

There could be discrepancies between different images of the same species and its verbal descriptions in canonical encyclopedias of the past, but the division of plants and animals into the discrete species of *Honzō kōmoku* was never substantially questioned before Itō Keisuke's adoption and adaptation of Linnaean classification. And even then, the Western notion of "species" and the classical notion of *shu* fused together, since both conceived of species not as epistemological categories but as real entities in and of themselves, as "natural kinds."[41]

Daston and Galison have persuasively shown that "objectivity"—a European notion that in the context of natural history was close to the concept of *shajitsu*—changed over time with the changes of scholars' "epistemic virtues," those norms shared by a scholarly community "that are internalized and enforced by appeal to ethical values, as well as to pragmatic efficacy in securing knowledge."[42] In the early modern period, for both the European natural philosopher and the Japanese *honzōgaku* specialist, ethical and aesthetic concerns were as important as cognitive norms in structuring and legitimating their research. The illustrations they produced were hence an objectification of a series of cognitive, aesthetic, and ethical operations. But precisely because they were so charged with paradigmatic epistemological value, these pictorial representations of natural species contributed to *create* species—that is, to transform an epistemological category into material, concrete objects to observe, manipulate, collect, and reproduce. In other words, the main function of *shashin* illustrations was to display, to make visible all those essential ontological characteristics of a *shu*, or a species. They were a kind of translation of invisible properties in visible appearances.

There were of course other kinds of illustrations of nature—each designed to satisfy different functions even within the narrow confines of *honzōgaku* practices—besides the drawings of natural species appearing in the 1736 survey—for example, in atlases like *Honzō zusetsu* or in albums like Hōbun's *Honzō shashin*.

Elaborate pictures that enriched eclectic albums and catalogs celebrating private collections were often accurate and aesthetically appealing representations of specimens of rare and exotic plants and animals owned by wealthy amateur collectors and naturalists. Others reported and chronicled specific events, like the capture of an impressive fish or bird never seen before (like the sunfish of fig. 11.6), the beaching of a rare species of otter, or the parading of rare animals along the Tōkaidō. But even then, the pictures were true to nature precisely because they epitomized the defining characteristics of vegetal and animal species, visually representing what made a species an unambiguous unit rather than capturing the idiosyncrasies and imperfection of a particular specimen. Even when a bird or a fish was caught in action or immersed in a natural landscape, like in many Audubon's or Tanshū's paintings, or when a plant was portrayed bent by strong winds or wet with the morning dew, their standardized shapes functioned even more successfully as paradigmatic and pedagogical models. Realism does not represent reality; it constructs it by teaching how to look and what to look at.

SHASHIN ILLUSTRATIONS AND THE OBJECTIFICATION OF SPECIES

The function of *shashin* pictures in atlases and catalogs was in fact not only to map the diversity of vegetal and animal diversity but also, and especially, to train the observing gaze of scholars and amateurs to recognize *species* when looking at actual specimens and to distinguish those characteristics peculiar to each species. As distillates of a species' essence, their role was to see the defining tracts of species, not the particularities of specimens. As Umberto Eco has argued, there is no such thing as a pure empirical experience, but our language, culture, education, professional training, and so on have preformatted our mind to see what we already know, to look for familiar cognitive patterns.[43] So just as Marco Polo was unable to acknowledge a new species when he saw a rhinoceros and preferred instead to correct the idea he previously held of what a unicorn looked like, Mizutani Hōbun, as we saw in chapter 10, did not hesitate to see a tufted puffin (*etohiruka*) in the horned puffin (*tsunomedori*) he had captured, even though the bird he portrayed was radically different from the verbal and pictorial representations of *etohiruka* he was familiar with. Illustrations in early modern European and Japanese atlases, hence, served to discipline and restrain empirical observation rather than merely register its results.

The objectivity of vegetal and animal species—that is, their existence as objects of study—was an effect of a series of cognitive practices performed by a community of professional scholars, often in institutions receiving state support and legitimation, via a protocol of authoritative linguistic and visual technologies. Such a conception of species is often at work even today in our common language. When I see a viper, I not only observe an individual snake with specific characteristics but perceive it as a member of a precise species indistinguishable from all other members of the same species. One might be tempted to argue, parodying Karl Marx, that natural species appear at first sight an extremely obvious, trivial thing, but that their analysis brings out that they are a very strange thing, abounding in metaphysical subtleties and theological niceties.[44]

Such a "fetishized" notion of species suggests the apparent paradox that vegetal and animal species are at the same time concrete and abstract. Indeed, they are material: we can touch them, smell them, taste them, watch them, and depict them. But at the same time they are also pure abstraction: first and foremost because what we really experience are *individual* specimens that our cognitively disciplined gaze transforms into representatives of the entirety of their *taxon*, singular manifestations of a general model deprived of any unique specificity; second, because as concrete objects, they are the final product of—and as such they internalize—a series of social practices (the collecting, growing, breeding, exchanging, drying, storing, cataloging, painting, describing, etc.) specifically designed to model individual specimens into fitting the ideal standard. These practices vanish as soon as the specimens are presented as paradigmatic illustrations in atlases, as descriptive entries in encyclopedias, or as objects displayed in public exhibitions: they ceased to be specimens and become paradigms of species.[45]

The commodification of nature in the form of material and intellectual resources rendered natural species flexible entities. Just like commodities that once thrown in the market lose the heterogeneity of their material forms and functions to become expression of exchange equivalences, plants and animals could be removed from their environment, isolated from their life-worlds, and abstracted from their concrete existence to become paradigmatic images. In the illustrated albums of products (*sanbutsu ezuchū*) collected during the 1734–36 survey or in Tanshū's albums, flowers, roots, fish, and birds floated in a blank space as abstract models of the species they paradigmatically represented, or, alternatively, they could be grouped together in fantastic ensembles, birds or insects from different ecosystems unnaturally clinging to the same branch or squeezed in the same space, like in Maria Sibylla Merian's engraving, Itō Jakuchū's surreal paintings, or catalogs of merchandise or of private collections and in the cages of many *kujakujaya* and *chinbutsujaya*, as shown in chapter 9.

To paraphrase Marx again, one might venture that the mysterious character of species lies in their reflection of the social characteristics of men's own labor as objective characteristics of the products of labor themselves, as the sociona-tural properties of these things.[46] As we have seen in the previous chapters, this intellectual labor, internalized and veiled by the objective materiality of natural "specimens-as-species representatives," always takes the form of a social network of intellectual and manual workers. In the case of *honzōgaku*, this would include gardeners, farmers, painters, amateurs, and scholars, as well as tools, verbal and pictorial technology, institutions, and state sponsorship. This labor is necessar-ily contingent on the conditions of its realization: the institutions that sustain it, the worldviews that legitimize it, and the particular targets that motivate it.

To conclude, natural species are epistemological categories, not entities in and of themselves—at least in a precladogenetic context. Their standardization in accordance with visually recognizable features was the outcome of conceptual and practical operations of various kinds performed by networks of intellec-tual and manual workers. The function of "realistic" illustrations in atlases and manuals as paradigms or archetypes of a species essence (*shin*), however, had the effect of concealing human interventions in their construction, of blurring the distinction between concrete specimens and abstract species, and of disciplin-ing the inquiring gaze of professional and amateur naturalists to see abstract and universal species of birds, herbs, or fish when they look at actual, individual birds, herbs, or fish. As a result, the composition and order of *Honzō kōmoku*'s species might not have been affected, but its metaphysical foundations simply ceased to have any relevance at all. Natural species assumed a somewhat plastic character: they could be removed from their natural environment, freed from any metaphysical necessity, and reconceptualized in accordance to different human needs. Plants and animals became a particular kind of material and in-tellectual commodity: they could be accumulated and collected, ordered and exchanged, transformed and deconstructed, and displayed and experimented with. Natural species became resources that could be mobilized for intellectual curiosity and aesthetic pleasure, for instrumental reason and utilitarian goals, for moral edification and convivial divertissement, and for economic growth and ideological indoctrination.

PART V *The Making of Japanese Nature*
The *Bakumatsu* Period

Natural History is the base for all Economics,
Commerce, Manufacture . . . because to want to
progress far in Economics without mature or sufficient
insight in Natural History is to want to act a dancing
master with only one leg.
— *Carl Linnaeus,* Bref och skrifvelser

The term *bakumatsu* is a historiographical category that refers to the last thirty to forty years of the Tokugawa bakufu, characterized by internal social, political, intellectual, and economic crises and the menace of Western imperial powers after the imposition of the first unequal treaties by Commodore Perry in 1854.

Having been reduced to a collection of discrete and exchangeable objects by practices aimed at satisfying disparate intellectual, cultural, aesthetic, medicinal, agronomical, entertaining, and economic needs, nature—a concept absent in both Chinese and Japanese languages until the late nineteenth century, as it has been argued in the introduction—had lost any meaning associated with religious beliefs or Neo-Confucian metaphysical order and had become intelligible to human understanding. This reification of nature—the source of its disenchantment—is evidenced by the names under which the myriad of natural objects (*banbutsu*) were classified. These names mobilized natural objects to satisfy different human needs: cognitive (*meibutsu*), medicinal (*yakubutsu*), economical (*sanbutsu*), and entertaining (*misemono*).

The construction of natural objects—in the forms of dried or embalmed specimens of plants and animals, pictorial representations (*shashin*), encyclopedic entries, collection items, catalog samples (*hyōhon*), commodities (*yakuhin*), and the like—was the result of intellectual as well as manual practices conducted in different institutional settings (private schools, cultural circles, institutes of medicine, apothecary shops, etc.) by amateurs and professionals of different social standings to satisfy different goals. The instrumentality of knowledge (*jitsugaku*) was the assumption behind these practices, sustained by an increasing commercialization and monetization of the economic life.

Such is the story of the development of *honzōgaku* that I have presented thus far. It began with the appropriation of the Chinese *bencao* tradition in late sixteenth- and early seventeenth-century pharmacological and Neo-Confucian circles and continued by tracing some processes of exaptation of that tradition in diverse scholarly fields of nature study that lost their connection with medicinal practices.[1] I argued how concerns with economic necessity, intellectual curiosity, and educated entertainment dominated this transformation and contributed to produce an ensemble of disciplines and practices that accumulated an impressive load of data on plants and animals as well as a sophisticated technology of observation, description, reproduction, and manipulation of specimens.

The last chapters of this book have the task of reconstructing how the multiplicity of styles, approaches, forms, and goals of Tokugawa-period *honzōgaku* gave space, in the course of the nineteenth century, to a more integrated and less diverse discipline. I argue that three factors contributed to this process of standardization: the collapse of the feudal social order that a completely monetized market economy stimulated; the economic crisis of the 1830s that called for a complete reformation of both the political sphere and the productive system; and the acceptance by a larger number of scholars and naturalists of the new Western sciences, thanks to the activities of the VOC surgeon Franz von Siebold.

Bakumatsu Honzōgaku
The End of Eclecticism?

THE VARIETIES OF *HONZŌGAKU* IN NINETEENTH-CENTURY JAPAN

A recurring theme of this study has been the eclecticism of *honzōgaku* practices and goals throughout the Tokugawa period. Heterogeneous ways of reading and using Chinese and, later, Japanese pharmacopoeias produced a variety of practices and disciplines that the narrow concept of *honzōgaku* could hardly do justice to. This trend continued through the beginning of the nineteenth century, when more traditional works of materia medica were accompanied by agronomical, gastronomical, natural historical, aesthetic, and artistic ones. At one extreme of the spectrum, Kuroda Suizan, a physician of the Kii Domain, stretched the lexicographical and encyclopedic tradition of *honzōgaku* to its greatest limits. Suizan studied natural history under Ohara Tōdō, himself a student of Ono Ranzan, and pursued nativist studies under Motoori Ōhira.[1] He enjoyed the patronage of Lord Tokugawa Harutomi of Kii, who placed him in charge of the pharmacological office (honzōkyoku) of the domainal Institute of Medicine (Igakukan) and its botanical garden.[2] Although Suizan hardly ever left his home province, he carried out an exhaustive field survey of the flora and fauna of the Kii peninsula, which later became the basis for a reform program of agricultural production. He recorded the results of his extensive work in two of the most comprehensive *honzōgaku* encyclopedias ever written in Tokugawa history. *Suizokushi* (Marine Life), in ten volumes, focused on marine life in the rivers and seashores of the peninsula. *Komeiroku* (Records of Old Names), consisting of eighty-five volumes, described a further 2,585 species of plants and animals on the basis of Chinese and Japanese sources, along with his personal observations in the field.

Throughout his life, Suizan kept his distance from the major networks of *honzōgaku* scholars. He corresponded regularly with an Osaka illustrator named Hotta Tatsunosuke, who specialized in naturalistic subjects. Tatsunosuke eventually became Suizan's disciple, together with Yamamoto Shinzaburō, son of Yamamoto Bōyō.[3] Suizan did not publish any of his works. For this reason, he remained virtually unknown until 1877, when Shishido Sakari, a bureaucrat of the Finance Ministry, spotted a manuscript copy of *Suizokushi* in a bookstore in Kōjimachi, Tokyo. After learning more about Suizan from Tatsunosuke, Sa-kari passed the text to Tanaka Yoshio, an influential scholar-bureaucrat in the emerging scientific world of Meiji Japan. Tanaka, in turn, acquired the publish-ing rights for all of Kuroda's writings and sponsored their publication.[4] It did not matter to Tanaka that Suizan adhered to the "anachronistic paradigm" of the Neo-Confucian *Honzō kōmoku*; Tanaka recognized that this attention to detail was of enduring scientific value.[5]

On the opposite end of the methodological spectrum was Udagawa Yōan, a professional Dutch translator turned naturalist. He was a committed proponent of the idea that Japan needed to reject the *honzōgaku* tradition in its entirely and adopt in its stead Western natural sciences. The Udagawa family, of which Yōan was a member of the third generation, was known for producing some of the finest translators of the Dutch language. The family also had a long history of engaging in intellectual exchange with other scholars in the late Tokugawa period. The first of the Udagawa to acquire notoriety was Genzui, the official physician of the Tsuyama Domain who studied Koihō medicine in Edo and then the Dutch language under the guidance of Maeno Ryōtaku, Katsurakawa Kuniakira, and Ishii Tsuneemon. Genzui achieved a degree of fame after his translation of Jan van Gorter's *Gezuiverde geneeskonst of kort onderwyks der meeste inwendige ziekten; ten nutte van chirugyns* (Accurate Medicine, or Brief Instructions to Many Internal Sicknesses, for the Use of Surgeons, Amsterdam, 1744) and a survey of Western surgical techniques in eighteen volumes, which he completed with the title *Seisetsu naika senyō* (A Survey of Western Surgery, 1793–1810).[6]

Genzui's main interest, however, was botany. He began by translating an ob-scure Dutch herbal manual, Petrus Nylandt's *Der Nederlandsche herbarius, of Kruydt-boeck, beschryvende de geslachten, plaetse, tijt, oeffening, aert, krachten en medicinael gebruyck* (The Netherlands Herbal, or a Herbal Describing Family, Place, Time, Exercise, Nature, Strength, and Medicinal Use, Amsterdam, 1670), under the title *Ensei yakukei* (Far Western Materia Medica). Despite his efforts, the translation was never published. The fact that most of these texts were not published does not mean that they were unsuccessful or did not attract schol-arly attention. On the contrary, large portions of Tokugawa scholarly production were never published and circulated in manuscript form through professional book lenders (*kashi hon'ya*).[7] Scholars often had their treatises copied by their students and sent to the various circles or schools of his networks. In this sense, the introduction of print at the end of the sixteenth century was perhaps less revolutionary in Japan than in Europe, in particular for specialized texts.[8]

Genzui went on to produce a survey of Western herbalist traditions in his *Ensei meibutsu kō* (On Western Semasiology), which also circulated in manuscript form. Most important, Genzui, in collaboration with Inamura Sanpaku, Dutch scholar and physician of the Tottori Domain, and his adoptive son Genshin, began the compilation of the first Dutch-Japanese dictionary, the *Haruma wage* (Explanation in Japanese of Halma's Dictionary), published in 1796. It was in this dictionary that the word *shizen* was first used to translate the Dutch word *natuur*.[9]

Genshin continued Genzui's translation projects of medical manuals on surgical techniques.[10] He acquired the reputation of being one of the most accomplished Dutch-to-Japanese translators. This led the shogunate to hire him for the translation of Noël Chomel's *Dictionnaire œconomique, contenant divers moyens d'augmenter son bien, et de conserver sa santé* (1767) and Matsudaira Sadanobu to entrust him with the complete translation of Dodoens's herbal.[11] Genshin also published two *honzōgaku* manuals based on Western sources, *Oranda yakukyō* (The Mirror of Dutch Pharmacopoeia, 1819) and *Ensei ihō meibutsu kō hoi* (Addenda to Reflections on Western Medical Terms, 1834).

It was the third Udagawa, Genshin's adopted son Yōan, who succeeded in establishing himself as both a gifted translator of Dutch botanical encyclopedias and a scholar of natural history in his own right. His first original work was a short text published in 1822 under the title *Botanika kyō* (The Sutra of Botany). Rather surprisingly, this text was written in the style of a Buddhist sutra, and it was the first work in Japan to use the Latin term *botanica* (botany). More than just a scholarly work, *Botanika kyō* could be described as a manifesto asserting the necessity of replacing traditional Japanese natural history with Western science, taught in Nagasaki by the German surgeon von Siebold in his own private academy. Yōan advocated substituting *honzōgaku* with the term *botanika*.[12]

The text opened with the Buddhist ritual formula "*nyoze gamon*," or "I have heard," probably to emphasize the foreign or, rather, the received nature of Western knowledge. He then explained that in the remote regions of the West, "great saints"—and again he used a Buddhist term *daishō*—"had developed a deeper working knowledge of the natural world than that of *honzōgaku*." Yōan's saints included Conrad Gessner, the Scottish botanist Robert Morison, the English botanist John Ray, the French naturalist Joseph de Tournefort, the Dutch botanist Paul Hermann, the Swiss botanist Kaspar Bauhin (author of *Pinax theatri botanici*, 1596), the Dutch physician Hermann Boerhaave, and Carl Linnaeus. This short text was less a manual of natural history than propaganda for an alternative way of studying nature. Yōan introduced and explained the breakdown of Linnaeus's classification of plants into twenty-four classes, which Yōan called *kei*. The choice of Buddhist terms and style to introduce Western botany in Japan was motivated, as Nishimura has explained, by Yōan's conception of the new paradigm as a gift from the West and as a received teaching. Thus he began the formula of Buddha's disciples that expressed their gratitude in receiving his teachings.[13] Yōan was not the only one to adopt this strategy. In the same year *Botanika kyō* was published, the Dutch interpreter and *rangaku* scholar Yoshio Nankō published a survey essay on Western astronomy titled *Seisetsu kanshō kyō* (The Sutra of Western Astronomical Observations), which also utilized a Buddhist terminology.

This strategy is puzzling. Not only did Yōan publish his first work in a Buddhist style, but he also shaved his head and wore clothes resembling those of a Buddhist monk, although he never actually took vows or ever claim to be a Buddhist believer. Respect and gratitude for the gift of a new corpus of knowledge might have motivated Yōan's and Nankō's choice of expression. But it might also have been a strategy to separate themselves from mainstream scholars, a visual and rhetorical stance to highlight their distance from the established traditions of *honzōgaku*, which since the Ming period had adopted a Neo-Confucian terminology. In a sense, his strategy was exactly the opposite of the fate that two centuries before Hayashi Razan had endured, forced by his patron Tokugawa Ieyasu to dress like a Buddhist monk in order to gain social acceptance as a "scholar" when such a profession did not yet exist.[14] Yōan's choice, in contrast, suggests a resistance against the now well-established socioprofessional identity of the *jusha*. As a translator, he was not recognized as a peer by other scholars, despite his efforts to establish himself in the field of *honzōgaku*. Having failed in his attempt, perhaps he was trying to invent a new socioprofessional position for himself by adopting the rhetoric of Buddhism.

After *Botanika kyō*, Yōan spent the following decade revising and expanding his adoptive father Genshin's works. It was in those years that Yōan completed a Dutch translation of Antoine-Laurent de Lavoisier's revolutionary works, which introduced the first notions of modern chemistry to Japan. In the expanded edition of Genshin's *Ensei ihō meibutsu kō hoi*, Yōan came up with a translation for the concepts of chemical element (*genso*), calorie (*onso*), light (*kōso*), gas (*gasu*), oxygen (*sanso*), nitrogen (*chisso*), hydrogen (*suiso*), carbon (*tanso*), and carbonic acid (*tansan*), among others, thus creating many of the words still in use today in Japanese chemistry.[15] The revised edition of *Ensei ihō mebutsu kō hoi* was published in 1834. That year also marked the beginning of Yōan's promotion of the new discipline of botany. *Shokugaku keigen* (Fundamentals of Botany), also published in 1834 in Chinese, was the first comprehensive treatment of Western botany in Japan. Plant organography, morphology, physiology, and phytochemistry replaced *honzōgaku* and its traditional taxonomical and descriptive modules. Yōan intended nothing less than to revolutionize the study of nature by rejecting the tradition of *Honzō kōmoku*. To do so, he developed a new technical language and a new terminology for naming plants and animals. He abandoned the usage of the old taxonomical categories (*kō, moku*, etc.) and replaced them with different terminology, often choosing words with strong Buddhist connotations. He abandoned the term *honzōgaku* first for *botanika* and then for *shokugaku* (literally, the study of plants), the ancestor of *shokubutsugaku*, which is the modern Japanese word for botany. As Yōan stated in his introduction to *Shokugaku keigen*, "I decided to translate what in the West is called '*botanika*' as '*shokugaku*.'"[16]

Yōan's manual was a challenging text for contemporary scholars to read, not to mention for amateur naturalists. Nonetheless, it made an impact on scholarly communities through the neologisms it introduced to describe the natural world, many of which are still in use today. Yōan ceased referring to plant and animal species as products (*sanbutsu*), medicinal herbs (*yakusō, honzō*), or names (*meibutsu*). He also ended the vague references to a multitude of natural objects (*banbutsu*), as well as the strings of unrelated objects, such as "herbs-trees-birds-beasts-insects-fish-metals-jewels-grounds-stones." Thanks to Yōan's new vocabulary, there were plants (*shokubutsu*) and animals (*dōbutsu*), two distinct kingdoms (*kai*) that, together with fungi (*kin*), constituted the three realms of all living things: *animalia* (*dōbutsukai*), *plantae* (*shokubutsukai*), and *fungi* (*kinkai*), which remain the terms used today.[17] The processes of reification of nature into discrete specimens bearing "objective" properties characterizing *honzōgaku* practices in the second half of the Tokugawa period were now sustained also by a terminology that excluded human intervention as did *sanbutsu, yakuhin*, or *meibutsu*.

A revolution in knowledge, for Yōan, required a revolution in language. In this respect, Yōan's epistemology followed almost literally Kuhn's notion that two scientific paradigms competing for the best explanation of natural phenomena seldom share a linguistic common denominator.[18] Yōan's natural history was consciously conceived in opposition to the established practices and notions of *honzōgaku*, and so he invented a completely different language that could express nature in its objectivity. At the same time, as Biagioli has argued in his study of Galileo Galilei, proposing a new epistemology in a new language is often a strategy to make a mark in the field of cultural production: "The adaptation and articulation of different worldviews are linked to the development and maintenance of socioprofessional identities."[19] In fact, Yōan was struggling to be recognized as something more than a simple translator—like Galileo, a mere mathematician, was struggling to be recognized as natural philosopher.[20] Being a translator, as Sugimoto has argued, did not necessarily provide credibility as a scholar, since translators occupied a subservient position in the field of cultural production.[21]

Seimi kaisō (The Foundation of Chemistry) is further evidence that Yōan was seeking to reform the study of nature and in doing so find acceptance as a scholar in a specific disciplinary field.[22] Published between 1837 and 1847, *Seimi kaisō* had the ambition to propose a new way to organize knowledge of the natural world. In it, Yōan divided all disciplines into three groups: *benbutsu* (the division of things), *kyūri* (the study of principle), and *seimi* (chemistry; the Chinese characters he used for their phonetic value can be literally translated as "the secret abode"). *Benbutsu* was the study of the morphology of natural species and their place in a taxonomical order according to their shape. Yōan explained that in the West, this field of study was called *hisutōri* (natural history). *Kyūri* was the study of the general laws governing natural phenomena and was therefore a more fundamental discipline than *benbutsu*. In the West, this field of study was known as *hishika* (physics). *Benbutsu* and *kyūri* are so different, Yōan argued, that it is difficult to see how they relate to one another. For that reason he elevated a third discipline, *seimi*, the phonetic rendition of the Dutch *chemie*, or chemistry, to the highest place because it provided the link between the laws of physics and the natural world as we perceive it. Yōan attempted to make sense of the natural world first by studying the process by which elements form molecules on the basis of the laws of *kyūri*, also known as physics, and then how these molecules constituted the plants and animals that are the objects of *benbutsu*, also known as natural history. As Yōan had already explained in *Shobutsu keigen*, "*Benbutsu* [natural history] clarifies the end of *kyūri* [physics], *kyūri* forms the foundation for *seimi* [chemistry]. *Benbutsu* is the threshold of learning, *seimi* is the inner sanctum of the laws of nature."[23] Yōan proposed rejecting the categories of the traditional knowledge and investigation of the natural world. In their place, he advocated the adoption of the Western scientific approaches introduced in his surveys and translations.

Yōan would surely have continued to promote the merits of Western science had he not died shortly after the publication of *Seimi kaisō* at forty-eight years of age. He never saw himself as the creator of this new science, just its "prophet." Perhaps Yōan adopted Buddhist style and terminology because he saw Western science as having the potential to transform Japanese culture in the same way Buddhism did when it was introduced more than a millennium earlier. In Yōan's eyes, Western science, like Buddhism, represented a corpus of knowledge and techniques that, if adopted, would produce a revolution in the life of Japan—an awe-inspiring event that perhaps motivated the religious tones he used to express it.

Between these two poles of the spectrum of *honzōgaku* production, there were numerous combinations of different practices and styles that often mixed together Western and Eastern techniques and knowledge. Ono Ranzan, as briefly described in part IV, maintained an overarching lexicographical approach at the same time that he engaged in herborizing expeditions for the shogunal Institute of Medicine. He was also earnest to absorb information from Western sources. An eclecticism of methods and scopes also characterized *honzōgaku* practices in various cultural circles, as we have seen. Even in the medical field—of which *honzōgaku* had been ancillary—physicians began to utilize traditional pharmacopoeias in more heterodox ways. A revealing example is Hanaoka Seishū, a physician who succeeded in mixing together Western surgical techniques and traditional *honzōgaku* knowledge.

The sixteenth day of the tenth month of 1804, the physician Hanaoka Seishū performed the first surgical operation using general anesthesia in the world, preceding the American Crawford Williamson Long, who is typically credited as the first physician to use a form of anesthesia in his surgeries, by thirty-eight years.[24] Seishū's operation was to remove a cancerous mass from a woman's breast, using only an herbal concoction he had formulated as a powerful anesthetic. The operation was illustrated in the 1829 book *Kikanzu* (Illustrations of a Rare Disease) by one of Seishū's students. The woman woke up three hours after the surgery to a manageable amount of pain. She recuperated well and was discharged after a short period of observation. Over the following thirty years, Hanaoka Seishū used general anesthesia in surgery on 154 patients, most of which were successful. By the time of his death, Seishū was probably the most famous physician in Japan, boasting more than 1,800 students personally trained in his private school, Shunrinken, in the village of Hirayama, Kii Province.

Seishū was an official physician in the employ of the Kii Domain. He had begun to study Koihō medicine in 1782 and was also trained in Western-style surgical techniques and *honzōgaku*. Seishū started experimenting with his concoction on dogs, but he only discovered a precise formula and dosage after he tried it on his mother and his wife. He practiced the most on his young wife Kae, putting her to sleep every other day for the general amusement of their fellow villagers. Her sacrifices and his determination paid off, and Seishū became famous, treating patients from more than sixty provinces. Seishū described in great detail how he obtained his concoction, which was based principally on *mandarage (Datura alba)*—a plant related to the poisonous thornapple, *Datura stramonium*, a hallucinogen often used by Native American populations in religious ceremonies. The efficacy of the recipe depended mainly upon atropine, a tropane alkaloid, and scopolamine, another alkaloid with powerful narcotic and sedative properties, which are the main chemical components of the *mandarage*.

THE ARRIVAL OF PHILIPP FRANZ VON SIEBOLD AND THE AFTERMATH

Stories like those of Yōan and Seishū reveal the growing attraction of Japanese scholars for Western learning in the last third of the Tokugawa period. The longest and most profitable Japanese contact with Western natural history came when a young German physician arrived at Deshima on the twelfth day of the eighth month of 1823. The energetic twenty-seven-year-old Philipp Franz Balthasar von Siebold, newly graduated from Würzburg medical school, arrived in Nagasaki with a copy of Thunberg's *Flora Japonica* and Humboldt's *Personal Narrative of Travels to the Equinoctial Regions of America, during the Year 1799– 1804*. He was seeking adventure and fame as naturalist. In those years, Deshima, like other Dutch stations in Southeast and East Asia, was experiencing a period of renewed ferment. Holland had reclaimed national autonomy after the Congress of Vienna of 1815 and was now the new Kingdom of the Netherlands. With the exception of Deshima, which between 1799 and 1815 had been the only place in the world outside Holland flying the Dutch flag, all colonies were returned to the Netherlands after periods of British and French administration. Deshima was put under the direct control of the Kingdom of the Netherlands after the Dutch East India Company was dissolved in 1800. This meant that the Dutch *kapitan* was de jure and de facto a representative of the Dutch state in Japan.

One of the first tasks Siebold was entrusted with was the production of a new intelligence report on the natural resources, productions, population, and geography of Japan. This was carried out with the help of his adjutant Heinrich Bürger and sent to Batavia, where it was published in 1824 with the title *De historiae naturalis in Japonia statu* (Natural History of Japan). As the title suggests, the report intended to establish the conditions of Japanese development in regard to the exploitation of natural riches. His early assessment described Japan as a land with magnificent richness in vegetation, which, however, awaited full exploitation owing to the backward knowledge of natural history of its population.[25] Siebold's suggestion was that it was in the interest of Dutch trade to bring Japanese scholars to a higher level of comprehension of Western scientific technologies and knowledge of the natural world.

Siebold began the instruction of his Japanese pupils right away. At first he utilized the operating room of the Deshima station, where he showed medical techniques to the translator Yoshio Gonnosuke and the Nagasaki physicians Narabayashi Eiken and his younger brother Sōken. As the number of scholars interested in Siebold's teachings on Western medicine, surgical techniques, natural history, pharmacology, and geography (taught by Heinrich Bürger) increased, he began lecturing at Narabayashi's house, until Siebold was able to build, in the sixth month of 1824, his own private academy on the outskirts of Nagasaki. He called it Narutaki-juku, from the name of its location.

The reputation of the school attracted students from all over Japan, since it gave Japanese scholars the opportunity to receive a thorough firsthand training in Western knowledge.[26] According to Rubinger, "Narutaki Juku was the first school where systematic and sustained teaching of the latest scientific knowledge and medical techniques was provided by a European to leading scholars in Japan."[27] The majority of Siebold's students were physicians and Dutch scholars (*rangakusha*), including Minato Chōan, Mima Junzō, Oka Kenkai, Itō Genboku, Ninomiya Keisaku, and Kō Ryōsai. All students, in addition to attending lectures, were required to write a report in Dutch, submitted directly to Siebold. In this report, they were asked to assess their level of expertise in different disciplines, including zoology, botany, pharmacology, economics, politics, geography, and ethnographical information about religions, festivals, and the like.[28] For the Japanese scholars, these reports were an occasion to practice their linguistic skills in Dutch. For Siebold, they created a database of information on Japanese affairs he would use once he returned to Europe. It was in fact through information supplied by his students that he published the ethnographic essay *Nippon* in 1832; *Flora Japonica* in 1835, in collaboration with the German botanist Joseph Gerhard Zuccarini; and *Fauna Japonica*, which he cowrote between 1833 and 1850 with the zoologists Coenraad Temminck, Hermann Schlegel, and Wilhem de Haan. As a matter of fact, Siebold was a rather mediocre naturalist who routinely relied on the skills of other scholars. He was able to exploit the favorable situation created by the newly formed Kingdom of Netherlands, which was eager to reestablish its control in parts of Asia. Japan, a country with a significant population of scholars starving for information from the West, was an ideal setting for Siebold to develop his authority.[29]

What Siebold lacked in scholarly talent he compensated for in enthusiasm, diplomatic entrepreneurship, and an ability to create networks of informants and scholars.[30] These social skills would prove beneficial on his only trip to Edo in the first month of 1826, when he was able to establish a number of relationships with Japanese scholars and domainal lords interested in *honzōgaku* like Shimazu Shigehide of Satsuma.[31] He also met twice with Mizutani Hōbun, the leading scholar of the Shōhyakusha circle in Owari, and convinced one young member of the group to join him in Nagasaki.[32] This young scholar was Itō Keisuke, who later became a pioneer in modern biology.

Siebold resided in Japan longer than any of his predecessors—a total of six years and five months.[33] He was not a scholar of the stature of Thunberg, especially considering the fifty intervening years of scientific developments in European botany and zoology. Nonetheless, his long stay, combined with the changing attitudes of Japanese scholars, which had become far more receptive to Western knowledge, made his presence extremely important in the development of Japanese natural studies. What happened in the fifty years separating Thunberg's and Siebold's stays to justify the different impact they had on Japanese scholars? Why did a first-rate intellectual like Thunberg fail to capture the attention of Japanese scholars with Linnaeus's botanical ideas, while a smart but moderately talented surgeon like Siebold was able to influence the future of Japanese natural history? The answer to these questions lies in the changes taking place in Japanese society in the first half of the nineteenth century.

ITŌ KEISUKE AND THE BIRTH OF JAPANESE SCIENCE

Honzōgaku maintained throughout its history a marked eclecticism, but toward the end of the Tokugawa period, and especially after the Tenpō crisis of the 1830s, its instrumental value for economic growth became a leading impulse for *honzōgaku* practices. This transformation in scopes and methods put in motion two distinct but interrelated dynamics. On the one hand, *honzōgaku* became more similar to Western natural history, in particular to that brand of naturalistic research sponsored by Linnaeus and his "apostles" that submitted natural knowledge to the necessities of national economic growth—in particular of its agricultural production.[34] After seventy years, Hiraga Gennai's appeal in the manifesto of the 1762 Edo exhibition was finally vindicated. On the other hand, the fusing together of *honzōgaku* and the new field of *keizaigaku*—a Confucian notion translatable as "ordering the realm and saving the people" that would soon transform into something similar to the European discipline of "political economy"—facilitated the conversion of late Tokugawa scholars into modern scientists in the early decades of the Meiji period. The cases of the Satsuma Domain and Itō Keisuke are paradigmatic examples of these dynamics.

The son of a town physician in Nagoya, Itō Keisuke studied medicine with his father and *honzōgaku* under Mizutani Hōbun, the leader of the Shōhyakusha circle of amateur naturalists.[35] After specializing in medicine and pharmacology, the eighteen-year-old Keisuke left home for a long herbalist trip throughout the provinces of central Japan. In Kyoto, he met Yamamoto Bōyō and Fujibayashi Taisuke, a physician who taught Keisuke the basics of Western medicine. Later that year, Keisuke had the opportunity to spend more than a month on a field expedition from Nikkō to Kiso with Udagawa Yōan, which was the beginning of a lifelong friendship.[36] The meeting that radically changed Keisuke's intellectual trajectory, however, happened on the twenty-ninth day of the third month of 1828 in Atsuta. It was there that von Siebold, on his way to Edo, stopped to meet Mizutani Hōbun, Ōkōchi Zonshin, and his younger brother Itō Keisuke.

Siebold had been conducting an epistolary exchange with members of the Shōhyakusha for some years by that point, but meeting with the German physician must have deeply inspired Keisuke, because the following year he left Nagoya to enter Siebold's Narutaki school in Nagasaki. The only book he brought with him was a copy of his master Hōbun's *Buppin shikimei* (Clarifications of the Names of Things, 1809), a short dictionary that listed species of plants, animals, and minerals under their Japanese names together with their translation in Chinese. Itō Keisuke and Siebold together decided to supplement Hōbun's text by adding the Latin scientific name for each species. In turn, Keisuke transliterated the Japanese pronunciation of the species of plants and animals into the Roman alphabet, thus enabling Siebold to pronounce them correctly.[37]

Keisuke spent a little more than six months in Siebold's school. In 1829, he left Nagasaki with a copy of Thunberg's *Flora Japonica*, which Siebold had given him as a gift, together with the frontispiece from the French edition of *Voyages de C. P. Thunberg au Japon, par le cap de Bonne-Espérance, les îles de la Sonde, etc.* (Paris, 1796), complete with a portrait of the Swedish naturalist. Keisuke brought both publications back to Owari and added them to the Shōhyakusha's library. In the same year, he completed the manuscript of *Taisei honzō meiso* (Annotations on Western Names of Herbs), in which he applied Linnaeus's classification to a number of indigenous species, closely following Thunberg's *Flora Japonica*. Yōan had already introduced Linnaean method to Japanese scholars in his unconventional botanical sutra. It is Keisuke, however, who is credited with putting the Linnaean system into practice for the first time in Japanese his-

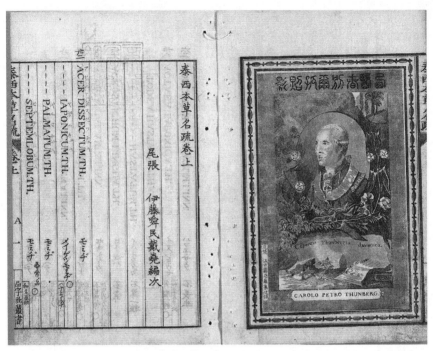

FIGURE 12.1. A page from Itō Keisuke's *Taisei honzō meiso* (1829). Tokyo: National Diet Library.

tory. In *Taisei honzō meiso*, the main body of the text consisted of a list of plant names in Latin arranged in alphabetical order—or *a-be-se*, as Keisuke put it phonetically in the introduction. The introduction also elucidated the principle of Linnaean taxonomy and its advantages for Japanese natural history in a far simpler and more accessible manner than Yōan did in his *Botanika kyō* (fig. 12.1).

Keisuke opened his text with a reproduction of the frontispieces of Thunberg's *Voyages* that Siebold had given to him in Nagasaki. The structure of Keisuke's 1829 *Taisei honzō meiso* resembles that of Hayashi Razan's *Tashikihen*, circulating nearly two hundred years earlier. Razan's text, now regarded as the first Japanese work of *honzōgaku*, was in fact a glossary of the Chinese *Honzō kōmoku*. But although similar in structure, the two books had a different impact in the cultural landscape of early seventeenth and early nineteenth centuries. The *Tashikihen* was used not only by the scholars interested in pharmacology but also especially by poets, painters, and Neo-Confucian scholars who used the *Tashikihen* as a dictionary when reading Chinese texts. In contrast, Keisuke's *Taisei honzō meiso* was appreciated and used almost exclusively by naturalists.

It resembled Razan's glossary only in the format, not the intent or the content, and the text continued to be important for Japanese scientists reading Western botanical manuals into the 1880s.[38] On the one hand, this difference reveals the degree of specialization that scholarly production had reached in its two-century-long development. On the other hand, it reveals a curious parallelism between two texts that were implicated in profound changes in the field of cultural production: Hayashi Razan struggled for a social recognition of scholarly activity, conceived in the language and form of Zhu Xi's reading of the Confucian tradition; Itō Keisuke worked for the professionalization of specialists of nature studies, conceived in the language and form of Western science—two fields imported from abroad that underwent a similar process of assimilation.

The importance of correct translation for the practical application of natural history is explained by Keisuke in the introduction of the printed version of *Taisei honzō meiso*:

> For those of us who are studying Western science, it is urgent that we collect products from across Japan and discuss them in the context of Western theories in order to correct their names and investigate their true nature [*shin*] so that we can take advantage of the knowledge in curing patients. However, there are only a few people who can do this and I have not seen a book that articulates the necessary process to carry out our tasks. The reason is that those who are translators are not familiar with *honzōgaku* and those who are interested in *honzōgaku* are too preoccupied with arguing the proper Chinese names and are not interested in Western theories. Because these two types do not converge, *honzōgaku* is not elucidated.[39]

The motivation behind the compilation of the glossary was not merely lexicographical. Razan's *Tashikihen* intended to translate in Japanese the Chinese names of minerals, plants, and animals in order to facilitate access to the information stored in *Honzō kōmoku*: its classificatory scheme, the names of species, their treatments to obtain medicinal substances, and so on. In contrast, *Taisei honzō meiso* did not simply to do that, but it also intended to introduce Linnaeus's taxonomical system. Through the binominal nomenclature developed by Linnaeus, in fact, one can already identify the genus of which each species was part just by looking at the first part of its name. Thus, although alphabetically arranged, one could already acknowledge the taxonomical relations of species that share the same genus name, an information that, for example, was not immediately available in those glossaries that, like Hōbun's *Buppin shikimei*, arranged species names in the Japanese *iroha* order. In other words, the Latin names of species, for Keisuke, had the advantage of expressing in it the taxonomical identification of a plant.

Maki Fukuoka rightly observes that Keisuke conceived of his work as having practical application in therapeutic practices: for him, the purpose of *honzōgaku* ultimately "resided in coordinating the name and efficacy of the plant world."[40] In contrast, Yōan aimed to understand the epistemological principles of Western botany in order to replace traditional *honzōgaku* with it. However, the choice of arranging the species in accordance with their Latin name already indicated Keisuke's intention of adopting Linnaean taxonomy to correct and improve the pharmacological information contained in traditional *honzōgaku* texts. Further proof of his reformist, rather than revolutionary, intents is the terminology he selected to introduce the binominal system in the last section of the printed version of *Taisei honzō meiso*.[41]

In the last section of the book, titled "Nijūyon kō kai" (Explanation of the Twenty-Four Classes, fig. 12.2), Keisuke described Linnaeus's taxonomical system of twenty-four classes and used it to classify all previously listed species of plants in orders and classes. His table was less schematic than Linnaeus's original diagram (fig. 12.3) and was probably Keisuke's original work.[42] It was likely based on Johann Mueller's *An Illustration of the Sexual System of Linnaeus*, a copy of which Siebold had almost certainly brought with him to Japan.[43]

Keisuke marked respectively with a "*me*" (メ) and a "*wo*" (ヲ) the female (*mesu*) and male (*osu*) reproductive organs (*shibe*) in flowers, to clarify the principle of sexual division. Most importantly, he rejected Yōan's translation of Linnaeus's categories of "class" and "order," with their heavy Buddhist connotations, and replaced them with the more familiar *kō* and *moku*, taken from *Honzō kōmoku* and still used today in taxonomical terminology.[44] Keisuke explained the principles of species classification in the introduction:

> Once a division in classes [*kō*] and orders [*moku*] is established, species should be further divided into genera [*rui*]. Genera are groups into which similar species are organized, and the genus appears as the first name in all species[45]. . . . Once classes and orders are distinguished, the various genera should be named and established on the basis of the proximity among species that results from a thorough investigation of their morphology. Furthermore, if it turns out, after consulting all existing botanical encyclopedias and manuals, that the species under investigation has never been studied before and it is impossible to establish a relationship [with other known species], a new genus should be established and named accordingly. . . . Moreover, even if the

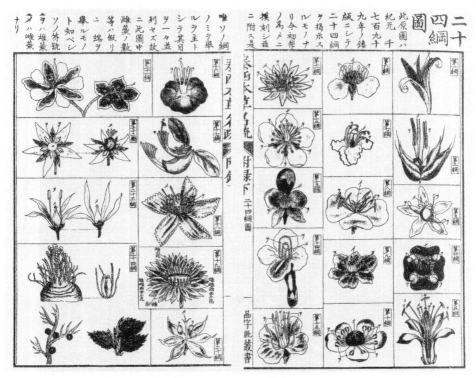

FIGURE 12.2. Linnaeus's twenty-four classes in Itō Keisuke's
Taisei honzō meiso (1829). Tokyo: National Diet Library.

identification of a new genus [*Neugeslacht*] has been conducted in a proper
manner, it should be carefully investigated if the species has never been analyzed before, and if it is not, it should be declared a new species and named
accordingly.[46]

Itō Keisuke's *Taisei honzō meiso* was a groundbreaking text, primarily because
it opened Japan to a new way of practicing and conceiving the study of nature.
In the years following its publication, Keisuke remained active in organizing
yearly meetings of professional and amateur *honzōgaku* scholars in Nagoya and
traveling on herbalist tours in various provinces of central Japan. In 1859, he
was nominated to be the Owari domainal physician and was ordered to organize classes on translation techniques from Dutch. Soon thereafter, however,
he moved to Edo, where in 1861 the shogunate recruited him as an instructor
(*kyōkan*) of the Bansho Shirabesho, a shogunal institute dedicated to research
and instruction in Western learning. There he was assigned to the newly estab-

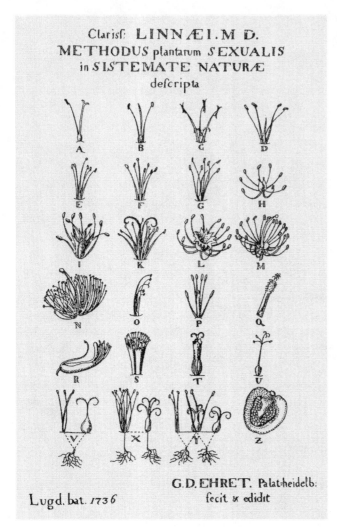

Clarisſ: LINNÆI. M D.
METHODUS plantarum SEXUALIS
in SISTEMATE NATURÆ
deſcripta

A B C D
E F G H
I K L M
N O P Q
R S T U
V X Y Z

Lugd. bat. 1736

G.D.EHRET. Palat-heidelb:
fecit & edidit

FIGURE 12.3.
Linnæi M.D.
methodus
plantarum sexualis
in Sistemate
Naturæ descripta
(Leiden, 1736),
with original
illustrations by
the painter Georg
Dionysius Ehret.

lished Bussankyoku (Office for Natural Products), the section of the Bansho Shirabesho in charge of natural studies. When, in the mid-1860s, the political situation in Edo became unstable, Keisuke returned to Nagoya and remained there for the duration of the Boshin War.

After 1868, the elderly Itō Keisuke was likely expecting a peaceful retirement of private study and herbalist tours in the Nagoya countryside.[47] However, in 1870 the newly formed Meiji government called him to lecture on botany at what was once the Bansho Shirabesho, renamed the Kaisei gakkō (School of Advanced Studies), the predecessor of Tokyo University. When Tokyo University was established in 1877, he was appointed first as adjunct professor (*buingai kyōju*) and in 1881 as professor of the department of science (*rigakubu kyōju*). In 1887, Keisuke was awarded the title of *hakase* of the Imperial University (Teikoku Daigaku, as the university was renamed in 1887), which at the time was the highest rank for a faculty member of the university.[48] He held this position until his death in 1901 at ninety-eight years of age.

Itō Keisuke succeeded in establishing himself as a pioneer naturalist in a crucial period of Japanese history. In the last decades of the Tokugawa period, *honzōgaku* networks accepted Keisuke as one of their leading scholars. He was renowned among professional and amateur natural historians at the same time that he worked for the shogunal Office for Western Studies. After the collapse of the shogunate and the establishment of a new central state, the Meiji government recruited him along with Western scientists to build a curriculum of natural sciences at the newly established University of Tokyo. Keisuke succeeded as a scholar in two different paradigms: traditional *honzōgaku* and modern natural sciences. Inspired by Linnaeus's classification and open to ideas and techniques arriving from the West, he succeeded among *honzōgaku* practitioners precisely because he encoded new methods into familiar and intelligible language and goals. As his use of the term *honzō* in the title of his *Taisei honzō meiso* suggests, Keisuke, unlike Udagawa Yōan, believed that Linnaean taxonomy and Western science were compatible with the traditional knowledge of *honzōgaku*. For him, Western botany shared with *honzōgaku* the common concern of creating a corpus of natural knowledge that could contribute to the pharmacological, agricultural, and economic welfare of the nation. In other words, the new biological science was a "Western *honzōgaku*," as the title of *Taisei honzō* (Western Materia Medica) can be rendered, and should be adopted because it constituted a more effective form than that of traditional *Honzō kōmoku*.

Keisuke was a scholar who straddled two paradigms. This explains the difficulty historians of science have had in understanding his scholarly production. Nishimura Saburō, for example, claimed that Itō Keisuke's work was not meaningful for the development of Japanese science, and Ueno Masuzō wondered to what extent Keisuke even understood the theoretical implications of Western science.[49] In fact, what Nishimura and Ueno, who were both trained as scientists before becoming historians of science, seem to be puzzled about is the fact that Keisuke navigated and mediated between two paradigms of natural knowledge that they deem as mutually incommensurable, as Kuhn would put it. Keisuke understood the language and methods of both *honzōgaku* and Western science. His "bilingualism"—both literally and, according to Biagioli,[50] in terms of his ability to move between two apparently incompatible explanatory paradigms—enabled him to adapt his own socioprofessional identity from that of *honzōgaku* scholar to modern scientist. In this sense, he translated the traditional concept of natural knowledge as a form of service to the welfare of one's domain into a form of service to the modern nation-state. The Neo-Confucian scholar-official thus became the scientist who strived to ensure technological, industrial, and agricultural advancement for his nation. The idea of science as an instrument of national progress was widespread in the nineteenth-century West and suited Japan's desires to become a modern nation-state and a "wealthy country with a strong army" (*fukoku kyōhei*). Science in the nineteenth-century world was not merely a learned pursuit but a matter of the most urgent national interest. In other words, Keisuke succeeded in bringing to fruition his "translation" of *honzōgaku* into science precisely because nineteenth-century *honzōgaku* already shared with Western science an investment in economic activities. After the Tenpō crisis of 1830s, in fact, the majority of domains had recruited *honzōgaku* specialists to assist their agricultural reforms.[51]

For Keisuke in particular, having grown up in a region affected by the agricultural crisis of the 1830s, the study of nature had always been a matter of public welfare. The motivation behind his early work on *honzōgaku* was precisely to prevent starvation and thereby prevent plagues and epidemics by strengthening the body's resistance, a precept shared by both the Goseihō and Koihō schools of medicine.[52] Keisuke stated this purpose clearly in the 1858 monograph *Ensei shōseki hen* (The Study of Saltpeter in the Far West), in which he argued that the study of Japan's natural resources was urgently necessary for the welfare of the nation and the prosperity of its people (*kokuri minfuku*).

In his practice, Keisuke followed the traditions of his masters Mizutani Hōbun and Ono Ranzan. He advocated serious study of manuals and encyclopedias and the application of knowledge acquired from field exploration in medicine and agriculture. The observation of live specimens and collection of herbs in the wild were always complemented by the study of multilingual texts. When Keisuke introduced the ink-rubbing technique he had learned from Siebold, he inserted it in the paratext of traditional albums.[53] The technique consisted of reproducing the texture of a leaf surface by placing a paper over it and rubbing it with an ink stone, a method that was fast, precise, and also economical, because it did not require drawing skills or the employment of professional painters (fig. 12.4). Some of Keisuke's contemporaries, many of whom were also caught between the two contrasting paradigms of natural knowledge, chose to stop relying on texts and received knowledge entirely and limited their work to observation and

FIGURE 12.4. A leaf of *haribuki* (*Oplopanax japonicus*) Itō Keisuke
had collected in the sixth month of 1826 during his herbal tour
at Nikkō with Udagawa Yōan. Tokyo: National Diet Library.

description of live specimens.[54] Keisuke succeeded because he consistently tried
to work within both paradigms. He was simultaneously a physician of samurai
origin and a scholar active in amateur and semiprofessional circles. Associated
with the major scholars of the period, he was able to introduce Linnaean taxonomy in an intelligible language. Keisuke acted as a "bilingual translator" who
prevented the clash of two epistemologically incommensurable paradigms.

In Renaissance Europe, the so-called Scientific Revolution—which, as Shapin has argued, was, strictly speaking, neither "scientific" nor "revolutionary"—was the product of two incommensurable worldviews (to oversimplify, Aristotelian physics and the quantitative methods of mathematical physics), which posed different problems in different "languages" and were supported by different groups of scholars with different socioprofessional identities.[55] But in late Tokugawa Japan, the new paradigm of Western science was introduced through neither a clash nor a revolution but through a work of purposeful translation and adaptation. In other words, Keisuke's efforts to launch Linnaeus's botany in as familiar a language as possible, and in as favorable and conciliatory a way as possible, was in part motivated by the fact that he did not need to create a different or subversive socioprofessional identity. And he could do that because—as I have been arguing thus far—*honzōgaku* had already transformed into a field in many ways analogous to Western biology, in terms of a similar professionalization of specialized scholars, the development of similar conceptions and representations of species via true-to-nature specimens embodying species-specific characteristics, and a similar tendency to conceive of nature knowledge in instrumentalist terms.

Linnaeus, as a representative of the Swedish Royal Academy of Science, envisioned his role as a scholar whose research was directly and practically implicated in improving the welfare of his native Sweden.[56] Keisuke inherited a similar understanding of scholarly activity, conceived as a form of service for the benefit of the shogunal or domainal administrations. The Meiji Restoration of 1868 and the establishment of the modern nation-state continued and actually reinforced the recruitment of scholars to the assistance of the nation-state. Intellectuals like Fukuzawa Yukichi encoded their thought in the language of civilization and enlightenment (*bunmei kaika*). And although former *honzōgaku* scholars began calling themselves scientists (*kagakusha*), biologists (*seibutsugakusha*), botanists (*shokubutsugakusha*), zoologists (*dōbutsugakusha*), and the like, they nonetheless all found themselves in the familiar role of scholar-officials in service to the state when called upon for their expertise in creating a modern wealthy nation with a strong army (*fukoku kyōhei*). After the Meiji Restoration of 1868, the name *honzōgaku* almost disappeared. It no longer indicated a recognized field of academic study, and its use was limited to a restricted number of amateurs and antiquarianists. Yet the immense amount of information it had accumulated throughout the Tokugawa period was transmitted intact to the new generation of scientists, and many of its practices and worldviews were preserved in what was now called *hakubutsugaku* (natural history). As a matter of fact, the study of nature now consisted of specialized disciplines like biology, geology, and organic chemistry but also of field observation, surveys, and classification that were entirely compatible with *honzōgaku* practices. Thus *honzōgaku* did not die. Only its name did, but the name, too, was already an anachronism, since many of the conceptions and practices of the materia medica of the seventeenth century had long since been transformed into "natural history."

13   *Nature as Accumulation Strategy*
Satō Nobuhiro and the Synthesis of
Honzōgaku and *Keizaigaku*

· · · · · · · · · · ·

It is a well-established cliché to describe modernity in terms of human dominion over nature. As Horkheimer and Adorno put it in their *Dialectic of Enlightenment*, "A philosophical interpretation of world history would have to show how the rational domination of nature comes increasingly to win the day, in spite of all deviations and resistance, and integrates all human characteristics."[1] The notion of human domination over nature can be found as far back as Francis Bacon. In *Novum organon* (1620), Bacon writes, "Let the human race recover that right over nature which belongs to it by divine bequest."[2] His *Instauratio magna* straightforwardly conceived of learning in general and of nature knowledge in particular as practical means to guide human action. Knowing nature through the experimental sciences that were set about precisely in those decades was not an aim in itself, a self-sufficient enterprise, but was engendered as a practical labor endowed with social utility for human redemption. Descartes's separation of extended matter (*res extensa*)—conceived as a dynamic mechanism to be explained through geometry—and mind (*res cogitans*) only reinforced that attitude. More than a century later, Carl Linnaeus explicitly linked his own scientific research on plants to economic growth: "Our own economy," he stated in a 1741 lecture at Uppsala University, "is nothing else but knowledge about nature adapted for man's needs."[3] As Peter Dear has rightly argued, the separation of pure theoretical sciences and applied practical ones is an academic distinction that originated in the second half of the nineteenth century.[4] This separation of pure and applied knowledge remains contentious even today.[5]

In the history of modern Europe, conceptions of nature and of nature knowledge seem to be inseparable from the development of industrial capitalism. As Neil Smith has argued, "For apologists and detractors alike, the global transformation of nature wrought by industrial capitalism dominates both the physical and intellectual consumption of nature."[6] The subsumption of nature under

A modified version of this chapter appeared in Federico Marcon, "Satō Nobuhiro and the Political Economy of Natural History in Nineteenth-Century Japan," *Japanese Studies* 34, no. 3 (2014), 265–287.

capitalism imbricated the already complex notion of "nature" in a dynamic and contradictory structural dualism of meanings that still survives today: nature as *external*, myriads of "objects and processes existing outside society ... waiting to be internalized in the process of social production"—what Adorno called reified nature[7]—and nature as *universal*, referring to a set of ahistorically conceived characteristics of human nature that situates the human species "as one among many in the totality of nature."[8]

This last chapter argues that analogous conceptions of nature as a resource for economic development arose in late Tokugawa Japan as a result of the recruitment of *honzōgaku* scholars in agricultural reform policies at both domainal and shogunal levels. The case of the Satsuma Domain is particularly illuminating, in terms of both practices and intellectual conceptions: on the one hand, in the aftermath of the economic crisis of the Tenpō era, the Shimazu leaders led a series of reforms that fostered the intensification of cash crop production and the commercialization of agriculture; on the other hand, one of the scholars involved in these reforms, Satō Nobuhiro, transformed the traditional notion of *kaibutsu*—meaning literally "the opening up of things" in the sense of "revealing the nature of things"[9]—into a concept akin to economic development that sustained practices leading to the exploitation, commodification, and accumulation of natural riches. These notions and practices survived in the Meiji period, where they were sustained, translated, and reconfigured into the language of Western science. The transformations of *bakumatsu honzōgaku* were at the core of the transformation of Japan into an industrialized nation-state.

SATSUMA'S "ECONOMIC MIRACLE"

When in 1829 the retired *daimyō* Shimazu Shigehide, Satsuma éminence grise, hired scholar Satō Nobuhiro to assist Zusho Hirosato, Satsuma's finance magistrate (*kanjō bugyō*), to carry on a plan of economic and administrative reforms, the domain was at the margins of the economic life of the Japanese archipelago. "In 1800," John H. Sagers argues, "Satsuma was at the periphery of the Japanese economy. It had several export commodities including rice, sugar, and processed fish. Yet, the domain could not solve its chronic trade deficits with the Tokugawa Shogunate-dominated market centers of Kyoto, Osaka, and Edo."[10] By 1840, only ten years after Nobuhiro's collaboration with Shigehide, Satsuma budget found itself with a surplus of 2.5 million *ryō*—an incredible success, considering that less than ten years earlier it was in deficit of five million *ryō*.[11] By the 1850s it had established one of the earliest industrial areas of Japan at Iso (the Shūseikan).[12]

Satsuma's economic recovery was surely impressive, but it was not the only domain that in the nineteenth century drastically reformed its administrative and economic structure, nor was it the only or the most successful one. True, we should pay attention to Satsuma's success because of the important role that it was soon to have in the downfall of the Tokugawa and because of the political hegemony that its elites enjoyed during the revolutionary decades of the Meiji period, when Japan underwent a process of radical modernization. But this is also a story of a surprising convergence. Nobuhiro's reform plan was sustained by a coherent reconceptualization of natural resources, agricultural production, labor relations, and state intervention in production that bore semblance, without being directly influenced, to similar notions that circulated in Europe at the same time. The philosophical foundations of this reform plan rested on a systematic human domination over nature that brings to mind similar notions developed in Europe and that unfolded in an analogous framework of exploitation of natural resources, labor reorganization, complete monetization of the economic life, and market-oriented productive activities. Nobuhiro invented a Japanese form of "political economy" that exercised, directly and indirectly, a considerable amount of influence among Meiji political oligarchs, economists, and intellectuals throughout its modernizing years, helping to redefine the human relations with the material environment and instrumentalizing knowledge of nature in the service of economic growth. Much of what he did independently mirrored European developments.

THE POLITICS OF NATURAL HISTORY: THE
SATSUMA DOMAIN AND *HONZŌGAKU*

The economic crisis of the Tenpō era had devastating consequences for the population of the Japanese archipelago between 1833 and 1837. Because of the magnitude of its impact, historians have often regarded it as an important watershed in the history of Japanese modernization, which played a pivotal role in the eventual collapse of the Tokugawa regime. It was during the crisis, it has often been claimed, that there emerged those strong ties between the economic elites among wealthy commoners and educated and idealist low-ranking samurai (*shishi*) that animated the struggle against the Tokugawa bakufu after the 1853 arrival of the US Commodore Matthew Perry's black ships that forced Japan to engage in unequal trading relations with Western powers.[13] But if we look at the strategies adopted, with success, by a domain like Satsuma—which played a leading role in the overthrow of the Tokugawa in 1868 and in the formation of the Meiji oligarchic government—then the Tenpō crisis was the event that put in motion new forms of economic policies that favored the emergence of transversal relations between *high*-ranking samurai and the economic elites.[14] These reforms, which directly engaged the samurai-led administration of Satsuma in the economic life of the domain, sprang from a recodification of Neo-Confucian political terminology, and they involved low-ranking samurai working in synergy with the domainal elites and with the economic powers of the region.

Crop failures—of rice, especially—after the unusually cold weather of 1833, followed in 1834 by Osaka brokers' speculations on the price of rice and bakufu mismanagements, produced one of the most deadly famines in Japanese history and the first modern economic crisis of Japan. Despite Susan Hanley and Kozo Yamamura's cautiousness in estimating the number of casualties, the figures easily passed one hundred thousand by 1836.[15] Riots and civil disturbances of various kinds followed, and attempts to cope with the social and economic crisis at the shogunal and domainal level were often uncertain.[16]

In domains like Satsuma, Hizen, Chōshū, and Saga, political reforms to cope with the consequences of the crisis consisted in the expansion of domainal control over agricultural production and over the nationwide commercialization of agricultural and manufactured commodities. The Shimazu leadership worked to consolidate and intensify the domainal monopoly of sugar, assuring the treasury a massive increase in profits. It also planned the expansion of both agricultural production and cottage industry, introducing "the production of silk, paper, indigo, saffron, sulfur, and medical herbs."[17] The reform plans carried over by the domainal lords Shimazu Narinobu (in office from 1787 to 1809), Narioki (in office from 1809 to 1851), and Nariakira (in office from 1851 to 1858) consisted in strengthening the *kokueki*, literally the prosperity of the "state" (meaning, actually, Satsuma), by assuring the agricultural self-sufficiency of the domain, cutting import expenses, acquiring zero-interest loans from local moneylenders to be repaid in 250 years, financing the development of new productions, and exploiting commercial opportunity by strengthening the trade of export goods both within Japan and, illegally, with Qing China via Okinawa.[18] After the Tenpō crisis, not only did Satsuma survive the economic slump as one of the wealthiest and most enterprising regions of the archipelago, but by the 1840s it "accumulated an enviable surplus of wealth."[19] Behind the success of Satsuma was the enterprising Lord Shimazu Shigehide, *honzōgaku* scholar and keen student of Dutch learning.[20]

Shigehide retired from the position of *daimyō* as early as 1787 but remained the behind-the-scenes mastermind of Satsuma politics until his death in 1833. In the long period of his leadership, he laid the foundation for the future agricultural reforms thanks to his own expertise in *honzōgaku* and his belief of its import for the economic development of the region. It was under his advice that Satsuma "invited a management consultant to give economic development advice in the late 1820s" (i.e., Satō Nobuhiro).[21] And it was under his suggestion that Satsuma leadership introduced new technology and knowledge from the West.[22] By conjugating together nature knowledge, administrative reorganization, and agricultural reforms, Shigehide's Satsuma remodeled its feudal political apparatus—prior to and more in depth than other domains—in an attempt to take control of an economy that was now almost systemically dominated by the coercive forces of the market. As a result, Satsuma became one of the first domains to put into practice the radically new ideas of "political economy" (*keizai*) that scholar Satō Nobuhiro developed precisely in those years. Also, Satsuma became the place where the eclectic field of nature studies was more drastically domesticated to the necessities of productive growth.

Satsuma's emergence as an economic power in the 1830s and 1840s, however, was not simply the long-term results of the enlightened administrative skills of Shigehide. There are, in fact, profound shifts in domainal economic policies from the late eighteenth century to the first decades of the nineteenth century. Shigehide himself, despite his wide range of interests, from Japanese poetry to Dutch studies and *honzōgaku*, initially implemented, while in charge as acting daimyo (1755–87), rather conventional administrative measures.[23] Quite in line with Confucian orthodoxy, he financed land reclamation projects, attempted to raise taxation (*reikin*) on trade licenses and monopolies, forcibly cancelled domainal debt with moneylenders, and imposed austerity measures to the population of the region with restrictive sumptuary laws; for example, in 1768 he prohibited the consumption of soup made with more than one vegetable. Diligence and frugality, the two keywords of Confucian morality, dominated the political language of Satsuma lawmakers. In practice, domainal authorities constantly attempted to squeeze more profit from agricultural surplus in a period characterized by sluggish productivity. In Satsuma, agricultural output had remained constant since the early eighteenth century and even started to fall during the Tenmei-era crisis (1782–87), despite the pressures from above to increase the production and trade of commodity crops like sugarcane (*satōkibi*). Sugarcane grew in the subtropical climate of the Ryūkyū kingdom and the Amami islands, both under Satsuma control after the military expedition of 1609. Since the mid-eighteenth century the Shimazu had pressured both directly and indirectly the population of those southern islands to develop a monocultural agriculture, to such an extent that by 1745, Ōshima peasants "had almost completely shifted from growing rice to raising sugar cane."[24] The cultivation of sugarcane was either forcefully coerced upon the local peasant populations or requested in return of monetary loans that Satsuma granted to the Ryūkyū kingdom.[25] The intensification of sugarcane production in the areas controlled by the Shimazu—especially in the Amami islands, closer to Satsuma and more tightly controlled than Okinawa by mainland samurai authority—was evidenced also by the local development of technological innovations to improve the extraction of sugar from the canes. In 1671, for example, a three-roller mill began to be used, "a unique case of an independent innovation within an area of strong Chinese influence."[26]

The inclusion of sugarcane into Japanese agriculture was still quite recent, dating back to the seventeenth century. Miyazaki Yasusada and his friend Kaibara Ekiken had both reported about the beneficial properties of sugar. Yasusada even advocated its large-scale production to strengthen the diet of Japanese peasants in his *Nōgyō zensho*.[27] But it was only in the first half of the eighteenth century that *honzōgaku* scholars began researching and experimenting with sugarcanes in medicinal gardens under Tokugawa Yoshimune's direct orders. Lord Shigehide's interest in cultivating the sweet grass derived from his knowledge of *honzōgaku* texts. From the second half of the eighteenth century, Satsuma officials began to buy sugarcanes in exchange for rice produced in the domain mainland. Each village had an allotted quota of sugarcane that was purchased at a price of 0.63 liters (0.35 *shō*) of rice for every *kin* (600 grams) of sugarcane. Anything over the quota was purchased for 0.72 liters of rice per *kin* of sugarcane, thus encouraging peasants to engage in monoculture production.[28]

The commercialization of agricultural production carried over in Satsuma in the last decades of the eighteenth century was a necessary condition for its economic success in the 1830s and 1840s, but it alone is not sufficient to explain it. Throughout the first two decades of the nineteenth century, in fact, Satsuma economy remained lethargic and its domainal leaders became increasingly indebted to local moneylenders. The two events that contributed the most to make of Satsuma one of the wealthiest domains of the country were the Tenpō crisis of the 1830s and the direct involvement of samurai authority in the management of production and commercialization of its agricultural riches.

The outbreak of the Tenpō crisis—which, as we have seen, not only had beaten hard the population of the Tōhoku regions but also disrupted the life of most provinces with 465 rural disputes, 445 peasant uprising, and 101 urban riots from 1834 to 1836[29]—was in a sense a stroke of luck for Satsuma, which survived the turmoil in much better condition, in part because its geographic location spared it from the worst of the bad weather in 1833 and in part because in the previous decades the domain authorities under Shimazu Shigehide had pursued an effective politics of agricultural self-sufficiency. Hence, by the end of the 1830s, Satsuma found itself in much better economic conditions than most domains. In addition, it had the chance to maximize profits from the sugar trade because, for the first time, it could evade the control of bakufu-licensed Osaka merchants, whose control over interdomainal trade had been weakened after the austerity and moralizing reforms that followed the 1837 revolt lead by Ōshio Heihachirō.

In a sense, then, Satsuma found itself at the right time and with the kind of state apparatus that allowed it to take advantage of the situation: it was a result of Shigehide's reform plan that the various stages of the economic circuit were tightly put under control of Satsuma samurai managers, who began to supervise every step of the process, from production to exchange and distribution to domain-chartered wholesale traders. The direct involvement of samurai in production management, discouraged by Tokugawa authorities since the *Buke shohatto*, had become not uncommon in many domains since the mid-eighteenth century.[30] The commercialization of agriculture, combined with the financial recession of the post-Genroku years and crop shortages in the Kyōhō era (1732), had convinced samurai authorities of the need for each domain to achieve economic self-sufficiency. Luke Roberts has described the rise of mercantilist practices among the samurai elites of Tosa.[31] Similar strategies were adopted in Chōshū, Hizen, Kii, and other domains and were similarly conceptualized as *kokueki* ("national prosperity").[32]

In this economic context, the combination of natural knowledge and mercantilist policies that Shigehide inaugurated had two origins. One was Shigehide himself, with his lifelong passion for *honzōgaku*, which he pursued with the help of his teacher, friend, confidante, and chancellor Sōhan. The other source was Satō Nobuhiro, a pioneer of political economy—soon to be known as *keizaigaku*—whom Shigehide hired as consultant for Zusho Hirosato, Satsuma's finance magistrate (*kanjō bugyō*) throughout the first half of the nineteenth century.[33] By 1848, the year of Zusho's death, Satsuma sent to Osaka merchants chartered by the bakufu more than half the sugar circulating annually in Japan and, as a result, "a surplus of about 62,400 *kan* had accumulated in the Satsuma treasury."[34]

Around 1830, precisely when he was working as consultant on economic affairs in Satsuma, Nobuhiro developed his own definition of *keizai* in a manuscript titled *Keizai yōryaku* (Essentials of Political Economy): "*Keizai* means managing the nation, developing its products, enriching the country and rescuing all its people from suffering. Thus the person who rules the country must be able to carry out his important task without relaxing his vigilance even for a single day. If this administration of *keizai* is neglected, the country will inevitable become weakened, and both rulers and people will lack the necessities of life."[35]

According to Nobuhiro, samurai leaders had the duty to orchestrate the productive life of their subjects. At the beginning of another manuscript, *Suitō hiroku* (Confidential Memorandum on Government), Nobuhiro had urged the domainal elites and the bakufu to take direct control over the economy through the creation of six governmental offices (*fu*), or "ministries" as Morris-Suzuki calls them: an office of basic affairs (i.e., agriculture, *Honjifu*), an office of development (*Kaibutsufu*), an office of manufacturing (*Seizōfu*), an office of circulation (*Fukutsūfu*), an office of the army (*Rikugunfu*), and an office of the navy (*Suigunfu*).[36] "Each ministry," Morris-Suzuki explains, "has substantial economic functions (the Ministry of the Navy, for example, is responsible for the shipbuilding and fishing industries), but it is the first four which are of particular interest from an economic point of view."[37] The office, or ministry, of basic affairs focused on the organization of agricultural production, including technological improvements, the coordination of labor, and the clearing up of new land. It was assisted by the office of manufacturing, whose main task was to develop tools, implements, vehicles, and the like to help maximize agricultural yield, whereas the office of development, supervising farmlands, forests, and mines, had the fundamental assignment of organizing surveys, experimenting with fertilizers and new alternative crops, exploring the land for new mineral resources, and improving extractive techniques. Lastly, the office of Circulation had the responsibility to "standardize the price of goods produced by each region, conduct trade with foreign countries, and apply the profits of this trade within the nation so that posterity may flourish."[38]

Nobuhiro's system of six ministries was never institutionally implemented in Satsuma or elsewhere and remained a purely theoretical construct. But the core of his plan was de facto put into effect when the Satsuma leaders directly engaged their samurai retainers in the various productive sectors. It involved the training and education of both samurai administrators and commoner laborers on new productive technologies aimed at improving agricultural cultivation, bettering extractive techniques in mines, and developing more efficient manufacturing tools. The first books of *Seikei zusetsu* (Illustrated Explanations of the Forms of Things), an encyclopedic work on agronomy edited by Sōhan and Shigehide, treated every aspect of agricultural life, from the organization of village life and the coordination of field labor to botanical information about grains and crop products as well as plants that could be utilized as fertilizers or insect repellants.

In *Satsuma keii ki* (On How to Administer Satsuma), Nobuhiro had laid out a strategic plan tailored for Satsuma. After emphasizing, in the introductory note, that "the study of agriculture and the development of products are fundamental to enriching the country and a vital duty for the rescue of the people,"[39] Nobuhiro proposed a detailed plan that included the organization of a new survey of Satsuma's natural riches and the development of research facilities (*yakuen*) to study the conditions for large-scale production of strategic agricultural and medicinal commodities. He assigned the domainal administration a central role in orchestrating the entire process, from the research stage to the final marketing of the products. A research team, composed of both samurai and commoners, focused on the development of techniques to better the quality of the sugar extracted from sugarcanes. Samurai inspectors were dispatched to various sugarcane-producing villages to oversee the processing of sugar, as the cases of *daikan* Higo Hachinoshin at Mishima and *daikan* Higo Yaemon at Ōshima well illustrate.[40] In essence, Nobuhiro envisioned a productive system that conjugated productive self-sufficiency with state planning inside and market relation outside the Satsuma Domain.

Managerial centralization under the tight supervision of Zusho Hirosato proved effective to consolidate Satsuma finances. Hirosato, who had served Shigehide since the late 1780s, was the executioner of the reform plan that Nobuhiro developed.[41] Certainly, these administrative and technological reforms played an important role in the economic restoration of Satsuma, but they were only one among a series of strategies that Hirosato devised. New forms of long-term loans with local moneylenders and, especially, the intensification of the illicit trade with Qing and French merchants via the Ryūkyū Islands—which would be fateful for Hirosato, when, in 1848, Shimazu Nariakira denounced him to the bakufu senior counselor Abe Masahiro—were also indispensable ingredients of Satsuma's success.[42]

The economic history of Satsuma under Shimazu Shigehide sheds new light on our understanding of Japan's modernization. It was there and as a result of precise forces that close collaborations of samurai elites with *honzōgaku* experts, leading merchants, and agricultural entrepreneurs became systematic. Botanists were employed to oversee agricultural production and the cultivation of medicinal herbs to improve output, the selection of pest resistant species, and the development of new processing techniques. Merchants received from the state monopoly rights to commercialize specific products. Shigehide actively favored the development of close collaborations between samurai retainers and the elite members of the artisan, agricultural, and merchant classes, inviting outside traders to relocate in Kagoshima and organizing cultural events to promote interclass exchanges.[43] Nineteenth-century Satsuma developed into a centralizing mercantilist state that orchestrated the activities of high-ranking elites, mid- and low-ranking samurai administrators and technocrats, and commoners. These collaborations created an economic dynamic that combined agricultural self-sufficiency in Satsuma mainland, protocolonial monocultural production in the southern islands it controlled, and interdomainal and international trade. It thus anticipated in many respects the active role that the Meiji state would have in the late nineteenth century in financing and directing Japan's industrialization and in creating semicolonial relations with the peripheral regions of Hokkaidō and the Ryūkyū Islands as well as full-fledged colonial exploitation in its East Asian colonies later on.

SHIMAZU SHIGEHIDE, SATŌ NOBUHIRO, AND THE INSTRUMENTALITY OF KNOWLEDGE

In a manuscript Nobuhiro completed in 1823, *Kondō hisaku* (Secret Strategy against Disorder), there is a passage celebrating the natural bounty of Japan:[44] "The climate [of Japan] is mild, the soil is fertile and many natural products grow in great abundance. Our country faces the ocean on four sides and is unsurpassed among the nations for its ease of sea transport to neighboring countries. Its people are so superior to those of other nations in their courage, and its natural conditions are outstanding amongst all nations, and thus the country is perfectly equipped to control and advance [*bentatsu*] nature and the world [*tennen uchū*]."[45] The passage revealed in condensed form Nobuhiro's proposal to synthesize economic policies (*keizai*) and the study of nature (*honzōgaku*), which he deemed essential to the development of the socioeconomic potential of the country: "The economic administration of a country must begin with a survey of its lands."[46] Investigating nature and accumulating information about its riches was the preliminary step toward its domination—"to control and advance nature and the world."

Shogun Tokugawa Yoshimune, a century earlier, had laid the foundations for the development of such a view by sponsoring the first surveys of natural riches.[47] Since then, the value of *honzōgaku* for the state (i.e., domain or bakufu) had been a constant element of the field. In that respect, there was nothing uncommon in Shimazu Shigehide's love of *honzōgaku*. He authored manuals and monographs that occupied an important place in the history of the field and sponsored herbal surveys of the Satsuma region and the Ryūkyū Islands. What distinguished Shigehide from the other *daimyō* was his conception that nature studies were inseparable from economic concerns. Shigehide would have indeed agreed with Carl Linnaeus, for whom "Natural History is the base for all Economics, Commerce, Manufacture . . . because to want to progress far in Economics without mature or sufficient insight in Natural History is to want to act a dancing master with only one leg."[48] With his decision to employ Satō Nobuhiro, he institutionally recruited *honzōgaku* as a fundamental element of the state's economic policies, both in practice and in theory.

Shigehide, born in 1745, became *daimyō* at the age of ten after his father Shigetoshi's death in 1755. He enjoyed a long life (he died at the age of eighty-eight) and dominated from behind the scenes Satsuma's political and cultural life. He retired early in 1787, and in 1804 he shaved his head and took Buddhist vows but indirectly continued to influence Satsuma politics until his death, right before the outbreak of the Tenpō crisis. Fluent in Chinese, Shigehide was connected to various scholars of the period. But it was *honzōgaku* that triggered his intellectual curiosity. In 1768, he organized a catalog of plants and animals of the Ryūkyū Islands, the largest survey in Japan after Niwa Shōhaku's country-wide inventory thirty years before. He sponsored the publication of the results in *Ryūkyū sanbutsu shi* (Flora and Fauna of Ryūkyū) in 1772. Under his reign, Satsuma had one of the most active Igakuin (Institute of Medicine) among the various domains and a celebrated medicinal garden, the Yoshino *yakuen*, which Shigehide built in the northern side of the Kagoshima castle.[49] In the garden, he charged the professional *honzōgaku* scholar Satō Nakaoka with the task to grow and experiment with plants he collected from another survey of the Satsuma region he organized in 1781. In *Seikei zusetsu* (Illustrated Explanations of the Forms of Things), an encyclopedia in thirty volumes, Shigehide recorded the results of his lifelong study of plants and animals in a distinctively utilitarian fashion. The scholar Sōhan edited the text. Shigehide was also renown for *Chōmei benkan* (Handbook on Bird Names), a useful guidebook on bird names, published in 1830, and *Nanzan zokugo kō* (Shigehide's Reflections in Simple Chinese),[50] a collection of anecdotes and information about fish, shellfish, birds, insects, herbs, animals, and food recipes in easy Chinese for the education of his retainers, published in 1812. After his retirement, Shigehide spent more time in Edo than in Kagoshima. In the shogunal capital, he could enjoy the company of scholars, naturalists, and artists in a salon he hosted in his retirement villa in Ebara, on the outskirts of Edo. He directly engaged in the cultivation of herbs and plants in a 1,600-square-meter garden attached to the villa. It was in the occasion of a trip to Edo that in the third month of 1826, Shigehide met Philipp Franz von Siebold.

On the ninth day of the first month of the ninth year of the Bunsei era (February 15, 1826, in the Western calendar), Siebold had left Nagasaki for the periodic visit of the Dutch mission to Edo. Siebold reported that the fourth day of the third month, after his arrival to Edo, he met Shigehide, together with his son (Narioki), and nephew (Nariakira), all of whom seemed interested in *rangaku*.[51] They discussed plants and animals, astronomy, and the Dutch names of insects and birds.[52] A few days later, Siebold and Shigehide met one more time in the house of the lord of the Nakatsu Domain Okudaira Akitaka, who was also an amateur *honzōgaku* scholar. The two met a third time a month later, on the twelfth day of the fourth month, in Shinagawa, during the Dutch envoys' return trip to Deshima.

In a collection of miscellaneous notes, *Gyōbō setsuroku* (Observational Notes), Shigehide confessed his lifelong passion for natural history: "Ever since I was young, I collected curiosities from the various provinces of our country and hunted for rare things from abroad. I cultivated herbs and trees and kept birds and animals. I have always desired to investigate the true nature (*shin*) of those things. This is why I decided to record my experience. I carry with me these notes, together with my books and poetry, as a pleasure for my tranquility."[53] Part of these "notes" he mentioned were posthumously collected in *Gyōbō setsuroku*; others were attached to many of the objects in Shigehide's private collection (Shuchin Sōko), which included art pieces and potteries, embalmed animals, rare insects, and dried plants; still others became part of the illustrated encyclopedia *Seikei zusetsu*, edited by Sōhan and printed in 1807.

Ueno Masuzō argued that Shigehide's commendation of *honzōgaku* in *Gyōbō setsuroku* is quite distinct from the customary celebration of the morally uplifting value of natural history as a pastime that we can read in the writings of other naturalist domainal lords—a sentiment that Shigehide too endorsed by defining it "a pleasure for my tranquility" (*seichū no ichiraku nari*).[54] Shigehide did not simply pay tribute to *honzōgaku* for nurturing the investigation of things (*kakubutsu chichi*) or broadening one's knowledge of the universe (*hakubutsu*), as conventional rhetoric dictated in many privately published albums and monographs. He unambiguously stressed that what led him to *honzōgaku* was a desire to know the true nature or essence of things (*ōku sono shin wo shiru koto wo hosshi*). By using a term like *shin*, he explicitly referred to a precise cognitive paradigm that included both encyclopedic exactness and true-to-nature representational practices (*shashin*). His familiarity with Dutch studies furthermore suggests that Shigehide shared with scholars like Hiraga Gennai, Ono Ranzan, and Itō Keisuke the belief that Western learning could contribute, along with traditional *honzōgaku*, to shed light on the innermost secrets of plants and animals. This attitude and intent pervaded the major work he sponsored, *Seikei zusetsu*, which devoted large sections to the rationalization of labor organization, the innovation of agricultural techniques, and the deepening of accurate knowledge of the essential properties (*shin*) of agricultural products.[55] But there is more. In a previous paragraph of *Gyōbō setsuroku*, Shigehide had explicitly stated that the "study of agriculture and its products as well as knowledge of medicinal herbs is indispensable for the welfare of the people [*tami no rieki*]."[56] Nature knowledge, for Shigehide, was not simply a private enterprise, however praiseworthy and uplifting, but a discipline of public interest deeply connected to government and the welfare of the subjects.

Seikei zusetsu—his most noteworthy achievement in the field of natural history—embraced and expressed the three meanings Shigehide attributed to the field of *honzōgaku*: it is a morally uplifting pastime that offers authoritative knowledge of natural objects, which, in turn, is a fundamental element of state-planned agricultural production. Previously collected under the title of *Seikei jitsuroku* (True Records of the Forms of Things, 1773) were more than thirty years of notes on rocks, plants, and animals that Shigehide jotted down either from encyclopedias or from observations in the field. In 1793, he ordered Sōhan, the nativist scholar Shirao Kunihashira, and the Confucian scholar Mukai Tomoaki, a graduate of the Shōheikō and administrator at the time of the Yoshino *yakuen*, to reedit the collection of field notes for publication in 1804 by the presses of the Kagoshima castle.[57] The thirty volumes that were printed out of the projected one hundred contained an encyclopedic explanation of the various crops, vegetables, herbs, and trees; their properties and characteristics; their utility for human beings; and the best system to cultivate them. As Sōhan put it in the introductory outline in the first book of *Seikei zusetsu*,

Our Lord has taught to his people [*tami*] first agriculture and sericulture and then established a medicinal garden to produce many useful medicinal substances. He investigated differences and similarities of [species] from various regions and reflected about their lifecycles through the seasons. It did all these things in order to get the most from each generation. He did not do it to satisfy his curiosity, but to make sure that people quickly followed nature's will [*ten'i*]. Of the herbs and trees that he grew in the medicinal garden, and the birds and animals that he kept as well as those that were brought from abroad, he produced true-to-nature illustrations [lit., he reproduced their essences, *sono shin wo utsushi*] and stored them away for their future utilization.[58]

The constellation of notions, information, pictures, techniques, and practices in *Seikei zusetsu*—with its "systematic arrangement [*ruiju*] of all species, all of which accurately depicted [*subete shin wo utsushite*]"[59]—all gravitated around the combination of a search for truthful knowledge (*shin*) of plants and animals and a concern for the applicability of this knowledge for the improvement of the economy of the domain and the life of its inhabitants. The accuracy of knowledge and its usefulness are never separated, but one required the other: only precise knowledge can serve the purpose to "save the people from salvation and despair";[60] conversely, only useful knowledge realizes the *shin* of things.

After the introductory outline, which continues by explaining textual sources and complaining about the ambiguities created by regional names, the first book (*maki*) of *Seikei zusetsu* is devoted to agricultural matters (*nōji*) and begins with an argument of strong nativist overtones about the creation of the world by the *kami*. It emphasizes the blessedness of "our country" (*mikuni*) for the people of Japan, who were instructed, by divine decreed of the sun goddess (Amaterasu Hinomikami), in how to correctly grow the herbs that nourished them generation after generation.[61] And just like *kami* provided the Japanese people with a correct way to peacefully rule the country, they also assisted them in naming correctly all things in the world and in cultivating the land in such a way (*tadashiku*) as to actualize its productive potential.[62] The passage, however mythical in tone, insisted that there was a correct way to agriculture (and sericulture) and that only by following it would the land actualize its productive potential: the deities had taught people the way of agriculture (*tatsukuri no michi*)—giving them useful animal like cows and horses to assist them—in order to prevent them from starving.

There is a tension, almost a contradiction, in this argument. On the one hand, the deities had blessed the Japanese land more than any other country with immense natural riches. On the other, nature, left to itself, is parsimonious in dispensing those riches, which requires human intervention. In other words, in *Seikei zusetsu* nature is a generous dispenser of bounties only on the condition that, as Nobuhiro had also put it, it is "controlled and advanced" (*bentatsu*) to actualize its potential, and that was what knowledge of the *shin* of natural species really consisted of.

At the beginning of *Kondō hisaku*, Nobuhiro had similarly stated that the natural riches of the imperial country of Japan (Sumera Ōmikuni) derived from the fact that it had been created by the gods (Musubi-no-kami).[63] But their creative act was not random: rather, it followed precise laws. Understanding those laws (*hōkyō*) was indispensable, because for Nobuhiro the abundance or scarcity of the myriads species of natural products (*manshu no sanbutsu*) depended exclusively on human capacity to develop techniques and institutions capable of utilizing and enhancing them (*bentatsu*).[64] The term that Nobuhiro chose to render the concept of extracting from nature its riches, which Morris-Suzuki translated with "control and advance," is *bentatsu*, which literally means "to whip with a cane," a violent metaphor that resonated with Bacon's notion that the sovereignty of man over nature entitled him to use violence, to rape nature of its riches. "Nature," for Bacon as for Nobuhiro, "takes orders from man and works under his authority."[65]

At the core of Shigehide's and Nobuhiro's thought rested the belief that the rationalization of production was grounded on mythological foundations: Nobuhiro's plan of developing state offices directly overseeing, organizing, and orchestrating in a systematic way the study, observation, experimentation, cultivation, technological development, and commercialization of natural resources allowed humans to enjoy and accumulate the inexhaustible reserve of resources that nature contained. Blending together myth and instrumental reason, divine powers and pragmatic rationalization of labor and production, Shigehide's Satsuma paved the way for the combination of natural knowledge and economic production that played a fundamental role in the subsequent Japanese modernization toward a "completely administered society," in Adornian terms, which combined economic growth and the exploitation of the material environment.

The two great famines of the Tenmei and Tenpō eras had certainly forced the Japanese population to face widespread shortages in an unprecedented scale, at least during the Tokugawa period. The 1830s had furthermore shown that the commodification and commercialization of agricultural products had the capacity to intensify rather than reduce the duration and extent of the crisis itself. It is therefore not surprising that in various domains as well as in Mizuno Tadakuni's shogunal reform of the 1830s, attempts had been made to control, more or less successfully, the system of production and distribution of agricultural goods.[66] In this context, Satō Nobuhiro's transformation of the traditional concept of *kaibutsu* functioned precisely as a theoretical justification of the necessity of human intervention to rationalize both knowledge and production.

Satō Nobuhiro was the first Japanese scholar who theorized and partly put into practice a system of human dominion over nature that consisted of the state control of production and commercialization of agricultural products, an instrumentalist conception of knowledge, a notion of the material environment as a potential treasure holder of resources exploitable for human needs, and the recruitment of scholars in economic activities. Nobuhiro explained the aims of his project at the very beginning of the "General Argument" (Sōron) of *Keizai yōryaku*: "I define *keizai* as the management [*keiei*] of the national land [*kokudo*], the development of its products [*bussan wo kaihatsu shi*], the enrichment of its provinces, the salvation of its people."[67] It is a responsibility of a ruler to organize and control production to create wealth and rescue his subjects from starvation. By "ruler," here, Nobuhiro meant *shutaru mono*, conceiving of power not in personalistic terms, such as the lord, the *shōgun*, or the emperor, but as an abstract role, a social function in an impersonal system of relations. The urgency of the ruler's duty was particularly felt in a period punctuated by crises: Nobuhiro, in fact, made clear that "if the economic policies are insufficient, the country will inevitably fall in decay and wither away and all people high and low will be deprived of the means of subsistence."[68]

He duly recognized that "all Japanese and Chinese classical texts had already discussed the principles of state administration," but he also emphatically stated that "these texts had treated these matters in a disorganized and vague manner without paying attention to the details."[69] And that was the reason, scholars (*jūshi*) in Japan "were unable to enrich the country and rescue its people."[70] What was needed now, he continued in the chapter, was a new organization of production (for him, the principal concern of *keizai*) that combined the knowledge of the natural world and the laws that the Musubi-no-kami embedded in it with a technology capable of activating the generative capacity inherent in things and forcing (*bentatsu*) nature to develop its full productive potential. Nobuhiro called the exploiting intervention of human beings *kaibutsu*, to which he dedicated the second chapter of *Keizai yōryaku*. *Kaibutsu* (*kaiwu*) is an ancient Chinese term connected to the *Yijing*. It literally means the "opening up of things," and, as Saigusa Hiroto and Tessa Morris-Suzuki have pointed out, it could mean either "revealing the essence of things" or "developing the potential of things" (in the sense of "making use of").[71] In Japan, the concept of *kaibutsu* was eclectically utilized throughout the Tokugawa period by various scholars either to emphasize the role of research to reveal the inner workings of principle (*ri*) in the context of the "investigation of things," as in the writings of Kaibara Ekiken,[72] or, as in the hermetic writings of eighteenth-century Confucian scholar Minakawa Kien, to reconstruct the original phonetic value of Chinese words for the sake of reactivating the metaphysical connections of words and things (given by their sharing the same universal force, *tenchi no ki*).[73] In Ming and Qing China, *kaiwu* was marginalized in the early modern period by other concepts practiced by the dominating philosophical systems of Zhu Xi and Wang Yangming. An exception was the *Tiangong kaiwu* (Development of the Works of Nature), an encyclopedic work compiled by scholar and provincial bureaucrat Song Yingxing. Despite Joseph Needham's characterization of Song Yingxing as the "Diderot of China,"[74] the *Tiangong kaiwu*, after its publication in 1637, was not widely known during the Qing period.[75] It had, however, far greater success in Japan, where, after its introduction in the late seventeenth century, it was reprinted twice in 1771 and 1830 in richly illustrated editions, and it influence scholars like Hiraga Gennai, Shiba Kōkan, and Satō Nobuhiro.[76] The *Tiangong kaiwu* focused on technological developments, covering the instruments and techniques developed in agriculture (from irrigation to hydraulic and milling engineering), sericulture, salt making, sugar making, ceramics, the metallurgy of bronze and iron, transportation (ships, carts), coal, vitriol, sulfur and arsenic mining, military technology, ink making, fermentation, pearls and jades, and so on. The overarching argument of the work emphasized the role of human creative and instrumental intervention in increasing nature's productivity.

The influence of the *Tiangong kaiwu* on Satō Nobuhiro's ideas is hard to ignore. The language of its long chapter on *kaibutsu* in *Keizai yōryaku* echoed many of the methods and terms used by Song Yingxing. Satō begins, as he usually did in all his writings, with a definition:

Kaibutsu consists in enriching the country [*kyōnai*][77] by developing its various natural products of the sea and the land, starting from grains and fruits. It means opening up mountains and valleys, rivers and seas for cultivation, turning over the ground in plains and valleys, nurturing grains and fruits to maturity, favoring the accumulation of wealth, and sustaining the expansion of the population of the country: it is the fundamental activity of the people of the land which follows the order of heaven. Although this principle has already been explained in texts on natural knowledge, I would like here to write a short treatise to elucidate the divine will of nature.[78]

Nobuhiro conceptualized *kaibutsu* as the development (*kaihatsu*) of the productive capacity of the land (together with rivers and seas) through strategically directed human intervention. *Kaibutsu* hence implied a reconfiguration of the natural environment into a treasure holder that awaits human exploitation. Like the French physiocrats of the eighteenth century and following the conventional Confucian view, Nobuhiro believed that the wealth of a nation derived primarily from agriculture, and it was therefore in the expansion, diversification, and maximization of agricultural production that the ruler should intervene (in terms of his notion of *keizai*). But unlike physiocrats, Nobuhiro did not believe in individual self-interest—and the consequent laissez-faire attitude of the government—as the leading motor of economic growth. Rather, as I have shown before, he maintained the necessity of a central coordinating authority administrating the various tasks of the economic life of the nation in order to ensure the welfare of the entire population. The agrarianist conceptions of the physiocrats and Nobuhiro converged in their shared belief that ultimately the source of wealth rested in the land, but the Japanese scholar remained true to his Confucian roots in emphasizing the communitarian well-being of a properly ordered society.[79]

Furthermore, he infused his writings with religious and mythological overtones. It was an axiomatic conviction of Nobuhiro that "in the beginning, the August Deity Takami Musubi had created the universe (*tenchi*), nurtured all things, and made the world a wealthy place."[80] It is therefore the duty of the ruler to follow the way of the deities and make proficient use of the wealth of nature to sustain his people. Although the religious implications of Nobuhiro's text echoed similar Christian conceptions of the earth as a garden of Eden providentially given to human beings to satisfy their needs, his belief in the inexhaustibility of nature as a resource of wealth—deriving from the constant intervention of the creative force of the two Musubi no kami—is an element that further distinguished Nobuhiro from the physiocrats.

The main task of the ruler that follows the proper way of *keizai*—that Nobuhiro understood as exercising control over productive activities to ensure a wealthy and ordered society—was to actualize a rationalization of labor in order to achieve a more effective exploitation of natural resources. For Nobuhiro, crises and famines were to be attributed not to shortages due to limits in nature's fertility but rather to the mismanagement of both production and distribution of agricultural goods as well as the failure of realizing the full potential of the soil.[81] He did not believe that there were limits in nature's productive potentiality or in the sustainability of intensive farming, because "Takami Musubi-no-kami has a great love of human beings."[82]

Like Ricardo and Malthus, he conceived that the potential fertility of the soil was naturally given, but unlike them he believed in its limitlessness, a notion he derived from his faith in the unceasing creative power of the Shinto deities—although he was well aware of the enriching activities of fertilizers and encouraged their use. Contrary to Marx—for whom nature productivity has precise "biological" limits that intensive cultivation risks exhausting[83]—Nobuhiro saw the intensification of human exploitation of nature and the expansion of the accumulation of its resources as the leading strategy to make nature produce more. Inherent in Nobuhiro's belief in the divine power of nature to create, its limitless capacity to generate, was the notion that the more you squeeze nature, the more you can get from it—a conception that has characterized the ideology of Japanese modernization since the Meiji period.[84]

Nobuhiro shifted the meaning of *kaibutsu* from understanding the essence of things to the right to exploit them for human needs. As he put it in *Keizai yōryaku*, "The exploitation (*kaibutsu*) of natural products is the first duty of the sovereign."[85] But that task was impossible to realize without a preliminary study of nature. Knowledge still played a fundamental role in Nobuhiro's system, but it was no longer confined to an abstract understanding of the essence of species in the tradition of the "investigation of things." Knowledge, to be of any value, had to concretely contribute to the enrichment of the country—*fukoku*, another important concept that Nobuhiro developed and was later adopted by the Meiji oligarchs, often combined with *kyōhei* (a strong army) to sustain the modernizing effort.[86] Knowing the life cycles, the habits, and the ecology of plants and animals was important to realize that seeds grow differently in accordance with climatic variations and the composition of the soil.[87] That is why, writes Nobuhiro, "my method of *kaibutsu* begins with an analysis of the nature of the soil, including astronomical calculations,[88] the measurements of the national land, and geographical investigations."[89] From this it followed that before starting the cultivation of a grain, one should determine whether "it is appropriate or not to the climatic conditions and the composition of the soil."[90] The combined knowledge of the territory and the individual species was the necessary condition of a successful harvest. Insofar as the individual species are concerned, Nobuhiro explains that "the number of products that the various deities had created with great effort is enormous, but they can all be divided into three groups: minerals, plants, and animals."[91] For him, in order to achieve a thorough knowledge of nature, it did not suffice to accumulate encyclopedic information of each species, as was conventional in *honzōgaku* texts, but one should rather understand how each species interacted with the others and with the environment. Ecological knowledge, for Nobuhiro, was clearly the key to the dominion of nature. In the remaining chapter on *kaibutsu*, he laid out a plan to investigate the properties of the various species of minerals, plants, and animals — from rare stones, jewels, pigments, metals, clays, salts, and sands to the various species of vegetables, herbs, grains, fruits, lichens, and so on and the various beasts of land and sea. He listed fifty-two fundamental species (*shu*) divided into three groups (*rui*), the investigation of which was a precondition for their development (*kaihatsu*).[92] This was the main task of the ruler if he wanted to rescue his people.[93]

The semantic transformations that Nobuhiro attributed to *kaibutsu* — in associations with the ideas of *keizai* and *fukoku* — thus allowed him to use this concept to justify his understanding of nature as a reservoir of riches to be exploited for the prosperity of the country and its people, his instrumental conception of knowledge as a means to maximize agricultural production, and his project of placing centralized economic planning at the core of the political life of the state.

CONCLUSIONS

Satsuma was one of the first domains to employ *honzōgaku* specialists to carry out a series of economic reforms that contributed to the transformation of the region in one of the wealthiest of late Tokugawa Japan. Originating in Shimazu Shigehide's activities as domainal lord and naturalist, Satsuma leaders adopted Satō Nobuhiro's ideas of political administration that, contrary to the received Confucian wisdom, emphasized the necessity for the ruler to keep an active control of the economy, from production and labor organization to distribution and commercialization of cash crops. This system favored the development of a new form of organically integrated society. In contrast to the received Confucian principle whereby members of different social classes had the duty to perform class-specific social functions — that is, samurai had monopoly of political power, while the commoners had the task of producing and distributing food and performing services that society needed — Nobuhiro developed a model of society that, while it did not question social hierarchy, it synergically engaged all classes in the joint task of enriching of the country. Nobuhiro, in other words, challenged the Neo-Confucian separation of the political from the economic and developed in its stead a "political economy" (*keizai*) that envisioned a centralized administration coordinating all productive activities to ensure the prosperity of its people.

Although his project of six ministries was not realized in a literal manner, the concepts and language he developed influenced the evolution of modern Japanese political economy. His vision of a central organizing power prefigured the model upon which the Meiji state developed. The subsumption of knowledge under production and the subsumption of scholars under a state apparatus that directed those productive activities were two steps that Nobuhiro saw as fundamental to ensure the wealth of the country. They survived in the Meiji period in the form of continuous collaborations of Japanese scientists with the government to sustain industrial and agricultural growth. But it was the consequential subsumption of nature under capital accumulation after 1868 that fully realized his paradigm of *kaibutsu*. These three processes became strictly interconnected after the Tenpō crisis and unleashed their full power only after the Meiji Restoration. It is therefore not surprising that the Meiji agriculturalist Edo Tekirei recognized in Satō Nobuhiro, alongside Andō Shōeki, Ninomiya Sontoku, and Tanaka Shōzō, one of the pioneers of modern Japanese economic thought.[94]

The dominion (*bentatsu*) of nature and the systematic accumulation of its riches that Nobuhiro encapsulated in the notion of *kaibutsu* prescribed "goals and purposes and means of striving for and attaining them," which differed from what was previously conceived of as the spontaneous realization of nature's inherent principle (*tenchi shizen no ri*).[95] This dominion was to be executed via a bureaucratization and rationalization of productive labor and, successively, via the commercialization of agricultural commodities. Nobuhiro's concepts can be easily understood in the light of Max Weber's idea that "the fate of our times is characterised by rationalisation and intellectualisation and, above all, by the 'disenchantment of the world.'"[96] The transformation of nature into a collection of objects to analyze, represent, manipulate, control, and produce that sprang from two and a half centuries of *honzōgaku* scholarship—of which Nobuhiro's writings should be considered a further contribution—had unquestionably favored the tendency to abandon the Neo-Confucian principle that nature consisted of an inherently meaningful order and to practice instead an epistemic stance that conceived of the myriad of objects that constituted it as intelligible and devoid of any metaphysical or sacred aura.[97] This transformation, I claim, constituted a surprising convergence with similar developments in early modern European natural history. In contrast to European mechanistic conceptions of nature, however, Nobuhiro's instrumentalist disenchantment was supported by his belief in a divine creative power sustaining an ever-expanding exploitability of natural resources. This was tantamount, I believe, to a paradoxical "re-enchantment" of nature that would later ideologically sustain the industrial expansion of Meiji Japan and the silencing of those who protested against the pollution and destruction of the environment.[98] Moreover, in the 1880s this re-enchantment contributed to create a new concept of *shizen* that on the one hand aimed to translate the Western notion of "nature" but on the other connoted it with an unquestionable veneer of Japanese uniqueness, which mythified both Japanese relations to nature and productive relations in Japanese society.[99]

Epilogue

In a short text he wrote a century ago, "Tsuka to mori no hanashi" (Discourse on Mountains and Forests, 1912), ethnographer Yanagita Kunio joined forces with biologist and folklorist Minakata Kumagusu in a harsh critique of the Meiji government, which at the time was destroying villages, forests, and old shrines to enhance Japan's agricultural and industrial infrastructures. The inhabitants of rural Japan, Yanagita claimed, had developed for centuries their communal identity around a precise symbiotic relation with the surrounding environment, symbolically expressed in myths and folktales that constituted their primeval (*genshiteki*) religious foundation. The destruction of the environment, to Yanagita, precipitated the erosion of Japanese communal spirit. The disappearance of *kappa, tengu, tanuki*, and other supernatural creatures from the woods, along with deer, bears, and boars, meant not only the destruction of the environment and the reduction of nature to exploitable resource but also, and in particular, the alienation of those communities from their own primeval ethos. "Without deep forest around a shrine," he wrote, "a feeling of worship will not be induced to us," because "as the place itself is appropriate for revering the gods, our worship of the gods originates not in some sacred body, or shrines, but in the land itself and the forest growing densely on the land."[1]

Yanagita conceived of Japanese identity as originating in a primeval metabolic relation with the environment that harmoniously contained a human domain of houses, villages, shrines, cultivated lands, and the like and a mythical, sacred space of inaccessible forests, populated and protected by gods and supernatural creatures. The two spaces were adjacent but separated by almost impermeable barriers. Stories about the unfortunate outcome of the breaking of that barrier sustained the cosmology of Japanese villagers, which the modernization of Japan was now threatening to obliterate.[2]

Yanagita was, to a certain degree, inventing a tradition that contributed to a heterogeneous multiplicity of discourses that jointly fabricated a sense of nationhood in modernizing Japan and, conversely, a reconfiguration of nature in nationalistic sense.[3] In fact, the history of the Japanese archipelago is punctuated, ever since the eighth century, by the continuous destruction of the environment that is at odds with his romantic idealization of the symbiotic relation of humans and nature.[4] The story of Yahazu—who violently cleared up for cultivation lands that were previously the abode of divine Yatsu no kami—that

299

opened this book symbolizes the dialectical and often violent relation of human communities and the natural environment throughout Japanese history. It was one of the goals of this book to demonstrate that various forms of dominion of nature have been thoroughly institutionalized and routinely performed since the seventeenth century; by the early nineteenth century the natural environment was explicitly conceptualized as a limitless reservoir of resources for the prosperity of the state (*fukoku*) and the study of nature was incorporated into political economy and state administration.

The myths and legends that Yanagita spent his life collecting had in reality served the purpose of familiarizing, and thus domesticating, the otherness of wilderness: rather than securing a harmonious separation of human and sacred (natural) spaces as Yanagita conceived, they favored the expansion of the human domain through pacifying ceremonies. But for Yanagita, who, as Harootunian has shown, was engaged in a struggle to counteract and overcome the alienating effects of Japan's rapid modernization, the invention of the harmonious coexistence of Japanese communities with nature was a necessary component of a "national life that was whole, unblemished by division and harmonious."[5] That is, the invention of a Japanese harmonious ethnical community was necessarily linked to the invention of a natural environment that was similarly conceived in terms of wholeness and harmoniousness. Yanagita's critique of the violent modernization of Japan carried out by the state and of the alienating effects it produced thus addressed both the destruction of Japanese forests and the obliteration of the essential customs of Japanese communities.

The presupposition at the core of Yanagita's project was the belief that science and technology were Western imports that the Meiji government had uncritically embraced since 1868. James Bartholomew summarized this view that saw "the paradigms in effect in the Tokugawa period ... all replaced by Western ones, and government policies restricting certain fields ... not surviv[ing] the demise of the shogunate."[6] This notion, however, presupposed the amnesia of two a half centuries and of *honzōgaku* history, which had developed, as I demonstrated, into a discipline similar to European natural history and organically imbricated the political economies of bakufu and domains.

The amnesia of *honzōgaku* was not exceptional. After the defeat of the Tokugawa and their allies in the Boshin War (1868–69) and the establishment of a new political order that restored power back to the emperor, Japanese politicians and scholars engaged in an effort of modernization that lasted for at least the first three decades of the Meiji period. Amnesia and erasure of the recent Tokugawa past were a vital prerequisite of this endeavor to "civilize and enlighten" (*bunmei kaika*) Japan and the Japanese.[7] This was—and often still is—the dominant master narrative describing the introduction of modern sciences from the West. In fact, if we follow the fate of *honzōgaku* in the transition from the Tokugawa to the Meiji period, a more complex and nuanced picture emerges. On the one hand, *honzōgaku* lost his name but transferred to the practitioners of the new scientific fields two centuries of knowledge, research, data, images, techniques, attitudes, styles, facilities, expertise, schools, social relations, books, and specimens. Scholars active in the transitional period of the 1850s to the 1880s were able to translate Western taxonomy and observational procedures into the familiar language of *honzōgaku*, so that the terminology developed in the Tokugawa period was adapted to convey Western concepts. On the other hand, *honzōgaku* became the name—associated with that of *kanpōyaku*, now referring to *traditional* (Chinese) medicine in antagonistic opposition to Western medicine—of a lost tradition, a forgotten wisdom, alternative practices, and a repository of Asian identity.

A glance at the names and biographies of some of the best-known scientists active in the first half of the Meiji period shows that, at least in the natural sciences, the Meiji Restoration of 1868 brought more continuity than rupture. Mori Risshi, physician of the Fukuyama Domain and strong advocate of the restoration of classical Chinese *honzōgaku*, became a high-ranking bureaucrat in the Finance Ministry (1869) and the Ministry of Education (1871) who was responsible for scientific education.[8] Yamamoto Keigu, son of Ono Ranzan's direct disciple Yamamoto Bōyō, continued his father's work in his private school in Osaka and spent his time organizing meetings that were regularly attended by both *honzōgaku* amateurs and Japanese biologists.[9] Ono Motoyoshi, son of Ono Ranzan's nephew Ono Mototaka, was another high-ranking bureaucrat in the Ministry of Education.[10] He was responsible, together with Tanaka Yoshio, for the administration and activities of the Office of Natural History, the Hakubutsukyoku. The office was established in the ninth month of 1871 as a substitute for the shogunal Bussankyoku (Office of Products) and acted as an interministerial agency connecting the Ministry of Education with the Ministry of Agriculture and Commerce and the Home Ministry, coordinating the collaboration of natural scientists with the government. Its first act was the establishment of the Tokyo National Museum of Natural History. Tanaka Yoshio, one of Itō Keisuke's many students, was one of its founders. Tashiro Yasusada, a pioneer in the study of East Asian tropical vegetation, received a formal education in *honzōgaku* in the Satsuma Domain before he became a member of the Office of Natural History after the Meiji Restoration of 1868.[11] When the University of Tokyo was founded in 1877, of the fifteen tenured professors in the new Department of Science, twelve were foreigners and the remaining three consisted of a professor of botany, Yatabe Ryōkichi; a professor of applied mathematics, Kikuchi Dairoku; and a professor of metallurgy, Imai Iwao, all of whom began their education in *honzōgaku* and Neo-Confucian studies in domainal schools and specialized, after 1868, in Western studies with periods of research abroad.

The composition of the early faculty of the science departments of the University of Tokyo reflected, in both style and method, the way science was being studied and taught in the West. Yet it was Itō Keisuke who inspired the first generation of Japanese natural scientists after his employment as adjunct professor shortly after the university was founded. This first scientific generation included Iijima Isao, son of a retainer of the Hamamatsu Domain, who graduated from Tokyo University under Itō Keisuke's supervision and became a pioneer zoologist.[12] The same was true for Miyoshi Manabu, a botanist born in a samurai family of the Noto Domain.[13] Another notable figure was the botanist Shirai Mitsutarō, whose graduating dissertation was highly praised by his advisor, Itō Keisuke, and who would become one of the first modern historians of *honzōgaku*.[14]

Both the social origins and the worldviews of Meiji natural scientists were more similar to the ones held by *honzōgaku* scholars before 1868 than the conventional historical narratives suggest. Tokugawa research practices were not abandoned. Indeed, the rise of modern natural sciences in Meiji Japan is better described as a transformation and adaptation of *honzōgaku* practices and theories to the language, methods, and aims of Western sciences. The classification of plants and animals changed as the scientific taxonomy derived from Linnaeus replaced the divisions established by Li Shizhen's *Bencao gangmu*. But the severity of this disruption was mitigated by the use of *honzōgaku* terminology to convey Western-derived biological concepts. For example, the names of the taxonomical divisions of the *Bencao gangmu* were carried over as the names of species (*shu*), genera (*rui*, later changed into *zoku*), families (*ka*), orders (*moku*), and classes (*kō*) of biological systematics. Chemistry and cellular biology coexisted with natural history (*hakubutsugaku*), which was still concerned with classification and fieldwork activities resembling *honzōgaku* herbalist expeditions.

This uniformity in attitudes was in part the result of the socially homogeneous background of the first Japanese scientists. Like their *honzōgaku* predecessors, they understood natural studies as a form of service for the welfare of the state. The majority of them collaborated with the Office of Natural History either as organizers of science curricula in the new national education system or as consultants in the development of agricultural reforms and industrial planning.[15] The activities of Tanaka Yoshio, Itō Keisuke's student before and after the Meiji Restoration of 1868, exemplify the connection between natural history and the government. He was one of the founders of the interministerial Office for Natural History (1870), the Ueno Zoological Garden (1882), the first Museum of Natural History of Japan (1873), and the Agricultural Association of Great Japan (1881).[16]

Tanaka was a scholar involved in various public activities, but even more research-oriented scientists like Shirai Mitsutarō could not avoid contacts with governmental offices. After his graduation from the Imperial University in 1886, he found employment as lecturer, thanks to Tanaka's recommendation, in the Tokyo School for Agricultural and Forest Studies (Tokyo Nōrin Gakkō) administered by the Ministry of Agriculture and Commerce. After the school was annexed to the Imperial University in 1893 as its Department of Agriculture, Shirai became an associate professor there. For the remainder of his career, he was involved in survey programs organized by the Ministry of Agriculture in various regions of the country. He is remembered today as the first historian of Japanese traditional natural history but also for having authored a number of works on Japanese forests, his favorite being a survey of Japanese trees he titled *Jumoku wamyō kō* (Reflections on the Japanese Names of Trees, published posthumously in 1933). This text was conceived and written in perfect *honzōgaku* style—as the title itself suggests.[17] Shirai summarized his decades of field expeditions sponsored by the Ministry of Agriculture, providing a list of the various regional names of each tree and giving information on its utilization in the Tokugawa period and how it could contribute to the economic growth of the modern nation-state. The text was written in the classical Japanese style used to translate Chinese throughout the Tokugawa period (*kundoku kakikudashi*) and is indistinguishable from an early modern *honzōgaku* text, except for the scientific Latin name attached to each species.

Hence *honzōgaku*, as a name for nature studies, disappeared in the Meiji period, but in part it survived in form and content under the new rubric of *hakubutsugaku*, or "natural history." Tokugawa scholars, called *jusha*, also disappeared. Yet their successors shared many of their inclinations, not least of which being the call to public service, though they were now known either by their professional title or generically as *gakusha* or *shikisha* ("scholars"), terms that had begun to circulate only toward the end of Tokugawa period, or as *hakase* ("scientists"), a term adopted from the old imperial school of the Daigakuryō of seventh-century Japan. Just as the *honzōgaku* spirit survived in modern *haku-*

butsugaku, the social role of scholar-officials of Tokugawa *jusha* continued to characterize the life of many intellectuals in modern Japan. As Andrew Barshay put it, "The vector of national service was very powerful; thinkers who were internally alienated could often be restored to the national community."[18] Many scientists were enlisted by the Meiji government in the service of modernization and directed their research activity to that purpose. They were now called biologists (*seibutsugakusha*), botanists (*shokubutsugakusha*), and zoologists (*dōbutsugakusha*), but their scholarly activities resembled those of their Tokugawa predecessors. This passage from *honzōgaku* to *hakubutsugaku*—just like the passage from *jusha* to *gakusha* and *hakase*—in the Meiji period suggests that although the names of the discipline and its practitioners changed, the Tokugawa legacy lived on and informed the modern names of nature.

At the same time and parallel to these developments, *honzōgaku* kept its name in explicitly antagonistic opposition to the modernization of medicine and pharmacology. This meant opposition to rapid institutional changes that privileged Western medicine and marginalized traditional medicine, as well as opposition to the rapid disavowal of the intellectual legacies of the Tokugawa period upon which *honzōgaku* was founded. For example, Mori Risshi, better known under his nom de plume Kien, promoted a revival of old *honzōgaku* precisely when the majority of Japanese botanists were translating their traditional knowledge in the language of Western biology. He rejected Li Shizhen's *Bencao gangmu* and dedicated his life to the recovery of the forms and meanings of ancient *bencao* tradition. Mori accused the *Bencao gangmu* of having abandoned the traditional conception of medicinal substances as being at the same time potentially healing and poisonous and of privileging the sorting out of natural species to the medicinal effects of their substances—that is, the object versus its subjective effects. In a quest to reconstruct the lost *Shennong bencao jing* (Shennong's Canon of Materia Medica, a text probably compiled around the second and the first centuries BCE and now lost), Mori dedicated his life to philological studies of Tao Hongjing's *Bencao jing jizhu* along with the *Xinxiu bencao* and Japanese classics of the Heian period like Fukane Sukehito's *Honzō wamyō*, Minamoto no Shitagō's *Wamyō ruijushō*, and Tanba no Yasuyori's *Ishinbō*. Mori's initiative was not merely philological nor was it motivated by a vane nostalgia of the past: not only did he want to restore the tripartite division of "princely," "ministerial," and "adjutant" drugs in the actual practice of traditional apothecaries and *kanpōyaku* physicians, but he also conceived of his work in open polemic against the modernizing fury of early Meiji Japan, so keen to reject its past.

Mori's struggle acquired public resonance precisely when the modernizing impulse of the early Meiji was making way for an identitarian discourse centered around the emperor, the national essence (*kokutai*), and the construction of a Japanese empire in East Asia, and it contributed to reinforce these ideas. However, that notoriety ensued independently from what he was defending of *kanpōyaku* and old-style *honzōgaku*. Indeed, I think that Kien's antiquarianism had nothing to do with identitarian constructions or nationalistic views. Instead, he defended a typically early modern East Asian cosmopolitanism—a belief in the Sinocentric cultural and intellectual sphere that deeply characterized early modern thought, especially in Japan—that was completely at odds with concepts like national essence (*kokutai*) and escape from Asia (*datsu-A*), the vision of a darkened, feudal backwardness of early modern East Asia, a lateness from which to escape that characterized the writings of the thinkers of the "Civilization and Enlightenment" movement. Returning to the original *honzōgaku*, just like preserving the ancient techniques of traditional Chinese medicine (*kanpōyaku*), meant for him to defend and preserve a sense of cultural unity that resisted the modern grammar of culturalism and national essentialism. An antiquarian in Nietzschean sense, Kien was indeed a surviving relic of a time far past.[19] But it was precisely for that reason that he could be later mobilized as a symbol of a past that needed to be brought back to life in order to recover the possibility of an alternative modernity, a modernity other than, and in opposition to, the West.

The survival of *honzōgaku* practices in modern life sciences, however, bears witness of the impossibility of that idea. As I have shown, *honzōgaku* had in fact already developed in the Tokugawa period into a discipline that conjugated knowledge of plants and animals with the reduction of nature to human needs. Indeed, *honzōgaku*'s history had to be forgotten or, in Mori Risshi's case, disavowed in order to maintain the delusional beliefs in a peculiar Japanese sensibility toward nature and in a harmonious coexistence of human communities and nature in traditional Japan that the writings of scholars like Yanagita Kunio, Orikuchi Shinobu, Watsuji Tetsurō, and many others disseminated at the same time that the rapid expansion of Japanese industry was progressively destroying the Japanese ecosystems.

Acknowledgments

One of the main points of *The Knowledge of Nature and the Nature of Knowledge* is that no idea is completely independent of the social environment in which it develops. Similarly, this book sprang not only from the research and philosophical obsessions of its author but also from serendipitous discoveries, institutional opportunities, and conversations with mentors, friends, and colleagues. I could not possibly remember all those who contributed to my project, but to all I extend my sincerest gratitude.

I would like to thank the Weatherhead East Asian Institute for supporting my book as a "WEAI Study" and, in particular, Carol Gluck and Daniel Rivero for their indefatigable help. A special thanks to Karen Merikangas Darling, Anita Samen, and Sophie Wereley of the University of Chicago Press and and Danny Constantino of Scribe Inc.: it was a pleasure working with you. My sincerest appreciation goes also to the two anonymous reviewers of the manuscript, without whom this book would have been much more imperfect. The Japan Foundation has graciously financed two research appointments at Waseda University and at the Institute for Advanced Studies on Asia of the University of Tokyo, which were fundamental to collect the primary material upon which this book is based. My gratitude goes also to the WEAI and the Princeton University Committee on Research in the Humanities and Social Sciences for their grants.

My project was conceived and nourished in different academic institutions. At Columbia University, Carol Gluck was the best mentor I could have. I am also deeply indebted to Matthew Jones and Elizabeth H. Lee for their intellectual, human, and editorial support. Many thanks to David Lurie, Henry D. Smith II, Greg Pflugfelder, Haruo Shirane, Pamela Smith, Donald Keene, David Moerman, and many other teachers who contributed to my intellectual growth, as well as to Mathew Thompson, Adam Clulow, Dennis Frost, Colin Jaundrill, Chelsea Foxwell, Eric Han, and Matt Augustine—among many others—for enduring endless conversations with me. I would like to thank professor Fukaya Katsumi of Waseda University for accepting me in his *komonjō* group and for his assistance in deciphering the intricate calligraphies of Tokugawa naturalists. Special thanks goes also to my friends of the Reischauer Institute for Japanese Studies of Harvard University: Hisa Kuriyama; Andrew Gordon; Stacie Matsumoto; Susan J. Pharr; Ted Gilman; Peter Bol; Stephen Shapin; Peter Galison; and the 2007–8 fellows Sam Perry, Hwansoo Kim, Gavin Whitelaw, and Matthew Marr for enduring my endless ranting. Many thanks to my former colleagues at the University of Virginia: Allan Megill, whose long conversations I will treasure forever; Brian Owensby; Bradley Reed; Duane Osheim; Ronald Dimberg; Cong Ellen Zhang; Karen Parshall; John K. Brown; W. Bernard Carlson; and many others.

This book would not have been nearly as good as I believe it is without the support and suggestions of my colleagues at Princeton University. I would like to thank them all and extend a special mentioning here to Benjamin Elman, an inexhaustible source of ideas and precious information; Sheldon Garon; Martin Kern; Anthony Grafton; Michael Gordin; Sue Naquin; Willard Peterson; Martin Collcutt; Joy Kim; Amy Borovoy; Angela Craeger; Erika Milam; Graham Burnett; Keith Wailoo; Daniel Garber; Bill Jordan; Michael Laffan; Philip Nord; and many others I cannot name here.

This project accompanied me across different continents and universities. I would like to acknowledge here only a few of the many scholars who contributed, directly or indirectly, to make it better: my friend and mentor Massimo Raveri, who taught me to always think from different points of views; Adriana Boscaro; and all my teachers at the University of Venice Ca' Foscari. I would also like to thank Bernardo Dipentina, my high school professor of philosophy and history, to whom I will forever be grateful for teaching me that philosophy is, first of all, rigorous thinking and for making me understand that, as Dionysius of Heraclea put it, "history is philosophy teaching by examples." Many thanks to David Howell, whom I consider a role model. And many thanks also to Kären Wigen, Carla Nappi, Ezra Rashkow, Mathias Vigouroux, Trent Maxey, Brett Walker, Morgan Pitelka, Paula Findlen, Susan Burns, James Bartholomew, all my friends at the Institute for Advanced Studies on Asia at the University of Tokyo, the stuff of the Rare Books Collection of Waseda University Library and of Tokyo National Diet Library, and anybody who graciously listened to my incessant questioning.

Last but not least, an especially heartfelt thanks to my intellectual comrades Ian J. Miller, Robert Stolz, and Julia Adeney Thomas, in the company of whom I have always felt like *Sanshirō*'s "stray sheep": this is also for you, my friends.

I want to dedicate this book to my family: to my parents and my mother-in-law for their support; to my wife, Elisa, for her honest critiques, intellectual stimuli, tireless encouragement, and patience; and to my children, Sofia and Leo, without whom I could have finished this book much sooner, but it surely would have been an incomplete book, without the intimate understanding of what it means to be natural, material beings that the experience of being their father gave me.

Japanese and Chinese Terms

Abe Masahiro 阿部正弘 (1819–57)
a-be-se 亜別泄
Abe Shōō 阿部将翁 (d. 1753)
ai 愛
Aichi 愛知
akaei 黄鱏
akai 赭い
Akita 秋田
Akizuki 秋月
amaboshi 甘干
Amami 奄美
Amaterasu ômikami 天照大神
ametsuchi 天地
Andô Shôeki 安藤昌益 (1703–62)
aobae 青蠅
aogiri 青桐
Aoki Kon'yō 青木昆陽 (1698–1769)
Aoyama 青山
Arai Hakuseki 新井白石 (1657–1725)
Arakida Hisakata 荒木田尚賢
  (1739–88)
araseitō アラセイトウ (紫羅欄花)
Asada Gōryū 麻田剛立 (1734–99)
asagao 朝顔
asagao mizukagami 朝顔水鏡
asahi 朝日
Asaka Naomitsu 浅香直光 (fl. 1835–38)
Asakusa 浅草
Asano Shundō 浅野春道 (d. 1840)
ashigaru 足軽
Ashikaga 足利
Atsuta 熱田
Azabu goyakuen 麻布御薬園
azarashi 海豹
azarashi sawagi 海豹騒
Baba Daisuke 馬場大助 (d. 1868)

bakufu 幕府
bakuhan seido 幕藩制度
bakumatsu 幕末
banbutsu 万物
banbutsu no ri 萬物之理
banji 万事
Bansha no goku 蕃者の獄
banshō 万象
Bansho Shirabesho 蕃書調所
Bansho Wage Goyōgakari 蕃書和解御
  用掛
Banshokō 蕃薯考
banyū 万有
bao 豹 (*hyô*)
beishoku 米食
benbutsu 弁物
bencao 本草 (*honzō*)
Bencao gangmu 本草綱目
Bencao jing jizhu 本草経集注
Bencao jiyao 本草集要
Bencao mengquan 本草蒙筌
bencaoxue 本草学 (*honzōgaku*)
benijukei 紅綬鶏
benisuzume 紅雀
benkeisō 弁慶草
bentatsu 鞭撻
betsu 鼈
betsuroku 別録
*Bizen no kuni Bitchū no kuni no nairyō
  sanbutsu ezuchō* 備前国備中国内領
  産物絵図帳
bonsai 盆栽
Boshin 戊辰
Botanika kyō 菩多尼訶経
bowu 博物 (*hakubutsu*)
Bowuzhi 博物誌

bōzushū 坊主衆
bu 部 (*bu*)
bugyō 奉行
buingai kyōju 部員外教授
Buke shohatto 武家諸法度
bunbu 文武
Bunchi seigenrei 分地制限令
bunchō 文鳥
bunjinga 文人画
Bunkairoku 文会録.
bunmei kaika 文明開化
Buppin shikimei 物品識名
bussan 物産
bussangaku 物産学
bussankai 物産会
Bussankyoku 物産局
Butsurui hinshitsu 物類品騭
cai 菜 (*sai*)
caomu 草木 (*sômoku*)
chadō 茶道
chen 臣 (*shin*)
Chen Jiamo 陈嘉谟 (1486–1570)
chiba 千葉
Chichibu 秩父
chiku 竹
Chikufu 竹譜
Chikuzen no kuni sanbutsu ezuchō 筑前
　国産物絵図帳
Chikuzen no kuni zoku fudoki 筑前国続
　風土記
chinbutsujaya 珍物茶屋
Chingan sodategusa 珍翫鼠育草
chisso 窒素
choku hikyaku 直飛脚
Chōmei benkan 鳥名便覧
chong 蟲 (*chū*)
chongshou 蟲獣 (*chūjū*)
chōsei karin shō 長生花林抄
chōsen ninjin 朝鮮人参
Chōshū 長州

Chosŏn 朝鮮
Chu 楚
chū 蟲
Chūfu zusetsu 蟲譜図説
Chūgoku 中国
chūjū 蟲獣
Chūka 中華
Chūrui ikiutsushi 虫類生写
cidian 詞典
da 蛇
dachō 駝鳥
dadian 大典
Daguan jingshi zhenglei beiyong bencao
　大観経史証類備急本草
Daigakuryō 大学寮
daihyō osso 代表越訴
daikan 代官
Daikan honzō 大観本草
daimyō 大名
daruma inko 達磨鸚哥
Date Munemura 伊達宗村 (1718–56)
Date Shigemura 伊達重村 (1742–1796)
datsu-A 脱亜
Dazai Shundai 太宰春台 (1680–1747)
Denpata eitai baibai kinshi rei 田畑永代
　売買禁止令
Dewa 出羽
difangzhi 地方誌
do 土
dōbutsu 動物
dōbutsugaku 動物学
dōbutsugakusha 動物学者
Dodoniyōsu koroitobokku どゝにやうす
　ころいとぼつく
dōgame どうがめ
Dōjima kome ichiba 堂島米市場
Dōjimon higo 童子問批語
Dōshun 道春 (*see* Hayashi Razan)
duoshi 多識 (*tashiki*)
Duoshibian 多識編

e 会

Ebara 江原

Echigoya 越後屋

Edo 江戸

Edo hanjōki 江戸繁昌記

Edo meisho zue 江戸名所図会

Edo Tekirei 江渡狄嶺 (1880–1944)

Edo-Tōkyō Hakubutsukan 江戸東京
博物館

Ehon kaga mitogi 絵本家賀御伽

Ehon saishikitsū 絵本彩色通

Engi shiki 延喜式

Enryakuji 延暦寺

Ensei Dodoneusu sōmoku fu 遠西独度涅
烏斯草木譜

Ensei ihō meibutsu kō hoi 遠西医方名物
考補遺

Ensei meibutsu kō 遠西名物考

Ensei shōseki hen 遠西硝石篇

Ensei yakukei 遠西薬経

ensō 園草

erekiteru エレキテル

Erya 爾雅

Esaki Hiromichi 江崎広道

Etchū 越中

etohiruka エトヒルカ

etopirika 花魁鳥

Ezo 蝦夷

faming 発明 (*hatsumei*)

feitouman 飛頭蛮

fengtu 風土 (*fūdo*)

fu 譜

fu 府

fudai 譜代

fūdo 風土

fufang 附方 (*fuhō*)

fuhō 附方

Fujibayashi Taisuke (Fuzan) 藤林泰助
(普山, 1781–1836)

Fujiwara 藤原

Fujiwara Seika 藤原惺窩 (1561–1619)

Fukane no Sukehito 深根輔仁 (fl. 900–
920)

Fukiage 吹上

fukoku kyōhei 富国強兵

Fukuoka 福岡

Fukuyama 福山

Fukuyama Tokujun 福山徳潤

Fukuzawa Yukichi 福沢諭吉 (1835–
1901)

furisode kaji 振袖火事

Fūryū Shidōken den 風流志道軒伝

Fūzan Bunko 楓山文庫

gaikokujin yatoikyōshi 外国人雇い教師

gakusha 学者

gang 綱 (*kō*)

Gang Hang 姜沆 (Kyokō, 1567–1618)

Ganjin 鑑真 (688–763)

gasu 瓦斯

ge 解

gehin nari 下品ナリ

Gejō kinmō zui 戯場訓蒙図彙

genki 元気

Genroku 元禄

genshiteki 原始的

genso 元素

Gensuke 源助

gesaku 戯作

gewu zhi xue 格物之学 (*kakubutsu no
gaku*)

gi 偽

Gifu 岐阜

ginmigakari 吟味掛

gōgakkō 郷学校

gogyō 五行

Go-Mizunō 後水尾 (1596–1680)

Gomō jigi 語孟字義

gōnō 豪農

Gonyūbu kyara onna 御入部伽羅女

Goseihō 後世方

gōshi 郷士
Gotō Gonzan 後藤艮山 (1659-1733)
goyakuen 御薬園
goyō toritsugi 御用取次
Go-Yōzei 後陽成 (1571-1617)
Gozan 五山
gundai 郡代
guo 果 (ka)
Gyōbō setsuroku 仰望節録
Gyofu 魚譜
gyoku 玉
gyokurōrei 玉瓏玲
gyokuseki 玉石
Gyūsan no ki zenshō kōgi 牛山之木全
　章講義
habubutsugaku 博物学
Hachinohe 八戸
Hachisuka Masauji 蜂須賀正氏
　(1903-53)
Hachisuka 蜂須賀
haikai 俳諧
hakase 博士
hakkan 白鶤
Hakone 箱根
hakubutsu 博物
hakubutsugaku 博物学
hakubutsugakusha 博物学者
Hakubutsukyoku 博物局
Hakura Yōkyū 羽倉用九
Hakurai chōjū zushi 舶来鳥獣図誌
hakusai mono 舶載物
Hakusan 白山
Hamamatsu 浜松
hana awase 花合
Hanabusa Iemon 花房伊右衛門
hananegi はなねぎ
Hanaoka Seishū 華岡青洲 (1760-1835)
hanashōbu 花菖蒲
hanazumō 花相撲
hankō 藩校

hanpon 板本
hanrei 凡例
Hanshu 漢書
haribuki 針蕗
Harima 播磨
Haruma wage 波留麻和解
hatamoto 旗本
Hatchōbori 八丁堀
hatsukanezumi 廿日鼠
hatsumei 発明
Hattori Nankaku 服部南郭 (1683-1759)
Hayashi Gahō (Shunsai) 鵞峰 (春斎,
　1618-80)
Hayashi Jussai 林述斎 (1768-1841)
Hayashi Morikatsu (Shuntoku) 林守勝
　(春徳)
Hayashi Nobuatsu (Hōkō) 林信篤 (林
　鳳岡, 1644-1732)
Hayashi Nobutoki 林信言 (1721-73)
Hayashi Razan 林羅山 (1583-1657)
Hayashi Ryōki 林良喜 (1695-1722)
hiden 秘伝
hidenju 秘伝授
Higo 肥後
Higo Hachinoshin 肥後八之進
Higo Yaemon 肥後八右衛門
hiiragi 柊
hikuidori 火食鶏
hikurage 火海月
hin 品
hinin 非人
Hiraga Gennai 平賀源内 (1728-79)
Hirakawa 平河
Hirayama 平山
Hiroshima 広島
hishika 費西加
hisutōri 斐斯多里
Hitachi no kuni fudoki 常陸国風土記
hitoezakura 一重櫻
hizakura 緋櫻

Hizen 肥前
hōgai mono 法外者
Hokkaidō 北海道
Hokuetsu seppu 北越雪譜
hōkyō 法教
honbyakushō 本百章
honchō no koseki 本朝之古籍
Hongō 本郷
Honjifu 本事府
Honkoku shokubutsugaku 翻刻植物学
honpō 本邦
hon'yakukan 翻訳官
honzō 本草
Honzō iroha shō 本草色葉抄
Honzō kōmoku 本草綱目
Honzō kōmoku jochū 本草綱目序註
Honzō kōmoku keimō 本草綱目啓蒙
Honzō kōmoku kibun 本草綱目記聞
Honzō shashin 本草写真
Honzō tsūkan shōzu 本草通串証図
Honzō tsūkan 本草通串
Honzō wamyō 本草和名
Honzō zusetsu 本草図説
honzōgaku 本草学
honzōgaku jidai 本草学時代
Honzōkyoku 本草局
Hori Kyōan 堀杏庵 (1585–1642)
horohorochō 珠鳥
hōsen 鳳仙
Hosokawa Shigekata 細川重賢
　(1718–85)
Hototogisu zusetsu ほとゝぎす図説
Hotta Masaatsu 堀田正敦 (1755–1832)
Hotta Tatsunosuke 堀田龍之助
　(1819–88)
hu 虎 (*tora*)
Huainan zi 淮南子
Hubei 湖北
hyakushō sōdai 百姓惣代
hyō 豹

hyōhon 標本
ibojiri 疣毟
Ichigaya 市谷
iemoto 家元
iemoto seido 家元制度
igaku 異学
igaku no kin 異学之禁
Igakukan 医学館
Igansai baihin 怡顔斎梅品
Igansai chikuhin 怡顔斎竹品
Igansai ōhin 怡顔斎桜品
Igansai ranpin 怡顔斎蘭品
Ihara Saikaku 井原西鶴 (1642–93)
ihōruto イホウルト
ihyō 遺表
Iijima Isao 飯島魁 (1861–1921)
Iimuro Masanobu 飯室昌栩 (b. 1789)
Iinuma Yokusai 飯沼慾斎 (1782–1865)
ikan 医官
Ike Taiga 池大雅 (1723–76)
Ikeda Michitaka 池田道隆
ikiutsushi 生写
ima anzuruni 今案
Imai Iwao 今井巌
Imaichi 今市村
imei 異名
in 陰
Inamura Sanpaku 稲村三伯 (1758–
　1811)
inko 鸚哥
Inō Jakusui 稲生若水 (1655–1715)
Inō Masatake 稲生正武 (1682–1747)
Inoue Tetsujirō 井上哲次郎 (1855–
　1944)
in-yō-zuhō 印葉図法
iogame いをがめ
Ippondō yakusen 一本堂薬選
Ise 伊勢
ishi 医師
Ishida Baigan 石田梅岩 (1685–1744)

Ishii Tsuneemon (Shōsuke) 石井常右衛
門 (庄助, b. 1743)
Ishinbō 医心方
Itō Genboku 伊藤玄朴 (1800–1871)
Itō Ihyōe Masatake 伊藤伊兵衛政武
(1676–1757)
Itō Ihyōe Sannojō 伊藤伊兵衛三之丞
(d. 1719)
Itō Jakuchū 伊藤若冲 (1716–1800)
Itō Jinsai 伊藤仁斎 (1627–1705)
Itō Keisuke 伊藤圭介 (1803–1901)
Itō Tōgai 伊藤東涯 (1670–1736)
Iwanaga Buntei 岩永文禎 (1802–66)
Iwasaki Tsunemasa (Kan'en) 岩崎常正
(灌園, 1786–1842)
Izu 伊豆
Izumo fudoki 出雲風土記
Jakōneko 麝香猫
Jia Sixie 賈思勰
Jiangxi 江西本
Jiga 爾雅
jijie 集解 (*shúkai*)
jijinsai 鎮神祭
Jikei 慈稽
jikken 実見
jin 仁
Jingshi zhenglei beiji bencao 経史証類備
急本草
Jinling ben 金陵本
Jinrin kinmō zui 人倫訓蒙図彙
jisharyō 寺社領
jisho 字書
jisshin 実心
jisshō 実証
jitchi 実知
jitō 地頭
jitsu 実
jitsugaku 実学
jitsuri 実理
jittoku 実徳

jiushu 灸術 (*kyūjutsu*)
jizamurai 地侍
Jō Izumi-no-kami Masamochi 城和泉
守昌茂
jō 丈
jō 情
jōhin nari 上品ナリ
jōi 攘夷
jōri 条理
jōruri 浄瑠璃
Jubustu mondō 儒仏問答
jūhō 重宝
jukan 儒官
Jumoku wamyō kō 樹木和名考
jun 君 (*kun*)
junkangaku 準官学
junsei hakubutsugaku jidai 純正博物学
時代
junsui 純粋
jusha 儒者
ka 果
ka 火
ka 花
kaboku 花木
kabu nakama 株仲間
kabutogani 鱟魚
kachi 徒士
Kada no Arimaro 荷田在満 (1706–51)
Kada no Azumamaro 荷田春満 (1669–
1736)
Kadan ji kinshō 花壇地錦抄
Kadan kōmoku 花壇綱目
kaede 楓
Kafu 花譜
kagakusha 科学者
Kagawa Shūan 香川修庵 (1683–1755)
Kagawa 香川
Kaga 加賀
Kagoshima 鹿児島
kahō 家法

kai 介
kai 会
Kai 甲斐
kai 界
Kai 花彙
kai 蟹
kaiawase 貝合
Kaibara Ekiken (Atsunobu) 貝原益軒
　(篤信, 1630–1714)
Kaibara Ekiken sensei den 貝原益軒先
　生伝
Kaibara Ekiken sensei nenpu 貝原益軒
　先生年譜
Kaibara Mototada (Sonzai) 貝原元端
　(存斎, 1622–95)
Kaibara Rakuken 貝原楽軒 (1625–1702)
Kaibara Toshisada 貝原利貞 (1597–
　1665)
Kaibara Yoshifuru 貝原好古 (1664–
　1700)
kaibutsu 開物
Kaibutsufu 開物府
kaidoriya 飼鳥屋
kaihatsu 開発
Kaihodō 会輔堂
kaisai 海菜
Kaisei gakkō 開成学校
Kaitai shinsho 解体新書
Kaitokudō 懐徳堂
kaiwu 開物 (kaibutsu)
kakekotoba 掛詞
kakubutsu chichi 格物致知
kakubutsu no gaku 格物之学
kakun 家訓
kamakiri 蟷螂
Kamakura 鎌倉
kame 亀
kami 神
Kamo no Mabuchi 賀茂真淵 (1697–
　1769)

kan 巻
Kan'ei 寛永
Kanagawa 神奈川
kanariya 金糸雀
Kanaseiri 仮名性理
Kanda 神田
kandai 寒鯛
kangaku 官学
kanibasakura カニバサクラ
kanjō bugyō 勘定奉行
kanmuribato 冠鳩
Kanō Kōkichi 狩野亨吉 (1865–1942)
Kanō Shigekata 狩野重賢
kanpōyaku 漢方薬
Kansei igaku no kin 寛政異学禁
Kansei 寛政
Kansho sensei 甘藷先生
Kanshoki 甘藷記
Kantokuan 管得庵
kapitan 甲比丹
kappa 河童
karatachibana 唐橘
karawatari 漢渡り
kashi hon'ya 貸本屋
Kashiragaki zōho Kinmō zui 頭書増補
　訓蒙図彙
Kasshi yawa 甲子夜話
kasō 花草
Katata 堅田
Katayama Hokkai 片山北海 (1723–90)
Katō Enao 加藤枝直 (1693–1785)
Katsuragawa Kuniakira (Hoshū) 桂川国
　瑞 (甫周, 1751–1809)
Katsushika Hokusai 葛飾北斎 (1760–
　1849)
Kawachi 河内
kawaraban 瓦版
kawatarō 河太郎
kawauso 獺
Kazuki no Amanoko 潜蜑子 (fl. 1690)

kei 経
Keian no furegaki 慶安触書
Keichō nikkenroku 慶長日件録
Keijō shoran 京城勝勝覧
keishitsu 形質
Keitai 継体天皇 (r. 507–31)
Keitekiin 啓迪院
keizaigaku 経済学
Keizai yōryaku 経済要録
keizugaku 系図学
Kenbō 妍芳
Kenkadō zasshi 兼葭堂雑誌
kenkon 乾坤
kenmon 権門
Kenninji 建仁寺
ki 気
ki ichigen ron 気一元論
Kidoku shomoku 既読書目
Kii 紀伊
Kikai zufu 奇貝図譜
Kikuchi Dairoku 菊池大麓 (1855–1917)
Kikuchi Kōsan 菊池黄山 (1697–1776)
kikuhi 菊婢
kikyō 桔梗
kimi 気味
Kimura Kōkyō (Kenkadō) 木村孔恭
　　(兼葭堂1736–1802)
kin 菌
kin 斤
kinhōke 金鳳花
Kinka gyofu 錦窠魚譜
kinkei 金鶏
Kinmō zui 訓蒙図彙
Kinoshita Jun'an 木下順庵 (1621–99)
kinsangindai 金盞銀臺
Kinshū makura 錦繍枕
kishitsu 気質
Kishū sanbutsu shi 紀州産物志
Kitano 北野
Kitsuki 杵築

kitsune 狐
Kiyohara Hidekata 清原秀賢 (1575–
　　1614)
kizukai 気遣
ki 気
Kō Ryōsai 高良斎 (1799–1846)
kō 綱
ko 苆
Kōbun'in 弘文院
kōbutsu 好物
Koga Tōan 古賀侗庵 (1788–1847)
Koga 古河
koganeyanagi 黄芩
kōgyo 鱴魚
kohonzō fukko 古本草復古
Koihō 古医方
Koishikawa goyakuen 小石川御薬園
Koishikawa yakuen 小石川薬園
Koishikawa 小石川
Kojiki 古事記
Kōjimachi 麹町
kōkibana 月季花
Kokinga sōgo hachi shu 古今画藪後
　　八種
koku 穀
kokueki 国益
kokugaku 国学
kokuri minfuku 国利民福
Kokuritsu Komonjokan Naikaku Bunko
　　国立古文書館内閣文庫
Kokusho sômokuroku 国書総目録
kokusho 国書
kokutai 国体
Komaba 駒場
Komagome 駒込
Komeiroku 古名録
Kōmō kinjū gyokaichū fu 紅毛禽獣魚
　　介虫譜
Kōmō zatsuwa 紅毛雑話
konando 小納戸

konchū hyōhon 昆虫標本
Kondō hisaku 混同秘策
Kondō Seisai (Morishige) 近藤正斎
　(守重, 1771–1829)
Koremune Tomotoshi 惟宗具俊
kōsaigusa かうさいぐさ
kōsaika 蒿菜花
Kōsei shinpen 厚生新編
Kōsho koji 好書故事
koshō 小姓
kōso 光素
Kotenseki sōgō mokuroku 古典籍総合
　目録
kotoriawase 小鳥合
Kōzu 高津
kū 空
kubiwa ōkōmori 首輪大蝙蝠
Kubo Sōkan 久保桑閑 (1710–82)
kujaku 孔雀
kujakujaya 孔雀茶屋
Kumamoto 熊本
Kumazawa Banzan 熊沢蕃山 (1619–91)
kumi 組
kumiai 組合
kun 君
kundoku kakikudashi 訓読書下
kundoku 訓読
Kunshikun 君子訓
kurabe 競
Kurama 鞍馬
Kurimoto Joun 栗本鋤雲 (1822–97)
Kurimoto Tanshū 栗本丹洲 (1756–
　1834)
Kurisaki Dōyū 栗崎道有 (1661–1726)
Kuroda Mitsuyuki 黒田光之 (1628–
　1707)
Kuroda Narikiyo 黒田斉清 (1795–1851)
Kuroda Tadayuki 黒田忠之 (1602–54)
Kuroda Tomoari (Suizan) 畔田伴存
　(翠山, 1792–1859)

Kuroda 黒田
Kurokawa Dōyū 黒川道祐 (d. 1691)
kusa awase 草合
kyōchikutō 夾竹桃
kyogaku 虚学
Kyōhō no kikin 享保の飢饉
Kyōhō 享保
kyōka 狂歌
Kyōka Nihon fudoki 狂歌日本風土記
kyōkan 教官
kyōkanchō 九官鳥
kyūjutsu 灸術
kyūri 究理
lang 莨
langdand 莨菪
leishu 類書
li 理 (ri)
Li Ai 李呆 (1180–1252)
Li Gong 李塨 (1659–1733)
Li Jianyuan 李建元
Li Shanlan 李善蘭 (1811–82)
Li Shizhen 李時珍 (1518–93)
lin 鱗, rin in Japanese
Lin Zhaoke 林兆珂
Liuyu yanyi 六諭衍義
Lunyu jizhu 論語集註 (Rongo shitchū)
machi isha 町医者
machi ishi 町医師
Maeda Ryōtaku 前野良沢 (1723–1803)
Maeda Toshiyasu 前田利保 (1800–59)
Maeda Tsunanori 前田綱紀 (1643–
　1724)
Makura no sōshi 枕草子
makura-kotoba 枕詞
Man'yōshū 万葉集
Manase Dōsan 曲直瀬道三 (1507–94)
Manase Gensaku 曲直瀬玄朔 (1549–
　1631)
Manase Shōjun 曲直瀬正純
manbō 翻車魚

mandarage 曼陀羅華
manga 漫画
manpōkei 万歩計
man'yōgana 万葉仮名
maō 麻黄
Maruyama Masao 丸山真男 (1914–96)
Maruyama Ōkyo 円山応挙 (1733–95)
Masuyama Masakata (Sessai) 増山正賢
　(雪斎, 1754–1819)
Matsu to ume taketori mongatari 松と梅
　竹取物語
Matsuda Chōgen 松田長元
Matsudaira Norisato 松平乗邑 (1686–
　1746)
Matsudaira Sadanobu 松平定信 (1758–
　1829)
Matsudaira Yoritaka 松平頼恭 (1711–
　71)
Matsumura Goshun 松村呉春 (1752–
　1811)
Matsunaga Sekigo 松永尺五 (1592–
　1657)
Matsunaga Teitoku 松永貞徳 (1571–
　1653)
Matsuo Bashō 松尾 芭蕉 (1644–94)
Matsuoka Izaemon 松岡伊左衛門
Matsuoka Joan 松岡恕庵 (1669–1747)
Matsushita Kenrin 松下見林 (1637–
　1703)
Matsuura Seizan 松浦静山 (1760–1841)
Matsuyama 松山
Matsuzaka 松坂
meibutsu 名物
meibutsugaku 名物学
Meiji 明治
Meiji ishin 明治維新
Meireki 明暦
meisho zue 名所図会
mesu 雌
metsuke 目付

mi 味
mi 実
michi 道
Mie 三重
Mikawa 三河
Mima Junzō 美馬順三 (1807–37)
Minagawa Kien 皆川淇園 (1734–1807)
Minakata Kumagusu 南方熊楠 (1867–
　1941)
Minamoto no Shitagō 源順 (911–83)
Minamoto Tarō みなもと太郎
Minato Chōan 湊長安 (d. 1838)
Ming 明
Mino 美濃
min'yō sōrui 民用草類
misemono 見世物
mishi 米食 (*beishoku*)
Miyazaki Yasusada 宮崎安貞 (1623–97)
Miyoshi Manabu 三好学 (1861–1939)
Mizuno Kōzan 水野皓山 (1791–1833)
Mizuno Motokatsu 水野元勝 (fl. 1680)
Mizuno Tadayuki 水野忠之 (1669–
　1731)
mizunomi hyakushō 水呑百姓
Mizutani Hōbun 水谷豊文 (1791–1833)
Mizutani shi tori fu 水谷氏禽譜
mō 毛
mokkōbana 木香花
moku 木
moku 目
Mokuhachi fu 目八譜
momiji 椛
Momijiyama Bunko 紅葉山文庫
mon 門
Mori Risshi (Kien) 森立之 (枳園,
　1808–85)
Morino Fujisuke 森野藤助 (1690–
　1767)
Morioka 盛岡
Morishima Chūryō 森島中良

Moriyama Takamori 森山孝盛 (1738–1815)

morokoshikibi 唐黍

Motogi Ryōei 本木良永 (1735–94)

Motogi Shōei 本木正栄 (1767–1822)

Motoori Norinaga 本居宣長 (1730–1801)

Motoori Ōhira 本居大平 (1756–1833)

mu 無

mu 目 (*moku*)

muchi 鞭

Mukai Genshō 向井元升 (1609–77)

Mukai Tomoaki 向井友章

mumei 無名

Murakami Sōshin 村上宗信

Muramatsu Hyōzaemon 村松標左衛門 (1762–1841)

Muraoka Tsunetsugu 村岡典嗣 (1884–1946)

Muro Kyūso 室鳩巣 (1658–1734)

Musashi Yoshitoki (Sekijun) 武蔵吉恵 (石寿, 1766–1860)

Musashi Yoshitoki 武蔵吉恵 (1766–1860)

Musashi 武蔵

Mutsu no kuni sanbutsuchō 陸奥国産物帳

Mutsu 陸奥

Myōbaru 女原

myōgakin 冥加金

myōgyō hakase 明経博士

myōshu 名主

Nagai 永井

Nagano 長野

Nagasaki 長崎

Nagasakiya 長崎屋

Nagashima 長島

Nagata Chōbei 永田調兵衛

Nagoya 名古屋

Nagoya Gen'i 名古屋玄医 (1628–96)

Naimushō 内務省

Nakae Tōju 中江藤樹 (1608–48)

Nakagami Kundo 中神君度

Nakagawa Jun'an 中川淳庵 (1739–86)

Nakai Riken 中井履軒 (1732–1817)

Nakai Shūan 中井甃庵 (1693–1758)

Nakamikado 中御門 (1701–37)

Nakamura Fumisuke 中村文輔 (1701–63)

Nakamura Tekisai 中村惕斎 (1629–1702)

Nakatsu 中津

Nakayama Zenzaemon 中山善左衛門

nakiawase 鳴合

Namikawa Tenmin 並河天民 (1679–1718)

nanban 南蛮

nanga 南画

nankinhaze 烏臼木

nanushi 名主

Nanzan zokugo kō 南山俗語考

Nara 奈良

Narabayashi Eiken 楢林栄建 (1800–1875)

Narutaki-juku 鳴滝塾

Nawa Kassho 那波活所 (1595–1648)

Negishi Yasumori 根岸鎮衛 (1737–1815)

Nenashi gusa 根南志具佐

Nihon no sakura 日本ノ櫻

Nihon shoki 日本書紀

Nihon sōezu 日本総絵図

Nijō Castle 二条城

Nijūyon kō kai 二十四綱解

nikawa 膠

Nikkō 日光

ningen 人間

ningyo 人魚

ninjin 人参

Ninjin kōsaku ki 人参耕作記

ninjinza 人参座

Ninomiya Keisaku 二宮敬作 (1804–62)

Ninomiya Sontoku 二宮尊徳 (1787–1856)

Nishinomaru 西丸

Niwa Shōhaku 丹羽正伯 (1700–1753)

niwaban 庭番

nō 能

nobemai 延米

Noda Yajiemon 野田弥次右衛門

Nōdoku 能毒

Nōgyō zensho 農業全書

nōji 農事

Nomura Kansai 野村観斎

Nomura Ritsuei 野村立栄 (1751–1828)

Nongshu 農書

Nongzheng quanji 農政全集

Norō Genjō 野呂元丈 (1693–1761)

Nōshōmushō 農商務省

Noto 能登

Nozuchi 野槌

Obama 小浜

Ochanomizu 水御茶ノ水

Ōdaka Motoyasu 大高元恭 (b. 1758)

ofuregaki 御触書

Ogata Tankō 尾形探香 (ca. 1812–68)

Ogiwara Shigehide 荻原重秀 (1658–1713)

Ogyū Sorai 荻生徂徠 (1666–1728)

ohanashishū 御咄衆

Ohara Tōdō 小原桃洞 (1746–1825)

Oka Kenkai 岡研介 (1799–1839)

Okada Atsuyuki 岡田淳之

ōkami 狼

Okamoto Rihee 岡本利兵衛

Okinawa 沖縄

Ōkōchi Zonshin 大河内存眞 (1796–1883)

okera 朮

okoze 虎魚

oku no miya 奥宮

Ōkubo Masaaki 大窪昌章 (1802–41)

Ōkubo Okaemon 大久保岡右衛門

Ōkubo Tadakata 大久保忠教 (1560–1639)

Ōkubo Tahei 大窪太兵衛 (1763–1824)

Ōkubo 大久保

Okudaira Akitaka 奥平昌高 (1781–1855)

omemie 御目見

Ōmi 近江

omoto 万年青

ōmu 鸚鵡.

onagabato 尾長鳩

ondokei 温度計

oniwaban 御庭番

Ono Motokata 小野職孝 (d. 1852)

Ono Motoyoshi 小野職愨 (1843–90)

Ono Ranzan 小野蘭山 (1729–1810)

onozukara 自然

onso 温素

Ōoka Kiyosuke 大岡清相 (1679–1717)

Oranda honzō goyō 阿蘭陀本草御用

Oranda honzō wage 阿蘭陀本草和解

Oranda honzō 阿蘭陀本草

Oranda kinjū chūgyo zu wage 阿蘭陀禽
獣虫魚図和解

Oranda yakukyō 和蘭薬鏡

ōren 黄連

ōrihare オオリハレ

Orikuchi Shinobu 折口信夫 (1887–1953)

oshidori 鴛鴦

Ōshima 大島

Ōshio Heihachirō 大塩平八郎 (1793–1837)

ōshokuki 黄蜀葵

osu 雄

Ōta Nanpō 大田南畝 (1749–1823)

otogizōshi 御伽草子

ōtsūji 大通詞

Ōtsuka goyakuen 大塚御薬園

Ōtsuki Genkan 大槻玄幹 (1785–1837)
Ōtsuki Gentaku 大槻玄沢 (1757–1827)
Ōuchi 大内
Owari meisho zue 尾張名所図会
Owari 尾張
ōyō hakubutsugaku jidai 応用博物学時代
Ozeki San'ei 小関三英 (1787–1839)
Ōzu 大津
pan'ya 古貝
pāruto パールト
pin 品 (*hin*)
pu 譜 (*fu*)
pulu 譜録
qi 気 (*ki*)
qi kaoshi zingli 其考釈性理
qiankun 乾坤
Qimin yaoshu 斉民要術
qing 情 (*jō*)
Qing 清
qiwei 気味 (*kimi*)
qizhi 気質 (*kishitsu*)
rakkasei 落花生
Rangaku kotohajime 蘭学事始
rangaku 蘭学
ranpōi 蘭方医
Razan Rin sensei gyōjō 羅山林先生行状 (1658)
Razan sensei nenpu 羅山先生年譜 (1659)
reikin 礼金
ren 連
renshen 人参 (*ninjin*)
ri 理
rigakubu kyōju 理学部教授
Rikugunfu 陸軍府
Rikuyu engi taii 六諭衍義大意
rin 鱗
Rinshi teihatsu juiben 林氏剃髪受位弁
Rinzai 臨済

Risshū 律宗
rōjū 老中
rokkasen 六歌仙
Rokuroku kaiawase waka 六々貝合和歌
rokurokubi 轆轤首
Rongo shitchū 論語集註
rui 類
ruizoku 類属
rurikazura ルリカヅラ
rusui 留守居
ryō 両
ryōshu 領主
Ryūkyū 琉球
Ryūkyū sanbutsu shi 琉球産物誌
Sabase Yoshiyori 佐橋佳依 (fl. 1835–40)
saboten 仙人掌
Saga 佐賀
Sagami 相模
Sagara 相良
sai 菜
saichō 犀鳥
Saifu 菜譜
Saikaku gohyakuin 西鶴五百韻
Sairan igen 采覧異言
Saitama 埼玉
Saiyaku dokudan 採薬独断
saiyakuki 採薬記
saiyakushi 採薬使
Sakamoto Juntaku 坂本純沢
Sakamoto Kōnen 坂本浩然 (1800–55)
sakoku seido 鎖国制度
sakui 作為
sakura 櫻
Sanbutsu Goyōdokoro 産物御用所
sanbutsu 産物
sanbutsuchō 産物帳
Sancai tuhui 三才図会
sanhuang 三皇
Sanjō Guest House 山上会館
sanka 山河

sankin kōtai 参勤交代
San 佐野
sanshuyu 山茱萸
sanso 酸素
Sansui 山水
Santō Kyōden 山東京伝 (1761–1816)
Sanuki 讃岐
sashi 佐使
Satake Yoshiatsu (Shozan) 佐竹義敦
　　(曙山, 1748–85)
Satō Nakaoka 佐藤中岡
Satō Naokata 佐藤直方 (1650–1719)
Satō Nobuhiro 佐藤信淵 (1769–1850)
sato no miya 里宮
satōkibi 甘蔗
satoyama 里山
satsuki 皐
Satsuma 薩摩
Satsuma imo 薩摩芋
Satsuma imo kōnōsho narabi ni tsukuriyō
　　no den 薩摩芋功能書並作様之伝
Satsuma keii ki 薩藩経緯記
sayakuki 採薬記
sei 性
sei 生
seibutsugakusha 生物学者
seigaku 正学
seigo 正誤
Seijukan 躋寿館
Seikei jitsuroku 成形実録
Seikei zusetsu 成形図説
Seikyō yōroku 聖教要録
seimei 正名
Seimi kaisō 舎密開宗
seimi 舎密
seiri 生理
Seisetsu kanshō kyō 西説観象経
Seisetsu naika senyō 西説内科撰要
seiyō honzōgaku 西洋本草学
Seiyō kibun 西洋紀聞

Seiyūki 西遊記
Seizōfu 製造府
sekai 世界
Seken munezan'yō 世間胸算用
seki 石
Sekigahara 関ヶ原
sekkyō 説教
Sen no Rikyū 千利休 (1522–91)
senbatsu 選抜
Senchūfu 千虫譜
Sendai 仙台
Senga Dōryū 千賀道隆 (1721–95)
sengoku daimyo 戦国大名
Sengoku period 戦国 (1467–1603)
senryū 川柳
seppō 説法
Setonaikai 瀬戸内海
setsuwa 説話
setsuyō 切要
Settsu meisho zue 摂津名所図会
sha 社
sha 写
Shabenkai hinbutsu ron teisan 赭鞭会品
　　物論定纂
Shabenkai 赭鞭会
Shajitsu 写実
shakyō 写経
shaku 尺
shakumei 釈名
shakuyaku 芍薬
Shanhaijing 山海経
shanhe 山河 (sanka)
shanshui 山水 (sansui)
sharei 謝礼
shasei 写生
Shasei gachō 写生画帖
shashin 写真
Shen Nanpin 沈南蘋
Shennong bencao jing jizhu 神農本草経
　　集注

Shennong bencao jing 神農本草経
Shennong 神農
shi 士
shi 獅 (*shi*)
Shiba Kōkan 司馬江漢 (1747–1818)
Shibano Ritsuzan 柴野栗山 (1736–
    1807)
shibe 蘗
Shidara Sadatomo 設楽貞丈 (b. 1785)
Shidoura 志度浦
shigai wuli 實該物理
Shiga 滋賀
Shige Shichirōzaemon 茂七郎左衛門
shihai zhi nei jie xiongdi ye 四海之内皆
    兄弟也
Shiiboruto jiken シーボルト事件
Shiiboruto sensei Narutaki jukusha no zu
    シーボルト先生鳴滝熟舎之図
Shiji no yukikai 四時交加
Shiji 史記 (*Shiki*)
Shijō kawara yūrakuzu byōbu 四条河原
    遊楽図屏風
shijuku 私塾
Shikaidōhō 四海同胞
Shiki 史記
Shikien 四季園
shikisha 識者
Shikishima 敷島
Shikitei Sanba 式亭三馬 (1776–1822)
Shimabara 島原
Shimada Mitsufusa 島田充房
Shimane 島根
Shimazu 島津
Shimazu Nariakira 島津斉彬 (1809–58)
Shimazu Narinobu 島津斉宣 (1774–
    1841)
Shimazu Narioki 島津斉興 (1791–1859)
Shimazu Shigehide 島津重豪 (1745–
    1833)
shiming 釈名 (*shakumei*)

Shimodera machi 下寺町
Shimokawabe Jūsui 下河辺拾水
Shimōsa 下総
Shimotsuke 下野
shin 心
shin 臣
shin 真
shin'ei 真影
Shin'i zukai 深衣図解
shin'i 深衣
Shinano 信濃
shinden kaihatsu 新田開発
shingaku 心学
Shinju narabi okonawarete aimotowazaru
    no ron 神儒並行而不相悖論
shinkeizu 真景図
Shinkōsei honzō kōmoku 新校正本草
    綱目
Shinpū Tokubetsu Kōgekitai 神風特別
    攻撃隊
shinrabanshō 森羅万象
Shinshiroku 慎思録
Shin-yū Oranda honzō 辛酉阿蘭陀本
    草
shinzan yūkoku 深山幽谷
Shirai Mitsutarō 白井光太郎 (1863–
    1932)
Shirao Kunihashira 白尾国柱 (1762–
    1821)
shiryō 私領
shishi 志士
Shishido Sakari 穴戸昌 (1841–1900)
shisho gokyō 四書五経
shiwu ru gewu zhi xue 実吾儒格物之学
shizen 自然
shizen no ri 自然之理
Shizen shin'eidō 自然真営道
Shizuki Tadao 志筑忠雄 (1760–1806)
Shizukidō Hōzan 指月堂芳山 (d. 1751)
Shizuoka 静岡

shō 升
shobutsu 庶物
shobutsu bugyō 書物奉行
Shobutsu ruisan 庶物類纂
Shōhaku sairai ikashomoku 商舶載来医
　家書目
Shōheikō 昌平校
Shōheizaka 昌平坂
Shōhyakusha 嘗百社
shōkō 小綱
shokubutsu 植物
shokubutsuen 植物園
shokubutsugaku 植物学
shokubutsugakusha 植物学者
Shokugaku keigen 植学啓原
shokugaku 植学
Shōma zusetsu 升麻図説
shōmoku 小目
shomotsu bugyō 書物奉行
Shosai chūchiku shomoku 書斎中蓄書目
shōtōkō 小桃紅
Shōtoku shinrei 正徳新例
shōya 庄屋
shu 種
shu 種
Shūhō gafu 衆芳画譜
shuihu 水虎
shuin 朱印
shuinsen bōeki 朱印船貿易
shuji 主治
shūji 修治
shujusho 種樹書
shūkai 集解
Shūkinzu 衆禽図
shuku 菽
shungiku 春菊
Shunkanshō 春鑑抄
shunran 春蘭
Shunrinken 春林軒
Shūrinzu 衆鱗図

shushigaku 朱子学
Sō 宗
sō 草
so 蔬
Sō Senshun 曽占春 (1758–1834)
Sō Shiseki 宋紫石 (1715–86)
sobana 薺苨
sōbyakushō ikki 惣百姓一揆
Sōhan 曽槃 (1758–1834)
Sōken 宗建 (1802–52)
sokukinsan 側金盞
Somei 染井
sōmoku 草木
sōmoku chōjū gyokai konchū kingyoku
　doseki 草木鳥獣魚介昆蟲金玉土石
sōmoku chūgyo kinjū 草木蟲魚禽獣
sōmoku-kinjū-chūgyo-kingyoku-doseki
　草木禽獣蟲魚金玉土石
Sōmoku shasei 草木写生
Sōmoku sodategusa 草木育種
Song Yingxing 宋應星 (1587–1666)
sonnō 尊皇
Sonsai okina zuihitsu 遜斎翁随筆
sōshaban 奏者番
Su Jing 蘇敬
Sugamo 巣鴨
Sugeno Kenzan 菅野兼山 (1652–1719)
Sugita Genpaku 杉田玄白 (1733–1817)
sui 水
Suigunfu 水軍府
Suiko jūni hin no zu 水虎十二品之図
Suiko kōryaku 水虎考略
suisai 水菜
suisen 水仙
suiso 水素
Suitō hiroku 垂統秘録
Suizokushi 水族志
Suminokura Ryōi 角倉了以 (1554–1614)
Sunpu 駿府
suppon 鼈

Suruga 駿河
Suwaraya Ichibei 須原屋市兵衛 (d. 1811)
Suzuki Bokushi 鈴木牧之 (1770–1842)
Suzuki Harunobu 鈴木春信 (1725–70)
Suzuki Sajiemon 鈴木左治衛門
Suzuki Shinkai 鈴木真海
ta no miya 田宮
tabako 煙草
Tachibana Kanzaemon 立花勘左衛門
Tachibana Nankei 橘南谿 (1753–1809)
Tai Yiyuan 太醫院
Taika 大化
Taionki 戴恩記
Taisei honzô meiso 泰西本草名疏
Takahashi Kageyasu 高橋景保 (1785–1829)
Takamatsu 高松藩
Takami Musubi 高御産巣
Takaoka 高岡
Takebe Katahiro 建部賢弘 (1664–1739)
Taki Genkō 多紀元孝 (1695–1766)
Tamaru Naonobu 田丸直暢 (fl. 1835–38)
tama shizume 霊鎮
Tamashiken 玉枝軒
Tamura Gen'yū 田村元雄
Tamura Ransui 田村藍水 (1717–76)
Tanaka Chōzaemon 田中長左衛門
Tanaka Shōzō 田中正造 (1841–1913)
Tanaka Yoshio 田中芳男 (1838–1916)
Tanba no Yasuyori 丹波康頼 (912–95)
tanben 單瓣
tanchōzuru 丹頂鶴
Tang Shenwei 唐慎微
Tani Bunchō 谷文晁 (1763–1841)
tansan 炭酸
tanso 炭素
tanuki 狸
Tanuma Okitsugu 田沼意次 (1719–88)
Tao Hongjing 陶弘景 (456–536)

taryū no sho 他流之書
tashiki 多識
Tashikihen 多識編
Tashiro Sanki 田代三喜 (1472–1544)
Tashiro Yasusada 田代安定 (1856–1928)
teien 庭園
Teikoku Daigaku 帝国大学
Ten'yakuryō 典薬寮
tenchi 天地
tenchi seibutsu 天地生物
tendai uyaku 天台烏薬
Tendai 天台
tengu 天狗
tenka 天下
Tenka gomen 天下御免
Tenka no i wo i suru 天下ノ医ヲ医スル
Tenkai 天海 (1536–1643)
Tenman 天満
Tenmei 天明
tenmongata 天文方
Tenpō 天保
tenri shizen no kotowari 天理自然之理
tenryō 天領
Ten'yakuryō 典薬寮
Terajima Ryōan 寺島良安 (fl. early seventeenth century)
Terakado Seiken 寺門静軒 (1796–1868)
terakoya 寺子屋
terumomeitoru テルモメイトル
Tesshoki 徹書記 (1381–1459)
tiandi 天地 (tenchi)
Tiangong kaiwu 天工開物
tianxia 天下 (tenka)
Tō Jakusui 稲若水 (see Inō Jakusui)
Tochigi 栃木
Toda Gyokuzan 戸田旭山 (1696–1769)
Tōdō Takayuki 藤堂高猷 (1815–95)
Tōga 東雅
Tōhoku 東北
tōhon 唐本

Tōkaidō 東海道
tōkakai 闘花会
Tōken 東軒 (1652–1713)
toki 鴇
Tokugawa Harutomi 徳川治寶 (1771–1853)
Tokugawa Hidetada 徳川秀忠 (1579–1632)
Tokugawa Ieharu 徳川家治 (1737–86)
Tokugawa Iemitsu 徳川家光 (1604–51)
Tokugawa Ienobu 徳川家宣 (1662–1712)
Tokugawa Ieshige 徳川家重 (1711–61)
Tokugawa Ietsugu 徳川家継 (1709–19)
Tokugawa Ietsuna 徳川家綱 (1641–80)
Tokugawa Ieyasu 徳川家康 (1542–1616)
Tokugawa jikki 徳川実記
Tokugawa Mitsusada 徳川光貞 (1627–1705)
Tokugawa Tsunayoshi 徳川綱吉 (1646–1709)
Tokugawa Yoshimune 徳川吉宗 (1684–1751)
Tokushima 徳島
Tōkyō Kokuritsu Hakubutsukan 東京国立博物館
Tōkyō Nōrin Gakkō 東京農林学校
Tōkyō Shinbun 東京新聞
Tōkyō Teikoku Daigaku 東京帝国大学
Toneri zusetsu 秦皮図説
tonkin nikukei 東京肉桂
Tono-sama hakubutsugaku 殿様博物学
tora 虎
torii 鳥居
Torimi Genryū 鳥海玄柳
toritsugidokoro 取次所
toriya 鳥屋
tororoaoi 秋葵
Tosa 土佐
Tōshōdaigongen 東照大権現

Tōto Yakuhinkai 東都薬品会
Tōtōmi 遠江
Tōtosaijiki 東都歳時記
Tottori 鳥取
Toyama 富山
Toyotomi Hideyoshi 豊臣秀吉 (1537–98)
tsubaki 椿
tsubo 坪
Tsuboiya Kichiemon 坪（壺）井屋吉右衛門
Tsuka to mori no hanashi 塚と森の話
tsunomedori 角目鳥
Tsurezuregusa 徒然草
Tsushima 対馬
tsutsuji 躑躅
Tsuyama 津山
Tsu 津
u 羽
uchi kowashi 打毀
uchū 宇宙
Udagawa Genshin 宇田川玄真 (1769–1834)
Udagawa Genzui 宇田川玄随 (1755–97)
Udagawa Yōan 宇田川榕庵 (1798–1846)
Ueda Akinari 上田秋成 (1734–1809)
uekiya 植木屋
Uemura Masakatsu (Saheiji) 植村政勝 (左平次, 1695–1777)
Uemura Saheiji 植村佐平次 (1695–1777)
Uemura Zenroku 植村善六
Ueno Masuzō (1900–1989)
ukiyoe 浮世絵
uma 馬
umagoyashi 首宿
uotchingu ウオッチング
Uragami Gyokudō 浦上玉堂 (1745–1820)
urushi 漆

ususakura 薄桜

Utagawa Toyokuni 歌川豊国 (1769–1825)

utsushi 写し

utsuwa 器

uyaku 烏薬

uzusakura 雲珠櫻

wahin 和品

Wajiga 和爾雅

waka 和歌

wakadoshiyori 若年寄

Wakan sansai zue 和漢三才圖会

(Wakan) shoseki mokuroku 和漢書籍目録

Wakayama 和歌山

Wamyō ruijushō 和名類聚抄

Wang Lu 王綸 (1488–1505)

Wang Qi 王圻 (1529–1612)

Wang Shizhen 王士禎 (1634–1711)

Wang Yangming 王陽明 (1472–1528)

Wang Zheng 王禎 (1271–1333)

wanwu zhi li 萬物之理 (*banbutsu no ri*)

Waseda 早稲田

Watanabe Tōgen 渡辺桃源 (1715–94)

Watsuji Tetsurō 和辻哲郎 (1889–1960)

Wayaku Aratame Kaisho 和薬改会所

wuming 無名 (*mumei*)

wuxing 五行 (*gogyō*)

xiang 象 (*zō*)

xianmu 小目 (*shōmoku*)

xiaogang 小綱 (*shōkō*)

xiesheng 写生 (*shasei*)

xin 心 (*shin*)

xing 性 (*sei*)

xingzhi 形質 (*keishitsu*)

Xinxiu bencao 新修本草

xiuzhi 修治 (*shūji*)

Xu Guangqi 徐光啓 (1562–1633)

Xunzi 荀子 (ca. 312–230 BCE)

Yaeyama kōmori 八重山蝙蝠

Yahazu-no-uji-Matachi 箭括氏麻多智

Yakko dako 奴師労之

yaku don'ya 薬問屋

yakubutsu 薬物

Yakuchō 薬徴

yakuen 薬園

yakuendai 薬園台

yakuhin 薬品

yakuhin-kai 薬品会

yakurui 薬類

Yakusei nōdoku 薬性能毒

yakusō 薬草

yakusō miwake 薬草見分

yama no miya 山宮

yamaarashi 豪猪

Yamaga Sokō 山鹿素行 (1622–85)

Yamaguchi 山口

Yamamoto Bōyō 山本亡羊 (1778–1859)

Yamamoto Keigu 山本渓愚 (1827–1903)

Yamamoto Shinzaburō 山本沈三郎 (1809–64)

Yamanashi 山梨

Yamaoka Shuzen 山岡守全

Yamashita Sōtaku 山下宗琢

Yamato 大和

Yamato honzō 大和本草

Yamawaki Tōyō 山脇東洋 (1705–63)

Yamazaki Ansai 山崎闇斎 (1618–82)

Yamazaki Kinbei 山崎金兵衛

yamazakura 山桜

Yanagisawa Kien 柳沢淇園 (1703–58)

Yanagita Kunio 柳田国男 (1875–1962)

yang 陽 (*yō*)

yangming 養命

yangxing 養性

Yao Hongjing 陶弘景 (456–536)

Yashima Gakutei 八島岳亭 (ca. 1786–1868)

Yatabe Ryōkichi 矢田部良吉 (1851–99)

Yatsu no kami 夜刀神

yibiao 遺表 (ihyō)
Yijing 易経
yin 陰 (in)
ying 櫻 (sakura)
yo no naka no yorozumono 世中の万物
yō 陽
Yodo 淀
yōga 洋画
yōgaku 洋学
Yōjōkun 養生訓
yomogi 蓬
Yoroichō 鎧蝶
Yosa Buson 与謝蕪村 (1716–83)
yose 寄席
Yoshida Haruyuki 吉田玄之 (1571–1632)
Yoshida Kyūichi (Seikyō) 吉田九市 (正恭, fl. 1800–1820)
Yoshida Takanori 吉田高憲 (1805–59)
Yoshimasu Tōdō 吉益東洞 (1702–73)
Yoshino 吉野
Yoshino yakuen 吉野薬園
Yoshio Gonnosuke 吉雄権之助 (1785–1831)
Yoshio Kōgyū 吉雄耕牛 (1724–1800)
Yoshio Nankō 吉雄南皋 (1787–1843)
Yoshio Shunzō 吉雄俊蔵 (1787–1843)
Yoshio Tōzaburō 吉雄藤三郎
youming weiyong 有名未用 (yūmei miyō)
Yūbun koji 右文故事
yūgaku 遊学
yūmei miyō 有名未用

yunibasaru jeniasu ユニバサル・ジェ ニアス
yushi 玉石 (gyokuseki)
Yushima 湯島
yuzhou 宇宙 (uchū)
Yūzūfu 融通府
Yuzuke Gansui 湯漬翫水
Zaikyō nikki 在京日記
zankuro 若榴
zaohua 像化 (zōke)
Zen 禅
Zhang Hua 張華 (232–300)
Zhenglei bencao 証類本草
zhengming 正名 (seimei)
zhengwu 正誤 (seigo)
Zhiwuxue 植物学
zhong 種 (shu)
zhongshushu 種樹書 (shujusho)
Zhu Xi 朱子 (1130–1200)
Zhu Zhenxiang 朱震享 (1281–1358)
zhuzhi 主治 (shuji)
zidian 字典
ziran 自然
zō 象
Zō no emakimono 象之絵巻物
Zō no mitsugi 象のみつぎ (1729)
(Zōho) Shoseki mokuroku 増補書籍目録
zōjō 造醸
zōke 造化
zoku 属
Zōshi 象志 (1729)
zuoshi 佐使 (sashi)
Zusho Hirosato 調所広郷 (1776–1849)

Notes

PROLOGUE

1. In today's prefectures of Ishikawa and Toyama, along the western shores of Honshū. The book follows the convention of writing Chinese and Japanese proper names: last name first, given name second. So, for example, Maeda is the surname and Toshitsune is the first name. Also, it follows the convention of referring to premodern Japanese renown personalities by their given name: so, for example, when referring to scholar Kaibara Ekiken, I use Ekiken (first name) rather than Kaibara (surname).

2. Natsume Sōseki, *Sanshiro*, trans. Jay Rubin (London: Penguin, 2009).

3. Max Horkheimer and Theodor W. Adorno, *Dialectic of Enlightenment*, trans. John Cumming (New York: Continuum, 1989), 3.

CHAPTER I

1. Akimoto Kichirō, "Hitachi no kuni fudoki," in *Fudoki, Nihon koten bungaku taikei*, vol. 2, ed. Akimoto Kichirō (Tokyo: Iwanami Shoten, 1976), 54–57. For a different, more literal English translation, see Mark C. Funke, "*Hitachi no Kuni Fudoki*," *Monumenta Nipponica* 49, no. 1 (Spring 1994): 16–17. From now on, unless specified otherwise, all translations from Japanese, Chinese, Latin, ancient Greek, German, French, and Italian are mine.

2. For an introductory essay, see Komatsu Kazuhiko, *Yōkaigaku shinkō: Yōkai kara miru Nihonjin no kokoro* (Tokyo: Shōgakukan, 1994).

3. See Komine Kazuaki, *Chūsei setsuwa no sekai wo yomu* (Tokyo: Iwanami Shoten, 1998).

4. See in particular Massimo Raveri, *Itinerari nel sacro: L'esperienza religiosa giapponese* (Venezia: Cafoscarina, 1984), 11–68; Yoneyama Toshinao, *Shōbonchi uchū to Nihon bunka* (Tokyo: Iwanami Shoten, 1989); and Hori Ichiro, *Folk Religion in Japan: Continuity and Change* (Chicago: University of Chicago Press, 1968).

5. The literature is vast. See in particular Nakao Sasuke, *Saibai shokubutsu to nōkō no kigen* (Tokyo: Iwanami Shoten, 1966); Sonoda Minoru, "Shinto and the Natural Environment," in *Shinto in History: Ways of the Kami*, ed. John Breen and Mark Teeuwen (London: Curzon, 2000), 32–46; Okatani Kōji, *Kami no mori, mori no kami* (Tokyo: Tokyo Shoseki, 1989). On the political meanings of Yanagita Kunio's ethnocentrism, see H. D. Harootunian, *Overcome by Modernity: History, Culture, and Community in Interwar Japan* (Princeton, NJ: Princeton University Press, 2000).

6. Raveri, *Itinerari nel sacro*; Yoneyama, *Shōbonchi uchū to Nihon bunka*.

7. The bibliography on Japanese environmental history is growing at a very rapid pace. See Brett Walker, Ian Miller, and Julia Adeney Thomas, eds., *Japan at Nature's Edge: The Environmental Origins of a Global Power* (Honolulu: University of Hawaii Press, 2013); Robert Stolz, *Bad Water: Nature, Pollution, and Politics in Japan, 1870–1950* (Durham: Duke University Press, 2014); Brett Walker, *Toxic Archipelago: A History of Industrial Disease in Japan* (Seattle: University of Washington Press, 2010); Conrad Totman's *A History of Japan* (Malden, MA: Blackwell, 2000) has long chapters on environmental history; William Kelly, *Water Control in Tokugawa Japan: Irrigation Organization in a Japanese River Basin, 1600–1870* (Ithaca, NY: Cornell University China-Japan Program, 1982); see also the two volumes of Kikuchi Kazuo, *Nihon no rekishi saigai* (Tokyo: Kokon Shoin, 1980–86).

8. Conrad Totman, *The Green Archipelago: Forestry in Pre-industrial Japan* (Berkeley: University of California Press, 1989).

9. As it will become clear, I do not conceive of space as neutral and empty, a preexisting Cartesian grid over which human events and natural phenomena unfold, but in the relational sense of being the product of a continuous process of social metabolism between human communities and natural environment. See Henri Lefebvre, *The Production of Space* (Oxford: Basil Blackwell, 1991); David Harvey, *Justice, Nature and the Geography of Difference* (Oxford: Basil Blackwell, 1996).

10. This description echoes Max Weber's notion of *Entzauberung*, or "disenchantment," which he adopted from Friedrich Schiller to describe the decline of the network of magical and symbolic correspondences characteristic of the premodern conceptions of the world as a result of the instrumental rationalization and bureaucratization of modern society. See Max Weber, "Science as a Vocation," in *The Vocation Lectures: "Science as a Vocation," "Politics as a Vocation,"* ed. D. S. Owen and T. B. Strong (Indianapolis: Hackett, 2004), 1–32.

11. This book does not offer a comprehensive survey of the variety and richness of *honzōgaku* practices, spanning from pharmacology to natural history and including culinary, landscaping, botany, agronomy, fishery, forestry, and so on. It offers but a cursory introduction of those scholars—like, for example, Ono Ranzan—who enjoyed great prominence in their own time but whose work is of marginal interest to my argument. For an overview on *honzōgaku*, see Yabe Ichirō, *Edo no honzō* (Tokyo: Saiensu Sha, 1984); Ueno Masuzō, *Nihon hakubutsugaku shi* (Tokyo: Kōdansha Gakujutsu Bunko, 1989); Kimura Yōjirō, *Nihon shizenshi no seiritsu: Rangaku to honzōgaku* (Tokyo: Chūōkōronsha, 1974); Sugimoto Tsutomu, *Edo no hakubutsugakushatachi* (Tokyo: Seidosha, 1985); Nishimura Saburō, *Bunmei no naka no hakubutsugaku: Seiō to Nihon*, 2 vols. (Tokyo: Kinokuniya Shoten, 1999); Sugimoto Tsutomu, *Nihon honzōgaku no sekai* (Tokyo: Yasaka Shobō, 2011).

12. Quoted in Ueno, *Nihon hakubutsugaku shi*, 66.

13. See Carla Nappi, *The Monkey and the Inkpot: Natural History and Its Transformations in Early Modern China* (Cambridge, MA: Harvard University Press, 2009).

14. See chapter 2.

15. Michel Foucault, *The Order of Things: An Archaeology of the Human Sciences* (New York: Vintage, 1994), 132–38.

16. The activities of Udagawa Yōan and of the naturalists of the Shōhyaku-sha circle in the nineteenth century are partial exceptions. See chapters 10 and 12. See also Maki Fukuoka, *The Premise of Fidelity: Science, Visuality, and Representing the Real in Nineteenth-Century Japan* (Stanford: Stanford University Press, 2012).

17. The expression comes from Karl Marx, *Capital: A Critique of Political Economy*, vol. 1, trans. Ben Fowkes (London: Penguin Classics, 1990), 167.

18. Modified from ibid., 165.

19. See Alexander Bird and Emma Tobin, "Natural Kind," in *Stanford Encyclopedia of Philosophy* (2008), accessed October 14, 2010, http://plato.stanford.edu/entries/natural-kinds; Marc Ereshefsky, "Species," in *Stanford Encyclopedia of Philosophy* (2010), accessed October 14, 2010, http://plato.stanford.edu/entries/species. See also John Dupré, "Natural Kinds and Biological Taxa," *Philosophical Review* 90, no. 1 (1981): 66–90; John Dupré, "In Defense of Classification," *Studies in History and Philosophy of Biological and Biomedical Sciences* 32, no. 2 (2001): 203–19; John Wilkins, *Species: The History of the Idea* (Berkeley: University of California Press, 2009).

20. See John Dupré, *The Disorder of Things: Metaphysical Foundations of the Disunity of Science* (Cambridge, MA: Harvard University Press, 1993); Wilkins, *Species*. See also Nishimura Saburō, *Rinne to sono shitotachi* (Tokyo: Asahi Sensho, 1997).

21. Hilary Putnam, "The Meaning of 'Meaning,'" in *Philosophical Papers: Mind, Language, and Reality*, vol. 2 (Cambridge: Cambridge University Press, 1975), 215–71.

22. Antonio Gramsci, *The Prison Notebooks: Selections* (New York: International Publishers, 1971), 9.

23. Jacques Le Goff pushed the beginning of the professionalization of scholars to the twelfth century. See Le Goff, *Les intellectuels au Moyen Age* (Paris: Editions du Seuil, 1957).

24. Bruno Latour developed a network of relations involving human actors and the institution they live in, the discursive apparatus they shared, and the objects they directly and indirectly engage with, all three factors having similar standing and, as "actants," being entwined with each other. See Bruno Latour, *Reassembling the Social: An Introduction to Actor-Network Theory* (Oxford: Oxford University Press, 2005).

25. Theodor W. Adorno, "The Idea of Natural-History," in *Things Beyond Resemblance: Collected Essays on Theodor W. Adorno*, ed. Robert Hullot-Kentor (New York: Columbia University Press, 2006), 260.

26. Deborah Cook, *Adorno on Nature* (Durham, UK: Acumen, 2011), 17.

27. Ibid., 11.

28. Two excellent studies on this subject are Harold J. Cook, *Matters of Exchange: Commerce, Medicine, and Science in the Dutch Golden Age* (New Haven: Yale University Press, 2007); Alix Cooper, *Inventing the Indigenous: Local Knowledge and Natural History in Early Modern Europe* (Cambridge: Cambridge University Press, 2007).

29. P. J. Crutzen and E. F. Stoermer, "The 'Anthropocene,'" *Global Change Newsletter* 41 (2000): 17–18.

30. Dipesh Chakrabarty, "The Climate of History: Four Theses," *Critical Inquiry* 35 (2009): 197–222.

31. Hilary Putnam, *Reason, Truth and History* (Cambridge: Cambridge University Press, 1981), xi.

32. This is a simplified treatment of a very contentious arena of philosophical debate that sees universalists opposed to relativists and realists opposed to antirealists. Introductory surveys of the problem abound: see, for example, Stuart Brock and Edwin Mares, *Realism and Anti-realism* (Montreal: McGill-Queen's University Press, 2007); William P. Alston, ed., *Realism and Antirealism* (Ithaca, NY: Cornell University Press, 2002); Patrick Greenough and Michael P. Lynch, eds., *Truth and Realism* (Oxford: Oxford University Press, 2006); Alfred I. Tauber, ed., *Science and the Quest for Reality* (Houndmills, UK: Macmillan, 1997). Some philosophers and historians of science have tried to mediate between the two contending positions. Ian Hacking, for example, has introduced the notion of "style of reasoning" to situate different epistemological attitudes in specific disciplinary contexts; see Hacking's "'Style' for Historians and Philosophers" and "Language, Truth, and Reason" in Ian Hacking, *Historical Ontology* (Cambridge: Cambridge University Press, 2002). See David Bloor, *Knowledge and Social Imagery* (Chicago: University of Chicago Press, 1991); Bruno Latour, *Science in Action: How to Follow Scientists and Engineers through Society* (Cambridge, MA: Harvard University Press, 1987); Bruno Latour, *We Have Never Been Modern* (Cambridge, MA: Harvard University Press, 1993); Richard Rorty, *Philosophy and the Mirror of Nature* (Princeton, NJ: Princeton University Press, 1980); Richard Rorty, *Objectivity, Relativism, and Truth: Philosophical Papers* (Cambridge: Cambridge University Press, 1981); Wolfgang Stegmüller, *The Structure and Dynamics of Theories* (New York: Springer-Verlag, 1976). See also the classic by Richard J. Bernstein, *Beyond Objectivism and Relativism: Science, Hermeneutics, and Praxis* (Philadelphia: University of Pennsylvania Press, 1983); Allan Megill, ed., *Rethinking Objectivity* (Durham: Duke University Press, 1994).

33. Raymond Williams, *Keywords* (Oxford: Oxford University Press, 1984), 219. Every time "nature" is written between quotation marks, it means the conception or idea

of nature, the focus being on the word qua signifier itself rather than on what it signified, what it refers to.

34. Arthur O. Lovejoy, "Some Meanings of Nature," in *Primitivism and Related Ideas in Antiquity*, by A. O. Lovejoy and George Boss (Baltimore: Johns Hopkins University Press, 1997), 447–56.

35. For a historical survey of the "idea" of nature, see Alfred North Whitehead, *The Concept of Nature* (Cambridge: Cambridge University Press, 1930); Adorno, "The Idea of Natural-History," 252–69; R. G. Collingwood, *The Idea of Nature* (Oxford: Oxford University Press, 1960); Maurice Merleau-Ponty's lectures on *Nature: Course Notes from the College de France*, trans. Robert Vallier (Evanston, IL: Northwestern University Press, 2003); Hans Blumenberg, *Die Lesbarkeit der Welt* (Frankfurt: Suhrkamp, 1983); Kate Soper, *What Is Nature?* (Oxford: Blackwell, 1995); Jean Ehrard, *L'Idée de nature en France dans la première moitié du XVIIIe siècle* (Paris: Albin Michel, 1994); Mario Alcaro, *Filosofie della natura* (Roma: Manifestolibri, 2006); Gianfranco Basti, *Filosofia della natura e della scienza* (Roma: Lateran University Press, 2002); Gianfranco Marrone, ed., *Semiotica della natura (Natura della semiotica)* (Milano: Mimesis, 2012); the magnificent Philippe Descola, *Beyond Nature and Culture*, trans. Janet Lloyd (Chicago: University of Chicago Press, 2013).

36. Yanabu Akira, *Hon'yaku no shisō: Shizen to neichā* (Tokyo: Heibonsha, 1977), 3–30. See also Yoshida Tadashi, "Shizen to kagaku," in *Kōza Nihon shisō*, vol. 1, *Shizen*, ed. Sagara Tōru, Bitō Masahide, and Akiyama Ken (Tokyo: Tōkyō Daigaku Shuppankai, 1983), 342.

37. See Terao Gorō, *"Shizen" gainen no keisei shi: Chūgoku, Nihon, Yōroppa* (Tokyo: Nōsangyoson Bunka Kyōkai, 2002) for a philological overview.

38. Graham Harman, *Guerrilla Metaphysics* (Peru, IL: Open Court, 2005), 251. See also Graham Harman, *The Quadruple Object* (Winchester, UK: Zero Books, 2011).

39. *"Quid est ergo tempus? Si nemo ex me quaerat, scio; si quaerenti explicare velim, nescio."* Augustine, *Confessionum libri*, XI-14-17.

40. See, for example, the photographs in the Wikipedia entry for "nature": http://en.wikipedia.org/wiki/Nature, last accessed October 2, 2012.

41. Sigmund Freud, *The Uncanny* (London: Penguin, 2003), 121.

42. Φύσις κρύπτεσθαι φιλεῖ—*Phýsis krýptesthai philei*—is Fragmentum B 123 in *Fragmente der Vorsokratiker*, ed. Hermann Diels (Berlin: Weidmann, 1903).

43. Galileo Galilei, *Il Saggiatore [The Assayer]*, quoted in Stillman Drake, ed., *The Controversy of the Comet of 1618* (Philadelphia: University of Pennsylvania Press, 1960), 183–84.

44. Alfred Tennyson, *In Memoriam A. H. H.*, Canto 56 (1850): "Who trusted God was love indeed / And love Creation's final law / Tho' Nature, red in tooth and claw / With ravine, shriek'd against his creed."

45. As in James Lovelock's Gaia hypothesis, first mentioned in James Lovelock and C. E. Giffin, "Planetary Atmospheres: Compositional and Other Changes Associated with the Presence of Life," *Advances in the Astronautical Sciences* 25 (1969): 179–93.

46. As in Baruch Spinoza's *deus sive natura*. See Hasana Sharp, *Spinoza and the Politics of Renaturalization* (Chicago: University of Chicago Press, 2011).

47. See Roger French, *Ancient Natural History* (London: Routledge, 1994); Geoffrey E. R. Lloyd, "Greek Antiquity: The Invention of Nature," in *The Concept of Nature*, ed. John Torrance (Oxford: Clarendon Press, 1992), 1–24; Charles E. Scott, *The Lives of Things* (Bloomington: Indiana University Press, 2002), 3–81; Giovanni Reale, *Storia della filosofia greca e romana*, vols. 1–4 (Milano: Bompiani, 2004).

48. For a critical analysis, see Bruno Latour, *Politics of Nature: How to Bring the Sciences into Democracy* (Cambridge, MA: Harvard University Press, 2004); Timothy Morton, *The Ecological Thought* (Cambridge, MA: Harvard University Press, 2010); and Steven Vogel, *Against Nature: The Concept of Nature in Critical Theory* (Albany: State University of New York Press, 1996).

49. Jean Baudrillard, *Simulacra and Simulation* (Ann Arbor: University of Michigan Press, 1994), 4–18; see also Jean Baudrillard, *The Perfect Crime* (London: Verso, 1996), 119.

50. See, for example, the "deep ecology" manifesto, Bill Devall and George Sessions, *Deep Ecology: Living as if Nature Mattered* (Salt Lake City, UT: Gibbs Smith, 1985). A harsh critique that is, however, completely enmeshed with it is Murray Bookchin, *The Philosophy of Social Ecology: Essays on Dialectical Naturalism* (Montreal: Black Rose Books, 1990).

51. Latour, *Politics of Nature*.

52. Theodor Adorno, "Musikpädagogische Musik: Brief an Ernst Krenek," in *Theodor W. Adorno und Ernst Krenek: Briefwechsel*, ed. Wolfgang Rogge (Frankfurt am Main: Suhrkamp Verlag, 1974), 219. English translation by Susan Buck-Morss, *The Origin of Negative Dialectic: Theodor W. Adorno, Walter Benjamin, and the Frankfurt Institute* (New York: Free Press, 1977), 228.

53. See Timothy Morton, *Ecology without Nature: Rethinking Environmental Aesthetics* (Cambridge, MA: Harvard University Press, 2007); Peter C. van Wyck, *Primitives in the Wilderness: Deep Ecology and the Missing Human Subject* (Albany: State University of New York Press, 1997); John Bellamy Foster, *Marx's Ecology: Materialism and Nature* (New York: Monthly Review Press, 2000); Bill McKibben, *The End of Nature* (New York: Random House, 2006); Robert Stolz, "Nature over Nation: Tanaka Shōzō's Fundamental River Law," *Japan Forum* 18, no. 3 (November 2006): 417–37.

54. Clearly, this is a paraphrase of Maximilien Robespierre's celebrated passage in the speech of December 3, 1792: "*Louis doit mourir parce qu'il faut que la patrie vive.*" My thanks go to Robert Stolz, the author of the paraphrase.

55. This is the case with only two exceptions: The first is the *Haruma wage*, a Dutch-Japanese dictionary compiled in 1796 by Inamura Sanpaku (1759–1811) with the collaboration of Udagawa Genzui (1756–98) and Okada Hosetsu and on the basis of the second edition of the Dutch-French dictionary by François Halma (1653–1722), *Nieuw Woordenboek der Nederduitsche en Freansche Taalen. Dictionnaire Nouveau Flamand & François*, printed in Amsterdam in 1722. The three translators were all disciples of the *rangaku* scholar and physician Ōtsuki Gentaku (1757–1827), who was active participant in *honzōgaku* intellectual circles. The *Haruma wage* is worth mentioning because, contrary to all other dictionaries compiled during the Tokugawa period, it was the only one that translated the Dutch *Natuur* as *shizen*. However, when in 1858 Katsurakawa Hoshū's (1751–1809) new edition of *Haruma wage* was finally printed with the title of *Waran jii*, the entry *Natuur* disappeared. The second exception is the works of rural thinker Andō Shōeki, whose manuscript *Shizen shin'eidō* treated *shizen* as a fundamental concept. Unfortunately, he remained largely unknown until 1899, when his manuscripts were discovered by Kanō Kōkichi (1865–1942). Shōeki's contribution in the philosophical debate of the Tokugawa period was virtually irrelevant. See Toshinobu Yoshinaka, *Andō Shōeki: Social and Ecological Philosopher in Eighteenth-Century Japan* (New York: Weatherhill, 1992); Ishiwata Hiroaki, *Andō Shōeki no sekai: Dokusōteki shisō wa ikani umareta ka* (Tokyo: Sōshisha, 2007).

56. Itō Jinsai, "Gomō jigi," in *Nihon shisō taikei*, vol. 33 (Tokyo: Iwanami Shoten, 1971), 116.

57. Ibid. For a different translation, see John Allen Tucker, *Itō Jinsai's* Gomō jigi *and the Philosophical Definition of Early Modern Japan* (Leiden: Brill, 1998), 73.

58. The following catalog is based upon Terao, *"Shizen" gainen no keiseishi*, 154–69.

59. Especially in the *Nihon shoki* (720) and the *Kaifūsō* (751). Terao, *"Shizen" gainen no keiseishi*, 157–58.

60. On Chinese gazetteers, see Peter K. Bol, "The Rise of Local History: History, Geography, and Culture in Southern Song and Yuan Wuzhou," *Harvard Journal of Asiatic Studies* 61, no. 1 (2001): 37–76; James M. Hargett, "Song Dynasty Local Gazetteers and Their Place in the History of Difangzhi Writing," *Harvard Journal of Asiatic Studies* 56, no. 2 (1996): 405–42. On Japanese *fudoki*, see Michiko Y. Aoki, *Records of Wind and Earth: A Translation of Fudoki, with Introduction and Commentaries* (Ann Arbor, MI: Association for Asian Studies, 1997). Modern philosopher Watsuji Tetsurō reconceptualized *fūdo* in terms of geocultural determinism in *Climate and Culture: A Philosophical Study* (Tokyo: Hakuseido, 1961). On Watsuji, see Harootunian, *Overcome by Modernity*, 202–92.

61. For a critical overview, see Julia Adeney Thomas, *Reconfiguring Modernity: Concepts of Nature in Japanese Political Ideology* (Berkeley: University of California Press, 2001).

62. This is conventional in the majority of Japanese sources.

63. Shirai Mitsutarō, *Nihon hakubutsugaku nenpyō* (Tokyo: Maruzen, 1891).

64. Steven Shapin, *The Scientific Revolution* (Chicago: University of Chicago Press, 1996), 1–14.

65. Simon Schaffer et al., eds., *The Brokered Word: Go-Betweens and Global Intelligence, 1770–1820* (Sagamore Beach, MA: Science History Publications, 2009).

66. Peter Dear, *The Intelligibility of Nature: How Science Makes Sense of the World* (Chicago: University of Chicago Press, 2006), 11.

67. I thus share Paul Boghossian's discontent with what he calls "postmodernist relativism"—that is, the belief that "there are many radically different, yet 'equally valid' ways of knowing the world, with science being just one of them." Paul Boghossian, *Fear of Knowledge: Against Relativism and Constructivism* (Oxford: Oxford University Press, 2006), 2.

68. In fact, the publications of Engelbert Kaempfer, Carl Peter Thunberg, and Franz von Siebold on Japanese flora and fauna in the eighteenth and nineteenth centuries were vehicles through which *honzōgaku* knowledge indirectly contributed to European natural history.

69. Satō Jin, *"Motazaru kuni" no shigenron: Jizoku kanōna kokudo wo meguru mō hitotsu no chi* (Tokyo: Tōkyō Daigaku Shuppankai, 2011).

70. Watsuji, *Climate and Culture*; see also Julia Adeney Thomas, "The Cage of Nature: Modernity's History in Japan," *History and Theory* 40, no. 1 (February 2001): 16–36; Thomas, *Reconfiguring Modernity*; Karatani Kōjin, *Origins of Modern Japanese Literature* (Durham: Duke University Press, 1993), 11–44.

CHAPTER 2

1. In fact, the Japanese archipelago was populated by a series of migratory waves no earlier than 35,000 years ago and continuing through the fourth century CE. Mark J. Hudson, *Ruins of Identity: Ethnogenesis in the Japanese Islands* (Honolulu: University of Hawaii Press, 1999).

2. This chapter offers a brief survey of *honzōgaku* in Japan before 1600. For this reason, I rely a great deal on research on Chinese *bencao*. Among them, Joseph Needham and Lu Gwei-djen, *Science and Civilization in China*, vol. 6, bk. 1, *Botany* (Cambridge: Cambridge University Press, 1986); Peter Unshuld, *Medicine in China: A History of Pharmaceutics* (Berkeley: University of California Press, 1986); Peter Unshuld, *Medicine in China: A History of Ideas* (1985; repr. Berkeley: University of California Press, 2010); Benjamin Elman, *On Their Own Terms: Science in China, 1550–1900* (Cambridge, MA: Harvard University Press, 2005); Benjamin Elman, *A Cultural History of Modern Science in China* (Cambridge, MA: Harvard University Press, 2006); Linda L. Barnes, *Needles, Herbs, Gods, and Ghosts: China, Healing, and the West to 1848* (Cambridge, MA: Harvard University Press, 2005); Vivienne Lo and Christopher Cullen, eds., *Medieval Chinese Medicine: The Dunhuang Medi-*

cal Manuscripts (London: Routledge Curzon, 2005); Elizabeth Hsu, ed., *Innovation in Chinese Medicine* (Cambridge: Cambridge University Press, 2001); Nishimura, *Bunmei no naka no hakubutsugaku*, vol. 1; Ishida Hidemi, *Chūgoku igaku shisōshi* (Tokyo: Tōkyō Daigaku Shuppankai, 1992); Okanishi Tameto, *Honzō gaisetsu* (Osaka: Sōgensha, 1977); Yamada Keiji, *Chūgoku kodai kagaku shiron* (Kyoto: Kyoto Daigaku Jinbunkagaku Kenkyūjo, 1989); Yamada Keiji, ed., *Higashi ajia no honzō to hakubutsugaku no sekai* (Tokyo: Shibunkaku Shuppan, 1995); Yamada, *Honzō to yume to tōkinjutsu*; Pan Jixing, "Tan 'Zhiwuxue' yi ci zai Zhingguo he Riben de youlai," *Daiziran tansuo* 3 (1984): 167–72; Zhongyi dacidian bianji wei-yuanhui, ed., *Zhongyi dacidian: yishi wenxian fence* (Beijing: Renmin Weisheng Chubanshe, 1981). On Li Shizhen, see Nappi, *The Monkey and the Inkpot*; Georges Métailié, "Des mots et des plantes dans le *Bencao gangmu* de Li Shizhen," *Extrême-Orient Extrême-Occident* 10 (1988): 27–43; Georges Métailié, "The *Bencao gangmu* of Li Shizhen: An Innovation in Natural History?," in *Innovation in Chinese Medicine*, ed. Elizabeth Hsu (Cambridge: Cambridge University Press, 2001), 221–61; Qian Yuanming, ed., *Li Shizhen yanjiu* (Guangzhou: Guangdong Keji Chubanshe, 1984); Lu Gwei-djen, "China's Greatest Naturalist: A Brief Biography of Li Shih-Chen," *Physis* 8, no. 4 (1966): 383–92.

3. Marius B. Jansen, *China in the Tokugawa World* (Cambridge, MA: Harvard University Press, 1992).

4. Unshuld, *Medicine in China*, 5, 14.

5. Literally, *bencaoxue* means the "study" (*xue*, *gaku* in Japanese) of the "fundamental" (*ben*, *hon*) "herbs" (*cao*, *sō*). See Yabe, *Edo no honzō*, 6.

6. Translation modified from Unshuld, *Medicine in China*, 113. See a different translation in Needham and Lu, *Botany*, 237.

7. Needham and Lu, *Botany*, 244.

8. Masayoshi Sugimoto and David Swain, *Science & Culture in Traditional Japan* (Rutland: Charles E. Tuttle, 1989), 88.

9. Ibid., 85.

10. For the early history of *honzōgaku* in Japan, see Yabe, *Edo no honzō*, 15–42; Ueno, *Nihon hakubutsugaku shi*, 24–63; Nishimura, *Bunmei no naka no hakubutsugaku*, vol. 1, 223–30.

11. Today the text is lost, but according to the eighth-century *Nihon shoki*, it was introduced twice from China and the Korean peninsula.

12. Needham and Lu, *Botany*, 242.

13. Nishimura, *Bunmei no naka no hakubutsugaku*, vol. 1, 194–98.

14. Métailié, "The *Bencao gangmu* of Li Shizhen," 224–25.

15. Needham and Lu, *Botany*, 266.

16. Fukane no Sukehito, *Honzō wamyō*, 2 vols. (Edo: Izumiya Shōjirō, 1796). Reprinted in Fukane no Sukehito, *Honzō wamyō*, ed. Yosano Hiroshi et al., 2 vols. (Tokyo:

Nihon Koten Zenshū Kankōkai, 1926). See also Mayanagi Makoto, "*Honzō wamyō* inyō shomei sakuin*," *Nippon ishigaku zasshi* 33, no. 3 (1986): 381–95.

17. Yabe, *Edo no honzō*, 15–24.

18. Koremune Tomotoshi, *Honzō iroha shō* (Tokyo: Naikaku Bunko, 1968). See also Needham and Lu, *Botany*, 282–83.

19. See Yabe, *Edo no honzō*, 28–31 for a complete list.

20. For Chinese medicine, see *Medicine in China: A History of Ideas* and Joseph Needham and Lu Gwei-djen, *Science and Civilization in China*, vol. 6, bk. 6, *Medicine* (Cambridge: Cambridge University Press, 2000). For Japanese medicine, see Sugimoto and Swain, *Science & Culture in Traditional Japan*; Hattori Toshirō, *Edo jidai igakushi no kenkyū* (Tokyo: Yoshikawa Kōbunkan, 1978).

21. Sugimoto and Swain, *Science & Culture in Traditional Japan*, 215.

22. See chapter 3.

23. Needham and Lu, *Botany*, 308. For a lively biographical portrait of Li Shizhen, see Nappi, *The Monkey and the Inkpot*, 12–49; see also Chen Xinqian, ed., *Li Shizhen yanjiu lunwen ji* (Wuhan: Hubei Kexue Jishu Chubanshe, 1985).

24. See Liu Hongyao, ed., *Lidai mingren yu Wudang, Wudang Zazhi Zengkan*, 1994. See also Nappi, *The Monkey and the Inkpot*, 136–49.

25. As Elman reminds us, Li failed the provincial examinations "like 95 percent of the candidates empirewide." Elman, *On Their Own Terms*, 30. On civil examinations, see Benjamin A. Elman, "Political, Social, and Cultural Reproduction via Civil Service Examinations in Late Imperial China," *Journal of Asian Studies* 50, no. 1 (February 1991): 7–28; Benjamin A. Elman, *A Cultural History of Civil Examinations in Late Imperial China* (Berkeley: University of California Press, 2000).

26. Needham and Lu, *Botany*, 310.

27. Quoted in ibid., 311.

28. Ibid., 316; Elman, *On Their Own Terms*, 32–33. Yabe counts 1,903 entries, in *Edo no honzō*, 47. The different numbers probably depend on which edition Needham and Yabe took into consideration for their analyses. While Needham bases his count on the 1596 first edition, Yabe considers the second edition of 1603, the so-called Jiangxi edition, which had a far wider diffusion thanks to the larger number of printed copies. At present, it is impossible to establish with any certainty whether the copy that Hayashi Razan purchased for Tokugawa Ieyasu in Nagasaki in 1607 was the first or the second edition, even though the majority of Japanese scholars favor the latter option. A third edition, the Hubei, followed in 1606, named as was usual in Chinese custom after the region where it was printed. As Elman notes, "The huge work sold out rapidly enough that eight reprints were issued in the seventeenth century alone." Elman, *On Their Own Terms*, 30–31.

29. Nanjing was at the time called Jinling, from which the *Bencao gangmu* was also known as *Jinling ben*.

30. Translated by Nappi, *The Monkey and the Inkpot*, 20.
31. Needham translates it as "The Great Pharmacopoeia" (Needham and Lu, *Botany*, 312), but a more literal rendition is "Pharmacopoeia Divided into Classes and Orders," or the more concise "Systematic Materia Medica," the translation I adopt here following Elman, *On Their Own Terms*, and Nappi, *The Monkey and the Inkpot*. Georges Métailié prefers "Classified Materia Medica," in Métailié, "The *Bencao gangmu* of Li Shizhen," 221.
32. Needham and Lu, *Botany*, 312.
33. In Nappi, *The Monkey and the Inkpot*, 20.
34. Métailié, "The *Bencao gangmu* of Li Shizhen," 241.
35. Needham and Lu, *Botany*, 312.
36. Li Shizhen, *Bencao gangmu* (Beijing: Renmin Weisheng Chubanshe, 1977–81). I use the modern Japanese edition *Kokuyaku Honzō kōmoku: Shinchū kōtei*, ed. Shirai Mitsutarō and Kimura Kōichi, trans. Suzuki Shinkai, 15 vols. (Tokyo: Shun'yōdō Shoten, 1973). For an English translation, see *Compendium of Materia Medica*, trans. and ann. Luo Xiwen, 6 vols. (Beijing: Foreign Languages Press, 2003).
37. Métailié argues that Li Shizhen might have been inspired by the title of *Tongjian gangmu*, published in 1189 by Zhu Xi. Given Li's explicit reverence for Zhu Xi, it is a persuasive hypothesis. Métailié, "The *Bencao gangmu* of Li Shizhen," 226.
38. Quoted in Métailié, "The *Bencao gangmu* of Li Shizhen," 226.
39. Quoted in Needham and Lu, *Botany*, 315. A more literal translation in Métailié, "The *Bencao gangmu* of Li Shizhen," 227.
40. See Nappi, *The Monkey and the Inkpot*, 50–68 for an analysis of the entry for "dragon" (*long*). See also Métailié, "The *Bencao gangmu* of Li Shizhen," 250–52.
41. Needham and Lu, *Botany*, 282. Known also in its abbreviated form of *Zhenglei Bencao*. The 1640 reprint of the encyclopedia displayed major changes in its illustration. The 1885 edition replaced the original illustrations with new ones based largely on the *Jiuhuang bencao* and Wu Qijun's *Zhiwu mingshi tukao* (1848). See Nappi, *The Monkey and the Inkpot*, 52–53; Xie Zongwan, "*Bencao gangmu* tuban de kaocha," in *Li Shizhen yanjiu lunwen ji* (Wuhan: Hubei Kexue Jishu Chubanshe, 1985), 145–99.
42. Brian W. Ogilvie, *The Science of Describing: Natural History in Renaissance Europe* (Chicago: University of Chicago Press, 2006), 219–21; Scott Atran, *Cognitive Foundations of Natural History: Towards an Anthropology of Science* (Cambridge: Cambridge University Press, 1990); Brent Berlin, *Ethnobiological Classification: Principles of Categorization of Plants and Animals in Traditional Societies* (Princeton, NJ: Princeton University Press, 1992).
43. Ernst Mayr, *The Growth of Biological Thought: Diversity, Evolution and Inheritance* (Cambridge, MA: Belknap, 1982), 147–48.
44. Mayr, *The Growth of Biological Thought*, 148.

45. Ibid.

46. Mayr, it seems to me, suggests that the first type of classification (identification schemes) leads to deductive identifications (aprioristic), while the second produces inductive inferences (scientifically empiricist), but this epistemological division is philosophically weak: to be purely inductive, a taxonomy should be prone to continuous adaptation not only of its structure but also of its inquiring principles on the basis of subsequent empirical discoveries, but this never actually happens, as the survival of Linnaeus's binominal nomenclature (a deductive identification scheme) in today's phylotaxonomies shows. See Mark Ridley, "Principles of Classification," in *Philosophy of Biology*, ed. Michael Ruse (Amherst, NY: Prometheus Books, 1998), 167–79; Ritvo, *The Platypus and the Mermaid*; Martin Mahner and Mario Bunge, *Foundations of Biophilosophy* (Berlin: Springer, 1997) for different overviews.

47. John Dupré, "Are Whales Fish?," in Medin and Atran, *Folkbiology*, 461–76.

48. See Umberto Eco, "Interpreting Animals," in *The Limits of Interpretations* (Bloomington: Indiana University Press, 1994), 111–22.

49. Also called phylogenetic systematics. See Willi Henning, *Phylogenetic Systematics*, trans. D. Dwight Davis and Rainer Zangerl (Urbana, IL: University of Illinois Press, 1999).

50. On the scientific debate, see Quentin Wheeler, *Species Concepts and Phylogenetic Theory: A Debate* (New York: Columbia University Press, 2000); David L. Hull, "The Limits of Cladism," *Systematic Zoology* 28 (1978): 416–40; Mayr, *The Growth of Biological Thought*; Claude Dupuis, "Willi Hennig's Impact on Taxonomic Thought," *Annual Review of Ecology and Systematics* 15 (1984): 1–24; Edward O. Wiley, D. Siegel-Causey, Daniel R. Brooks, and V. A. Funk, *The Compleat Cladist: A Primer of Philogenetic Procedures* (Lawrence: University of Kansas Museum of Natural History Special Publication No. 19, 1991); Ian J. Kitching, Peter L. Forey, Christopher J. Humphries, and David M. Williams, eds., *Cladistics: Theory and Practice of Parsimony Analysis* (Oxford: Oxford University Press, 1998). On the philosophical debate, see David L. Hull, "The Ontological Status of Species as Evolutionary Units," in Ruse, *Philosophy of Biology*, 146–55; David L. Hull and Michael Ruse, eds., *The Philosophy of Biology* (Oxford: Oxford University Press, 1998), 295–347; Mahner and Bunge, *Foundations of Biophilosophy*, 213–70. For a popular introduction, see Carol Kaesuk Yoon, *Naming Nature: The Clash between Instinct and Science* (New York: W. W. Norton, 2009).

51. Jorge Luis Borges, "John Wilkins' Analytical Language," in *Jorge Luis Borges: Selected Non-fictions*, ed. Eliot Weinberger (New York: Penguin Books, 1999), 231. Borges plays with the taxonomy of a "certain Chinese encyclopedia" that divided animals into "(a) belonging to the Emperor, (b) embalmed, (c) tame, (d) sucking pigs, (e) sirens, (f) fabulous, (g) stray dogs, (h) included in the present classification, (i) frenzied, (j) innumerable, (k) drawn with a very fine camelhair brush, (l) *et*

cetera, (m) having just broken the water pitcher, (n) that from a long way off look like flies." The passage is quoted also in Foucault, *The Order of Things*, xv.

52. See Geoffrey C. Bowler and Susan Leigh Star, *Sorting Things Out: Classification and Its Consequences* (Cambridge, MA: MIT Press, 1999).

53. Ian Hacking, "Biopower and the Avalanche of Printed Number," *Humanities in Society* 5 (1982): 280.

54. Foucault, *The Order of Things*, xv–xvi.

55. Ritvo, *The Platypus and the Mermaid*.

56. Marcel Proust, *Swann's Way*, trans. Lydia David (London: Penguin, 2004), 403.

57. I here use "encyclopedia" in metaphorical sense to mean the semantic competency of a community of speakers. Umberto Eco, *Kant and the Platypus: Essays on Language and Cognition* (New York: Harcourt Brace, 2000).

58. Edward Sapir, *Selected Writing of Edward Sapir on Language, Culture, and Personality* (Berkeley: University of California Press, 1949).

59. Martin Heidegger, *Poetry, Language, Thought* (New York: Harper & Row, 1971), 146.

60. *Correlationism* is a term coined by French philosopher Quentin Meillassoux. "By 'correlation,'" he explains, "we mean the idea according to which we only ever have access to the correlation between thinking and being, and never to either term considered apart from the other." Quentin Meillassoux, *After Finitude: An Essay on the Necessity of Contingency*, trans. Ray Brassier (London: Continuum, 2008), 5.

61. Theodor W. Adorno, *Metaphysics: Concepts and Problems*, trans. E. F. N. Jephcott (Stanford: Stanford University Press, 2001), 68.

62. Since "the peculiarity of the concept of ὕλη, or matter, is that we are using a concept ... which, by its meaning, refers to something which is not a concept or principle." Adorno, *Metaphysics*, 67.

63. Cook, *Adorno on Nature*, 11.

64. See Elman, *On Their Own Terms*, 24–60 for a treatment of Neo-Confucian theory of knowledge and materia medica.

65. Quoted in Needham and Lu, *Botany*, 320–21.

66. For Zhu Xi's metaphysics, see Yamada Keiji, *Shushi no shizengaku* (Tokyo: Iwanami Shoten, 1978); Shimada Kenji, *Shushigaku to yōmeigaku* (Tokyo: Iwanami Shinsho, 2000); Yasuda Jirō, *Chūgoku kinsei shisō kenkyū* (Tokyo: Chikuma Shobō, 1976); Fung Yu-Lan, *A History of Chinese Philosophy*, vol. 2 (Princeton, NJ: Princeton University Press, 1953), 533–672. For the Japanese understanding of Zhu Xi's cosmology, see Wajima Yoshio, *Nihon Sōgakushi no kenkyū*, rev. ed. (Tokyo: Yoshikawa Kōbunkan, 1988); Iwasaki Chikatsugu, *Nihon kinsei shisōshi josetsu*, vol. 1 (Tokyo: Shin Nihon Shuppansha, 1997), 116–76. "Heterodox" interpretations of Japanese Zhu Xi Neo-Confucianism can be found in Koyasu Nobukuni, *Edo shisōshi kōgi* (Tokyo: Iwanami Shoten, 1998); Koyasu Nobukuni, *Hōhō toshite no Edo: Nihon*

shisōshi to hihanteki shiza (Tokyo: Perikansha, 2000); Kurozumi Makoto, *Kinsei Nihon shakai to jukyō* (Tokyo: Perikansha, 2003). See also Robert Bellah, *Tokugawa Religion: The Cultural Roots of Modern Japan* (London: Free Press, 1985); Wm Theodore de Bary, *The Unfolding of Neo-Confucianism* (New York: Columbia University Press, 1975); Benjamin Elman, John Duncan, and Herman Ooms, eds., *Rethinking Confucianism: Past and Present in China, Japan, Korea, and Vietnam* (Los Angeles: UCLA, 2002); Maruyama Masao, *Studies in the Intellectual History of Tokugawa Japan*, ed. Mikiso Hane (Tokyo: University of Tokyo Press, 1974); Tetsuo Najita, *Vision of Virtue in Tokugawa Japan: The Kaitokudō Merchant Academy of Osaka* (Honolulu: University of Hawaii Press, 1987); Tetsuo Najita and Irwin Scheiner, eds., *Japanese Thought in the Tokugawa Period, 1600–1868: Methods and Metaphors* (Chicago: University of Chicago Press, 1968); Peter Nosco, ed., *Confucianism and Tokugawa Culture* (Honolulu: University of Hawaii Press, 1984); Herman Ooms, *Tokugawa Ideology: Early Constructs, 1570–1680* (Princeton, NJ: Princeton University Press, 1985); Samuel Hideo Yamashita, "Compasses and Carpenter's Squares: A Study of Itō Jinsai (1627–1705) and Ogyū Sorai (1666–1728)" (PhD diss., University of Michigan, 1981).

67. In Japan, Ishida Baigan—who founded the *shingaku* school of Neo-Confucianism—and Satō Naokata are two scholars that emphasized introspective practices. See Bellah, *Tokugawa Religion*; Paolo Beonio-Brocchieri, *Religiosità e ideologia alle origini del Giappone moderno* (Milano: Ispi, 1965); John Allen Tucker, "Quiet-Sitting and Political Activism: The Thought and Practice of Satō Naokata," *Japanese Journal of Religious Studies* 29, nos. 1–2 (2002): 107–46.

68. Neo-Confucian scholars mainly carried on the "scientific" movement in Edo Japan. On this regard, see Karatani Kōjin, *Kotoba to higeki* (Tokyo: Kōdansha Gakujutsu Bunko, 1993), 161–84; Tsuji Tetsuo, *Nihon no kagaku shisō* (Tokyo: Chūkō Shinsho, 1973), 25–61; Sugimoto and Swain, *Science & Culture in Traditional Japan*, 291–395; Itō Shuntarō and Murakami Yōichirō, eds., *Nihon kagakushi no shatei*, Kōza kagakushi 4 (Tokyo: Baifūkan, 1989), 64–89, 121–41.

69. Needham and Lu, *Botany*, 315.

70. Elman, *On Their Own Terms*, 5.

71. Métailié, "The *Bencao gangmu* of Li Shizhen," 233.

72. Ibid., 234.

73. *Bencao gangmu, juan* 34, 1911. Translated in Métailié, "The *Bencao gangmu* of Li Shizhen," 236.

74. Needham and Lu, *Botany*, 177.

75. Métailié, "The *Bencao gangmu* of Li Shizhen," 238. "Folk classification" in Métailié is taken from Scott Atran, "Origin of the Species and Genus Concepts: An Anthropological Perspective," *Journal of the History of Biology* 20, no. 2 (1987): 195–279. See also Atran, *Cognitive Foundations of Natural History*.

76. The Neoplatonic philosopher Porphyry explained Aristotle's logical-ontological model of *eidos-genos* with a hierarchical schema that revealed the logical, ontological, and empirically observable structure of reality. Roughly, it consisted in dividing what Aristotle called *genos* (genera) into *eidos* (species) in accordance with a hierarchical system of differences (διαφορα) that shows how species shared the essence of a genus and each individual of one species shared the essence of the species. The system was supposed to construct a treelike structure (*arbor porphyriana*) that visualized the hierarchical series of logical and ontological differences that allowed distinguishing all things in the universe. See Umberto Eco, "Dall'albero al labirinto," in *Dall'albero al labirinto. Studi storici sul segno e l'interpretazione* (Milano: Bompiani, 2007), 13–96.

77. See Wilkins, *Species: A History of the Idea* (Berkeley: University of California Press, 2009), 47–96.

78. See Wheeler, *Species Concepts and Phylogenetic Theory*.

79. See Aristotle, *De generation et corruptione* (On Generation and Corruption). See an English translation by H. H. Joachim at http://ebooks.adelaide.edu.au/a/aristotle/corruption (accessed October 21, 2010).

80. Ever since Parmenides, Greek philosophy was caught in the paradox of being and change. Democritus's atomism was an attempt to circumvent the problem of a being identical with itself and change, but Aristotle rejected it, and his deductive system became the dominant paradigm of Western philosophical tradition until the early modern period. For a juxtaposition of Greek and Chinese systems of thought in regard to the human body, see Shigehisa Kuriyama, *The Expressiveness of the Body, and the Divergence of Greek and Chinese Medicine* (New York: Zone Books, 2002).

81. See Nappi, *The Monkey and the Inkpot*, 69–135.

82. Ibid., 140.

83. Elman, *On Their Own Terms*, 5.

84. Métailié, "The *Bencao gangmu* of Li Shizhen," 241.

85. Elman, *On Their Own Terms*, 37. See also, John Makeham, *Name and Actuality in Early Chinese Thought* (Albany: State University of New York Press, 1994); and Fung, *History of Chinese Philosophy*, vol. 1, 59–62, 302–11.

86. *Luyun, pian* 13, *zhang* 3.

87. *Luyun, pian* 12, *zhang* 17.

88. Burton Watson, *Xunzi: Basic Writings* (New York: Columbia University Press, 2003), 154.

89. Dan Robins, "Xunzi," *Stanford Encyclopedia of Philosophy*, accessed October 21, 2010, http://plato.stanford.edu/entries/xunzi.

90. Fung, *History of Chinese Philosophy*, vol. 2, 631–36.

91. Métailié, "The *Bencao gangmu* of Li Shizhen," 241.

92. Nappi, *The Monkey and the Inkpot*, 140.

93. A similar belief in the homology of names and things also characterized European natural history of the Renaissance period. Foucault, *The Order of Things*, 129. See also William B. Ashworth Jr., "Emblematic Natural History of the Renaissance," in *Cultures of Natural History*, ed. Nicholas Jardine, James A. Secord, and Emma C. Spary (Cambridge: Cambridge University Press, 1996), 23.

94. Note, for example, that the terms to classify things into classes (*gang*) and orders (*mu*) were also utilized to name entries (*gang*) and chapters (*mu*).

95. William Shakespeare, *Romeo and Juliet*, II, ii, 1–2, accessed August 14, 2014, http://shakespeare.mit.edu/romeo_juliet/full.html.

96. Bernard of Cluny, *De Contemptu Mundi*, in *Bernard le Clunisien: De contemptu mundi. Une vision du monde vers 1144*, ed. André Cresson (Turnhout: Brepols, 2009), 126. More famous today is Umberto Eco's variation: *Stat rosa pristina nomine, nomina nuda tenemus*, in *Il nome della rosa* (Milano: Bompiani, 1980).

PART II

1. See Peter Kornicki, *The Book in Japan: A Cultural History from the Beginnings to the Nineteenth Century* (Honolulu: University of Hawaii Press, 2001), 277–305. See also Lucille Chia, *Printing for Profit: The Commercial Publishers of Jianyang, Fujian (11th-17th Centuries)* (Cambridge, MA: Harvard University Asia Center, 2002); Lucille Chia, "Of Three Mountains Street: The Commercial Publishers of Ming Nanjing," in *Printing and Book Culture in Late Imperial China*, ed. Cynthia Brokaw and Kai-wing Chow (Berkeley: University of California Press, 2005), 107–51.

2. Fukuo Takeichirō, *Ōuchi Yoshitaka* (Tokyo: Yoshikawa Kōbunkan, 1989), 93–115.

3. See chapter 10.

CHAPTER 3

1. Yabe, *Edo no honzō*, 43; Ueno, *Nihon hakubutsugaku shi*, 65; Kimura, *Nihon shizenshi no seiritsu*, 33; Sugimoto, *Edo no hakubutsugakushatachi*, 38; Nishimura, *Bunmei no naka no hakubutsugaku*, vol. 1, 106.

2. The debate over the ideological role of Neo-Confucianism in the foundation of the Tokugawa shogunate is immense. See in particular Wm. Theodore de Bary, *The Unfolding of Neo-Confucianism* (New York: Columbia University Press, 1975); Kurozumi, *Kinsei Nihon shakai to jukyō*; Yamada, *Shushi no shizengaku*; Shimada, *Shushigaku to yōmeigaku*; Yasuda, *Chūgoku kinsei shisō kenkyū*; Wajima Yoshio, *Nihon Sōgakushi no kenkyū*; Iwasaki, *Nihon kinsei shisōshi josetsu*, vol. 2; Robert Bellah, *Tokugawa Religion: The Cultural Roots of Modern Japan* (London: Free Press, 1985); Maruyama, *Studies in the Intellectual History of Tokugawa Japan*; Tetsuo and Scheiner, *Japanese Thought in the Tokugawa Period, 1600–1868*; Kate Wildman Nakai, "The Naturalization of Confucianism in Tokugawa Japan: The Problem of Sinocentrism," *Harvard Journal of Asiatic Studies* 40, no. 1 (June 1980): 157–99;

Nosco, *Confucianism and Tokugawa Culture*; Samuel Hideo Yamashita, *Compasses and Carpenter's Squares: A Study of Itō Jinsai (1627–1705) and Ogyū Sorai (1666–1728)* (Ann Arbor, MI: University Microfilms International, 1981); Watanabe Hiroshi, *A History of Japanese Political Thought, 1600–1901*, trans. David Noble (Tokyo: International House of Japan, 2012).

3. Yabe, *Edo no honzō*, 44; Ueno, *Nihon hakubutsugaku shi*, 65; Kimura, *Nihon shizenshi no seiritsu*, 34; Sugimoto, *Edo no hakubutsugakushatachi*, 38–46; Nishimura, *Bunmei no naka no hakubutsugaku*, vol. 1, 106.

4. See Robert L. Backus, "The Relationship of Confucianism to the Tokugawa Bakufu as Revealed in the Kansei Educational Reform," *Harvard Journal of Asiatic Studies* 34 (1974): 97–162; Robert L. Backus, "The Kansei Prohibition of Heterodoxy and Its Effects on Education," *Harvard Journal of Asiatic Studies* 39, no. 1 (June 1979): 55–106; Robert L. Backus, "The Motivation of Confucian Orthodoxy in Tokugawa Japan," *Harvard Journal of Asiatic Studies* 39, no. 2 (December 1979): 275–338.

5. For example, Inoue Tetsujirō and Muraoka Tsunetsugu. Maruyama Masao, *Studies in the Intellectual History of Tokugawa Japan*, xv–xxxvii. Published in Japanese as *Nihon seiji shisōshi kenkyū* (Tokyo: Tokyo Daigaku Shuppankai, 1952).

6. To justify his claim, Maruyama quoted the *Tokugawa jikki*: "Ieyasu had conquered the nation on horseback, but being an enlightened and wise man, realized early that the land could not be governed from a horse. He had always respected and believed in the way of the sages. He wisely decided that in order to govern the land and follow the path proper to man, he must pursue the path of learning. Therefore, from the beginning he encouraged learning." Maruyama, *Studies in the Intellectual History of Tokugawa Japan*, 15.

7. On the one hand, I respect Ooms's diminution of the role Neo-Confucianism played in the ideological foundation of the Tokugawa shogunate. On the other, I recognize that a Neo-Confucian worldview pervaded the study of nature throughout the period. See Kurozumi, *Kinsei Nihon shakai to jukyō*; W. J. Boot, *Keizers en Shōgun: Een Geschiedenis van Japan tot 1868* (Amsterdam: Amsterdam University Press/Salomé, 2001).

8. Ooms, *Tokugawa Ideology*, 73.

9. Information about Hayashi Razan may be found not only in the *Tokugawa jikki* but also in two manuscripts written shortly after his death by two of his sons: *Razan Rin sensei gyōjō* (Eulogy of Master Hayashi Razan, 1658) and *Razan sensei nenpu* (Biography of Master Razan, 1659); Kyoto Shiseki Kai, ed., *Razan sensei bunshū*, 2 vols. (Kyoto: Heian Kōko Gakkai, 1918). Hayashi Morikatsu's (Shuntoku) *Razan Rin sensei gyōjō*, in one volume, is an eulogy Shuntoku wrote the year after his father's death. *Razan sensei nenpu* is a biographical account of Razan's life divided into entries organized chronologically by year. It was compiled two years after Razan's death by his third-born son Shunsai, better known as Hayashi Gahō.

Modern references for Razan's life and work are Hori Isao, *Hayashi Razan* (Tokyo: Yoshikawa Kōbunkan, 1964); Suzuki Ken'ichi, *Hayashi Razan sensei nenpu kō* (Tokyo: Perikansha, 1999).

10. Mayanagi Makoto, "*Honzō kōmoku* no denrai to kinryōbon," *Nihon ishigaku zasshi* 37, no. 2 (1991): 41–43; Isono, *Nihon hakubutsushi nenpyō*; Nishimura, *Bunmei no naka no hakubutsugaku*, vol. 1, 106–8.

11. Mayanagi, "*Honzô kômoku* no denrai to kinryôbon," 108.

12. *Razan Rin sensei gyōjō* and *Razan sensei nenpu*, in *Razan sensei bunshū*; also Hori, *Hayashi Razan*; Suzuki, *Hayashi Razan sensei nenpukō*.

13. Medieval yeoman warriors who retained power as local magnates not in service of the Shogunate. Their origins are associated with the decline of shogunal power late in the Kamakura period, when landholders (known as *myōshu*) started loosening their ties to the central government and seizing actual ruling power in the villages. Many of them organized militarized units for self-protection. In the warring state period, the most powerful of these *jizamurai* gradually took military control of an entire region. Their retainers were usually called *gōshi*, or "village warriors." The background of the Hayashi family, before they moved to Osaka and then to Kyoto, is that of minor yeomen whose holdings were nonetheless enough to grant them the possibility of a comfortable urban life in the capital. On *jizamurai*, see Pierre François Souyri, *The World Turned Upside Down: Medieval Japanese Society*, trans. Käthe Roth (New York: Columbia University Press, 2001).

14. Quoted in Hori, *Hayashi Razan*, 9.

15. Martin Collcutt, *Five Mountains: The Rinzai Zen Monastic Institution in Medieval Japan* (Cambridge, MA: Harvard University Press, 1981).

16. For Zhu Xi's metaphysics, Yamada, *Shushi no shizengaku*; Shimada, *Shushigaku to yōmeigaku*; Yasuda, *Chūgoku kinsei shisō kenkyū*. For the Japanese understanding of Zhu Xi's cosmology, see Wajima, *Nihon Sōgakushi no kenkyū*; Iwasaki, *Nihon kinsei shisōshi josetsu*, vol. 1, 116–76. For "heterodox" interpretations of Japanese *shushigaku*, see Koyasu Nobukuni, *Edo shisōshi kōgi* (Tokyo: Iwanami Shoten, 1998); Kurozumi, *Kinsei Nihon shakai to jukyō*. See also Bellah, *Tokugawa Religion*; de Bary, *The Unfolding of Neo-Confucianism*; Elman, Duncan, and Ooms, *Rethinking Confucianism*; Maruyama, *Studies in the Intellectual History of Tokugawa Japan*; Nosco, *Confucianism and Tokugawa Culture*; Ooms, *Tokugawa Ideology*.

17. For example, the case of the so-called Harima and Tosa schools of Neo-Confucianism. See Wajima, *Nihon Sōgakushi no kenkyū*.

18. The Tokugawa attempted to link their family to the Minamoto; see Nakamura Kōya, *Tokugawa ke: Ieyasu wo shūshin ni* (Tokyo: Shibundō, 1961). After their deaths, Toyotomi Hideyoshi and Tokugawa Ieyasu were transformed into Buddhist avatars; see Mary Elizabeth Berry, *Hideyoshi* (Cambridge, MA: Harvard University

Press, 1982); Ooms, *Tokugawa Ideology*. For culture as a legitimating instrument, see William H. Coaldrake, "Edo Architecture and Tokugawa Law," *Monumenta Nipponica* 36, no. 3 (Autumn 1981): 235–84; William H. Coaldrake, *Architecture and Authority in Japan* (London: Routledge, 1996); Peter F. Kornicki, *The Book in Japan*; Timon Screech, *The Shogun's Painted Culture: Fear and Creativity in the Japanese States, 1760–1829* (London: Reaktion Books, 2000).

19. Hori, *Hayashi Razan*, 18.

20. Ibid., 20.

21. Ibid.

22. In the 1599 entry of *Razan sensei nenpu*, also quoted in Hori, *Hayashi Razan*, 27.

23. Ibid., 25–38.

24. In Ming and Qing China, "the late imperial examination system," Elman explains, "was not a premodern anachronism or an anti-modern monolith. Classical examinations were an effective cultural, social, political, and educational construction that met the needs of the Ming-Qing bureaucracy while simultaneously supporting the late imperial social structure whose elite gentry and merchant status groups were defined in part by examination degree credentials." Elman, *A Cultural History of Civil Examinations in Late Imperial China*, xx. According to Elman, "local elites and the imperial court continually influenced the government to reexamine and adjust the classical curriculum and to entertain new ways to improve the institutional system for selecting those candidates who were eligible to become officials.... Civil examinations, as a test of educational merit, ... were a cultural arena within which diverse political and social interests contested each other and were balanced." Elman, *A Cultural History of Civil Examinations*, xxiii–xxiv.

25. Ooms, *Tokugawa Ideology*, 73.

26. Ibid.

27. On Fujiwara Seika, see Ōta Seikyū, *Fujiwara Seika* (Tokyo: Yoshikawa Kōbunkan, 1985).

28. See Nakai Riken, *Shin'i zukai* (Illustrated Explanation of *shin'i*, 1795), electronic copy at http://wsv.library.osaka-u.ac.jp/tenji/kaitokudo/kaitoko9.htm (accessed June 2007). Nakai Riken was a scholar of the Kaitokudō academy of Confucian studies. See also Najita, *Vision of Virtue in Tokugawa Japan*, 186–220 for further details on Nakai Riken.

29. In the 1603 entry of *Razan sensei nenpu*, quoted also in Hori, *Hayashi Razan*, 38. Zhu Xi's commentary on the *Analects* is read in Japanese *Rongo shitchū*.

30. Kondō Seisai, *Kōsho koji*, quoted in Hori, *Hayashi Razan*, 222–23.

31. Sermons were called *seppō*, *sekkyō*, or *dangi*. See Sekiyama Kazuo, *Sekkyō no rekishi: Bukkyō to wagei* (Tokyo: Hakusuisha, 1992); Nakamura Yukihiko, "Taiheiki no kōshakushi-tachi," in *Nihon no koten: Taiheki* (Tokyo: Sekai Bunkasha, 1975);

Hyōdō Hiromi, *Taiheiki "yomi" no kanōsei: Rekishi to iu monogatari* (Tokyo: Kōdansha, 1995); Wakao Masaki, *"Taiheiki yomi" no jidai: Kinsei seiji shisōshi no kōsō* (Tokyo: Heibonsha, 1999).

32. Hori, *Hayashi Razan*, 40.

33. Ooms, *Tokugawa Ideology*, 75. If we compare the 200 *ryō* and the 5,340 *tsubo* of land (little less than four acres) to the 50,000 *ryō* that Tenkai received to build his temple, we can appreciate how trivial the grant was. Tenkai, a monk of the Tendai school of Buddhism, enjoyed Ieyasu's patronage and was involved in the political organization of the bakufu. Also, he actively participated from his temple, the Enryakuji, and later from the Tokugawa mausoleum in Nikkō, in the divinization of Tokugawa Ieyasu after his death as Tōshōdaigongen.

34. For further details on anti-Buddhist rhetoric in early modern Japan, see Tamamuro Fumio, *Edo bakufu no shūkyō tōsei* (Tokyo: Hyōronsha, 1971); Ōkuwa Hitoshi, *Nihon kinsei no shisō to bukkyō* (Kyoto: Hōzōkan, 1989); Ōkuwa Hitoshi, *Nihon bukkyō no kinsei* (Kyoto: Hōzōkan, 2003).

35. In particular of Go-Yōzei and Go-Mizunō.

36. *Keichō nikkenroku* (Diary of the Keichō Era, 1596–1615), also quoted in Hori, *Hayashi Razan*, 41.

37. Wajima Yoshio, "Kinsei shoki jugakushi ni okeru ni-san no mondai," *Ōtemae Joshi Daigaku ronshū* 7 (1973): 92–94.

38. Since, according to Japanese measures of age, Razan was twenty-one in 1603, the events he is narrating took place in 1604 and 1605. Moreover, we know from *Razan sensei nenpu* that Razan adopted the custom of dressing in a *shin'i* from Fujiwara Seika, whom he met the eighth month of 1604.

39. Quoted in Hori, *Hayashi Razan*, 41–42. The versions proposed in *Razan sensei nenpu*, *Razan Rin sensei gyōjō*, *Tokugawa jikki* and Razan's direct recollection of the incident in *Nozuchi*, a commentary on the Japanese classic *Tsurezuregusa* he wrote in 1621, differ in some details. According to *Razan Rin sensei gyōjō* and *Tokugawa jikki*, the Razan-Hidekata's dispute started in 1603. They omit mentioning Razan's attempts to gain Hidekata's patronage—which would have been contradictory with the subsequent struggle against the court monopoly over the interpretation of Confucian classics. Instead, they emphasize Ieyasu's plea for freedom of intellectual pursuit, which would eventually be granted to Razan. Moreover, an overt support from Ieyasu—who, it is worth remembering, at the time had yet to meet the young scholar—would have been contradictory with Ieyasu's appointment of Hidekata as bakufu consultant in matters of court procedure. For further details, see Hori, *Hayashi Razan*, 40–48.

40. Quoted in ibid., 43.

41. Ibid., 49. Mitigated by Ieyasu's comment to Hidekata that "he should not be concerned by Razan's lecturing any more than he would be by a housefly (*aobae*)."

Quoted in ibid., 44, from a letter Seika wrote to Razan referring to the result of Ieyasu's deliberation, as heard by the master of shogunal ceremony (*sōshaban*) Jō Izumi-no-kami Masamochi, who became in early 1605 Razan's student at Seika's suggestion.

42. Ibid.; Wajima, "Kinsei shoki jugakushi ni okeru ni-san no mondai"; Iwasaki, *Nihon kinsei shisōshi josetsu*, vol. 1.

43. Ooms, *Tokugawa Ideology*, 74.

44. Hiromi, *Taiheiki "yomi" no kanōsei*.

45. Mikael S. Adolphson, *The Gates of Power: Monks, Courtiers, and Warriors in Premodern Japan* (Honolulu: University of Hawaii Press, 2000); Souyri, *The World Turned Upside Down*; Kuroda Toshio, *Kenmon taiseiron*, in *Kuroda Toshio chosakushū*, vol. 1 (Kyoto: Hōzōkan, 1994).

46. In the 1607 entry of *Razan sensei nenpu*, quoted also in Hori, *Hayashi Razan*, 128–29.

47. Later generations of Confucian scholars like Nakae Tōju harshly criticized Razan's obedience to Ieyasu's order, describing him as a Confucian scholar only in manner (in *Rinshi teihatsu juiben*). Jean-François Soum, *Nakae Tōju, 1608–1648, et Kumazawa Banzan, 1619–1691: Deux penseurs de l'époque d'Edo* (Paris: De Boccard, Diffusion, 2000). It should also be remembered that in the Hayashi school, the Kōbun'in, teachers and students alike shaved their heads and wore black Buddhist robes until 1690. For further details, see Wajima Yoshio, *Shōheikō to hangaku* (Tokyo: Shibundō, 1962).

48. Nakai, "The Naturalization of Confucianism in Tokugawa Japan," 157.

49. As Kondō Morishige, head librarian of Shogunal library of Momijiyama, wrote in a 1817 entry of his *Yūbun koji*. Quoted in Hori, *Hayashi Razan*, 133.

50. Mayanagi, "*Honzō kōmoku* no denrai to kinryōbon."

51. *Meibutsugaku* is a Confucian discipline originating in China in the Han period, inspired by Confucius's precept of "rectifying names" (thirteenth book of the *Analects*). See Sueki Takehiro, *Tōyō no gōri shisō* (Kyoto: Hōzōkan, 2001).

52. *Honzō kōmoku jochū* (The Annotated Introduction to *Honzō kōmoku*), printed posthumously in 1666, consisted of an annotated edition of Li Shizhen's original introduction and the 1603 introduction added to the Jiangzhi edition.

53. There are six versions of *Tashikihen* surviving, a manuscript and five printed editions (1630, 1631, 1649, plus other two with no date). The earliest printed edition of 1630 consists of two small folio volumes with a total of 141 leaves. The 1631 edition is the revised edition titled *Shinkan Tashikihen*, published by Kyoto booksellers Murakami Sōshin and Tanaka Chōzaemon, and consists of three block-printed *hanpon* volumes, a copy of which is now at Waseda University and which I consulted.

54. *Man'yōgana* is a writing system based on Chinese characters taken only for their phonetic values to represent Japanese sounds. Used since the seventh century, it takes its name from the poetic anthology *Man'yōshū*, compiled around 759.

55. The *furigana*—the Japanese reading of a Chinese compound written in phonetic characters—is attached to *man'yōgana* only in the first pages of the text and is soon abandoned after the implied reader has familiarized with Razan's usage.

56. *Diospyros kaki.*

57. *Diospyros eriantha.*

58. *Lanshi*, a kind of persimmon jam.

59. A convergence with early modern European encyclopedias up to Buffon's *Histoire naturelle, générale et particulière* (1749–88).

60. *Haemopis marmorata.*

61. The term *tabako* derives from the Portuguese word *tabaco* after the introduction of the plant sometime between 1596 and 1614 in the Satsuma Province. Its introduction in East Asia is usually dated to 1571, when Spanish merchants brought it into the Philippines from Cuba.

62. To complicate matters further, scientists today identify *lang* with *Scopalia Japonica maxima*, which is not at all related to *H. niger* but to tobacco, being *Nicotiana tabacum*, like *lang*, also a species of the Solanaceae family.

63. The lexicographical approach informed all the entries of *Tashikihen*, including those on "human beings." In this section, there is the earliest attempt to compare Japanese and Chinese monsters. The last entry of the section, for example, was the following:

飛頭蛮　呂久呂久毘゛
Feitouman: rokurokubi

This seems to be the earliest extant source that translated the Chinese *feitouman* (*hitōban* in Japanese pronunciation)—a ghostly creature consisting only of the head of a diseased person, believed to have haunted abandoned houses—with the Japanese *rokurokubi*, identifying a gruesome creature able to stretch its neck at will haunting uninhabited houses.

64. John Knoblock, ed., *Xunzi: A Translation and Study of the Complete Works*, vol. 3, books 17–32 (Stanford: Stanford University Press, 1994), 129.

65. Elman, *A Cultural History of Civil Examinations*; Elman, *From Philosophy to Philology: Intellectual and Social Aspects of Change in Late Imperial China*, 2nd rev. ed. (Los Angeles: UCLA Asian Pacific Monograph Series, 2001).

66. Nakai, "The Naturalization of Confucianism in Tokugawa Japan," 159.

67. Fujimura Tsukuru, ed., *Nihon bungaku daijiten*, 8 vols. (Tokyo: Shinchōsha, 1932–35), vol. 2.

68. Yet the image of *Tashikihen* eventually changed from that of a dictionary to the earliest text of natural history written in Japan. In the 1930s, Shirai Mitsutarō, biology professor of Tokyo Imperial University, repeated this interpretation on the publication of the complete translation of *Honzō kōmoku* into Japanese. He criticized

the treatment of *Tashikihen* in Kameda Jirō's encyclopedia and stressed instead its status as natural history. See Shirai Mitsutarō, *Shirai Mitsutarō chosakushū*, ed. Kimura Yōjirō, 6 vols. (Tokyo: Kagaku Shoin, 1985–90), 6:369. Shirai's interpretation has never been repudiated, and contemporary historians of science continue to classify *Tashikihen* as natural history. For example, Sugimoto Tsutomu's comments in "Kaisetsu," Hayashi Razan, *Shinkan Tashikihen*, ed. B. H. Nihongo Kenkyū Gurupu (Tokyo: Bunka Shobō Hakubunsha, 1973), 1.

CHAPTER 4

1. Quoted in Daniel Rosenberg, "Early Modern Information Overload," *Journal of the History of Ideas* 64, no. 1 (January 2003): 1.

2. It is therefore not surprising that totalitarian regimes always exercised a tight control over the production, editing, and dissemination of encyclopedias.

3. *Wikipedia*, in this sense, is an ever-lasting work-in-progress that challenges this notion of the modern encyclopedia. See Erik Olin Wright, *Envisioning Real Utopias* (London: Verso, 2010), ch. 7.

4. Jorge Luis Borges, "On Exactitude in Science," in *Collected Fictions*, trans. Andrew Hurley (New York: Penguin, 1999), 325.

5. I here use the term *encyclopedia* quite improperly—certainly not in the sense given to the term by Enlightenment thinkers—to group together different genres of books that introduced various topics in a systematic manner. On the printing industry of Chinese encyclopedias, see Chia, *Printing for Profit*; Joseph P. McDermott, *A Social History of the Chinese Book: Books and Literati Culture in Late Imperial China* (Hong Kong: Hong Kong University Press, 2006). See also Craig Clunas, *Superfluous Things: Material Culture and Social Status in Early Modern China*, 2nd ed. (Honolulu: University of Hawaii Press, 2004).

6. See Carol Gluck, "The Fine Folly of the Encyclopedists," in *Currents in Japanese Culture: Translations and Transformations*, ed. Amy Vladeck Heinrich (New York: Columbia University Press, 1996), 223–51; Yamada, *Honzō to yume to renkinjutsu to*.

7. On the meaning of "architext", see Gérard Genette, *The Architext: An Introduction*, trans. Jane E. Lewin (Berkeley: University of California Press, 1992).

8. Chinese texts were not translated into Japanese but were edited with a system of markings and phonetic annotations that allowed Japanese to read the text in classical Japanese (*kundoku*).

9. The attribution, not unanimously shared by Japanese scholars, derives from the analysis of an explanatory note attached at the end of every volume, in which appeared detailed speculations on the correspondences between Chinese and Japanese terms similar to the ones Ekiken wrote in his *Yamato honzō*. See Yabe, *Edo no honzō*, 50 for further details.

10. It is difficult to track down the history of Japanese editions of Chinese books,

since, as Kornicki reminds us, bibliographical works like the *Kokusho sōmokuroku* (Complete Index of Japanese Books) and its sequel *Kotenseki sōgō mokuroku* (Comprehensive Index of the Classics) usually record only "Japanese books" (*kokusho*), excluding all Chinese books reprinted in Japan. See Kornicki, *The Book in Japan*, 1–2. For Japanese editions of *Honzō kōmoku*, Mayanagi, "*Honzō kōmoku* no denrai to kinryōbon"; Nishimura, *Bunmei no naka no hakubutsugaku*, vol. 1; Yabe, *Edo no honzō*; Isono, *Nihon hakubutsushi nenpyō*; Isono Naohide, ed., *Egakareta dōbutsu shokubutsu: Edo jidai no hakubutsushi* (Tokyo: Kinokuniya Shoten, 2005).

11. Despite the fact that they did not organize themselves into guilds (*za*), their social organization resembled that of specialized artisans and merchants.

12. Mary Elizabeth Berry, *Japan in Print: Information and Nation in the Early Modern Period* (Berkeley: University of California Press, 2006).

13. Mayanagi Makoto counts more than four hundred titles of manuscripts and printed books on medicine and *honzōgaku* published by Japanese scholars in the seventeenth century. See Mayanagi Makoto, "Nihon no iyaku, hakubutsu chojutsu nenpyō" (2006), accessed June 2007, http://www.hum.ibaraki.ac.jp/mayanagi/paper01/ChronoTabJpMed.html.

14. Brian W. Ogilvie, "The Many Books of Nature: Renaissance Naturalists and Information Overload," *Journal of the History of Ideas* 64, no. 1 (January 2003): 39.

15. Ōba Osamu, *Edo jidai ni okeru Karafune mochiwatarisho no kenkyū* (Suita: Kansai Daigaku Tōzai Gakujustsu Kenkyūjo, 1967).

16. Leibniz is quoted in Richard Yeo, "A Solution to the Multitude of Books: Ephraim Chambers's *Cyclopaedia* (1728) as 'the Best Book in the Universe,'" *Journal of the History of Ideas* 64, no. 1 (January 2003): 61–72. Gessner, *Bibliotheca universalis* (Zurich, Switzerland: Christoph Froschauer, 1545), sig. *3^V is quoted in Ann Blair, "Reading Strategies for Coping with Information Overload ca. 1550–1700," *Journal of the History of Ideas* 64, no. 1 (January 2003): 11–28. See also Blair, *Too Much to Know: Managing Scholarly Information Before the Modern Age* (New Haven: Yale University Press, 2010).

17. See Jansen, *China in the Tokugawa World*; for a "post-modern" interpretation, David Pollack, *The Fracture of Meaning: Japan's Synthesis of China from the Eighth through the Eighteenth Centuries* (Princeton, NJ: Princeton University Press, 1986). For a more thorough study of the reception of Chinese culture in Tokugawa Japan, see Ōba Osamu, *Edo jidai ni okeru Chūgoku bunka juyō no kenkyū* (Kyoto: Dōhōsha Shuppan, 1984). For a comparison with Europe, see Anthony Grafton, *Bring Out Your Dead: The Past as Revelation* (Cambridge, MA: Harvard University Press, 2001).

18. Scholars such as Nicolò Leoniceno, for example, based their knowledge of the natural world almost entirely upon ancient Greek and Roman texts. A professor of medicine and moral philosophy at the University of Ferrara, Leoniceno possessed

one of the largest collection of books on natural history and materia medica of his time. The majority of the texts he owned were Greek and Latin classics, from Aristotle's Περὶ Τὰ Ζῷα Ἱστορίαι ("Inquiries of Animals" in Greek, or *Historia animalium* in Latin translation) to Theophrastus's Περὶ φυτῶν ιστορία (*De historia plantarum*) and Pliny's *Historia naturalis*. He owned a total of 482 texts, as Daniela Mugnai Carrara gathered from a posthumous inventory; see Daniela Mugnai Carrara, *La biblioteca di Nicolò Leoniceno. Tra Aristotele e Galeno: Cultura e libri di un medico umanista* (Firenze: Olschki, 1991), 44. See also Atran, *Cognitive Foundations of Natural History*, 83–122; Ogilvie, *The Science of Describing*, 31.

19. Ogilvie, *The Science of Describing*, 28.

20. On *pulu*, see Martina Siebert, *Pulu: "Abhandlungen und Auflistungen" zu materieller Kultur und Naturkunde im traditionellen China*, Opera Sinologica 17 (Wiesbaden: Harrassowitz Verlag, 2006).

21. The two characters of the title are of difficult interpretation. It has been translated variously as "The Literary Expositor," "The Ready Rectifier," "Progress toward Correctness," or "Approaching Elegance/Refinement."

22. On *Erya*, see W. South Coblin, "*Erh ya*," in *Early Chinese Texts: A Bibliographic Guide*, ed. Michael Loewe (Berkeley: Society for the Study of Early China, 1993); Elman, *On Their Own Terms*, 38–41.

23. The study of *Erya* was in earlier centuries conducted in the Daigakuryō. The text, since its introduction into Japan, inspired the compilation of similar anthologies of essays and commentaries. The most famous was probably Minamoto no Shitagō's *Wamyō ruijushō* (Thematic Dictionary of Japanese Terms), compiled between 931 and 938 in *man'yōgana*. Five manuscript versions of *Wamyō ruijushō* circulated during the Edo period, until it was printed in 1801 and 1821.

24. Bernhard Karlgren, "The Early History of the *Chou Li* and *Tso Chuan* Texts," *Bulletin of the Museum of Far Eastern Antiquities* 3 (1931): 1–59.

25. Dividing them into the categories of herbs, trees, insects, fish, birds, and animals. See Roel Sterckx, *The Animal and the Daemon in Early China* (Albany: State University of New York Press, 2002), 30–32.

26. Yoshifuru's *Wajiga* is notable as the first East Asian source to mention peanuts, imported at the end of the Genroku era (1688–1704) from western Africa by Dutch merchants and henceforth identified by the Japanese term *rakkasei* (*Arachis hypogea*). *Wajiga* was also the first Japanese text to quote and briefly describe the cactus, a similarly exotic plant imported into Japan with the name *saboten* (a subspecies of *Opuntia maxima*). The same is true for sedum, called *benkeisō*, and sorghum, named *morokoshikibi* (*Hylotelephium erythrostictum*). A copy of the first edition of *Wajiga* by Kyoto bookseller Tamashiken is conserved at Waseda University. See Kaibara Yoshifuru (Chiken), *Wajiga* (Manuscript, 1694), http://www.wul.waseda.ac.jp/kotenseki/html/ho02/ho02_04324/index.html (accessed June 2007).

27. Arai Hakuseki, *Tōga: Eiin, honkoku*, ed. Sugimoto Tsutomu (Tokyo: Waseda Daigaku Shuppanbu, 1994); Arai Hakuseki, *Tōga: Kaitai, sakuin*, ed. Sugimoto Tsutomu (Tokyo: Waseda Daigaku Shuppanbu, 1994). *Tōga* was influential among mid-Tokugawa *kokugaku* scholars Kamo no Mabuchi and Motoori Norinaga.

28. See Harold Bolitho, "Travelers' Tales: Three Eighteenth-Century Travel Journals," *Harvard Journal of Asiatic Studies* 50, no. 2 (December 1990): 485–504; Constantine N. Vaporis, *Breaking Barriers: Travel and the State in Early Modern Japan* (Cambridge, MA: Harvard University Press, 1994); Marcia Yonemoto, *Mapping Early Modern Japan: Space, Place, and Culture in the Tokugawa Period, 1603–1867* (Berkeley: University of California Press, 2003); Berry, *Japan in Print*.

29. On *Shanhaijing* see Richard E. Strassberg, *A Chinese Bestiary: Strange Creatures from the Guideways through Mountains and Seas* (Berkeley: University of California Press, 2002).

30. Kagawa Masanobu, *Edo no yōkai kakumei* (Tokyo: Kawade Shobō Shinsha, 2005).

31. Yamada Keiji, *Honzō to yume to renkinjutsu to*, 27–101. See also Gluck, "The Fine Folly of the Encyclopedists."

32. Nishimura, *Bunmei no naka no hakubutsugaku*, vol. 1, 216.

33. The term was not in large use in Japan until the modern period.

34. Nishimura, *Bunmei no naka no hakubutsugaku*, vol. 1, 218.

35. See part IV.

36. Indeed, the 1895 entries of the *Honzō kōmoku*—excluding synonyms and vernacular names—far outnumbered the species named and described by sixteenth-century European naturalists of that day. Nicolò Leoniceno and Valerius Cordus, perhaps the most knowledgeable naturalists of their generation, classified no more than five hundred species in their texts, while Carolus Clusius (Charles de l'Écluse) described seven hundred. Ogilvie, *The Science of Describing*, 52.

37. "Introduction" in Nakamura Tekisai, *Kinmō zui*, ed. Sugimoto Tsutomu (Tokyo: Waseda Daigaku Shuppanbu, 1975), 1.

38. Of the four books treating animals, book twelve listed 64 images of animals, book thirteen had 76 images of birds, book fourteen had 64 illustrations of fish and amphibians, and book fifteen had 108 pictures of insects. Five books were dedicated to plants: book sixteen had 36 illustrations of various grains, book seventeen had 56 pictures of vegetables, book eighteen listed 52 illustrated entries of fruits and fruit trees, book nineteen focused on 84 species of trees and bamboos, and book twenty had 128 pictures of flowers and herbs.

39. Yabe, *Edo no honzō*, 57–60; Nishimura, *Bunmei no naka no hakubutsugaku*, vol. 1, 112–15. Interestingly, a recent new edition of *Kinmō zui* is subtitled "The First Illustrated Encyclopedia of Japan." Kobayashi Yōjirō, ed., *Edo no irasuto jiten: Kinmōzui, waga kuni hatsu no irasuto hyakka* (Tokyo: Bensho Shuppan, 2012).

40. In order, *Impatiens balsamica, Narcissus tazetta, Abelmoschus manihot* (improperly referred to also as *Hibiscus japonicus*), and *Leucanthemum coronarium*.

41. *Artemisia vulgaris.*

42. The "three realms" (*sancai, sansai* in Japanese) are heaven, earth, and human beings (culture).

43. Sixteenth- and seventeenth-century European naturalists similarly relied upon information provided by workers and commoners not involved in scholarly activity. See Pamela H. Smith and Paula Findlen, eds., *Merchants & Marvels: Commerce, Science, and Art in Early Modern Europe* (New York: Routledge, 2002); Ogilvie, *The Science of Describing*. In the eighteenth and nineteenth centuries, scholars like Cuvier, Linnaeus, and Darwin also relied upon information gathered from workers.

44. The fact that he did not name names reveals the status of illustrators in the seventeenth century. The situation changed in the following century. In a later edition of *Kinmō zui* (Kyoto, 1789), the illustrations were edited by Shimokawabe Jūsui, active in the eighteenth century.

45. *Kabutogani* (*Tachypleus tridentatus*). See also Kajishima Takao, *Shiryō Nihon dōbutsushi* (Tokyo: Yasaka Shobō, 2002), 194–95.

46. The freshwater turtle is the *Amyda japonica*, which Tekisai named *iogame* or *dōgame* but today is usually called *suppon*.

47. The last "working" edition was Terajima Ryōan, *Wakan sansai zue*, 2 vols. (Tokyo: Nihon Zuihitsu Taisei Kankōkai, 1929).

48. See Ronald Dore, *Education in Tokugawa Japan* (Ann Arbor: University of Michigan Press, 1985); Richard Rubinger, *Private Academies of Tokugawa Japan* (Princeton, NJ: Princeton University Press, 1982); and especially Richard Rubinger, *Popular Literacy in Early Modern Japan* (Honolulu: University of Hawaii Press, 2007). See also Umihara Tōru, *Kinsei no gakkō to kyōiku* (Tokyo: Shibunkaku Shuppan, 1988); Tsujimoto Masashi, *Kinsei kyōiku shisōshi no kenkyū: Nihon ni okeru "kōkyōiku" shisō no genryū* (Kyoto: Shibunkaku Shuppan, 1990).

49. Yabe, *Edo no honzō*, 59.

50. Ibid.

51. Ibid., 60.

52. Rubinger, *Popular Literacy in Early Modern Japan*; Ishikawa Matsutarō, *Hankō to terakoya* (Tokyo: Kyōikusha, 1978).

53. The original English title was *The History of Japan, Giving an Account of the Ancient and Present State and Government of That Empire; of Its Temples, Palaces, Castles and Other Buildings; of Its Metals, Minerals, Trees, Plants, Animals, Birds and Fishes; of the Chronology and Succession of the Emperors, Ecclesiastical and Secular; of the Original Descent, Religions, Customs, and Manufactures of the Natives, and of*

Their Trade and Commerce with the Dutch and Chinese. Together with a Description of the Kingdom of Siam.

54. Nakamura's success in life and obscurity in modern historiography reminds us of the similar case of Nicolas-Claude Fabri de Peiresc (1580–1637), one of the most well-known intellectuals in his lifetime but soon forgotten after his death. See Peter Miller, *Peiresc's Europe: Learning and Virtue in the Seventeenth Century* (New Haven: Yale University Press, 2000).

55. See Ihara Saikaku, *Saikaku gohyakuin* (Tokyo: Benseisha, 1976).

56. See http://www2.ntj.jac.go.jp/dglib/ebook01/mainmenu.html (accessed June 2007).

57. The *rokuroku* at the beginning of the title suggests that the anthology was compiled in the *rokkasen* tradition, which refers to the six great poets of the *Kokinshū*, a poetic anthology compiled in the early tenth century. *Rokuroku* may be also a pun for an adjective often used with the meaning of "ordinary," "common," "tiny," usually associated with small objects.

58. Emma C. Spary, "Rococo Readings of the Book of Nature," in: *Books and the Sciences in History*, ed. Marina Frasca-Spada and Nick Jardine (Cambridge: Cambridge University Press, 2000), 255–75. See also Gould, *The Hedgehog, the Fox, and the Magister's Pox*, 157–63.

59. Spary, "Rococo Readings of the Book of Nature," 260. For the connection of social status and truth telling in early modern England, see Shapin, *A Social History of Truth*. The epistemological separation of aesthetic concerns and scientific enterprise is a modern development, according to Spary. I will argue about the interconnection of artistic ideals and the knowledge of the natural world in part IV, where I focus on private collectors and amateur naturalists. From a different perspective, Eiko Ikegami has focused on the bonds of civility that characterized early modern Japan—that is, the formation of aesthetic networks and communities that, in the course of the Tokugawa era, formed of themselves an identity that would have, at the dawn of the modern world, political implications. See Eiko Ikegami, *Bonds of Civility: Aesthetic Networks and the Political Origins of Japanese Culture* (Cambridge: Cambridge University Press, 2005).

60. Kornicki, *The Book in Japan*; Mayanagi, "Nihon no iyaku, hakubutsu chojutsu nenpyō" for *honzōgaku*.

CHAPTER 5

1. Information on Ekiken's life is taken from *Kaibara Ekiken sensei nenpu* and *Kaibara Ekiken sensei den*, both in Kaibara Ekiken, *Ekiken zenshū*, vol. 1, ed. Ekikenkai (Tokyo: Ekiken Zenshū Kankōbu, 1910–11). See Inoue Tadashi, *Kaibara Ekiken* (Tokyo: Yoshikawa Kōbunkan, 1989); Okada Takehiko, *Kaibara Ekiken* (Tokyo: Meitoku Shuppan, 2012) for secondary sources. In English, see Tucker, *Moral and Spiritual Cultivation in Japanese Neo-Confucianism*, ch. 2.

2. *Kaibara Ekiken sensei den*, in *Ekiken zenshū*, vol. 1, 10.

3. Okada, *Kaibara Ekiken*, 19.

4. *Ekiken zenshū*, vol. 1, 3.

5. On Mototada, see Inoue, *Kaibara Ekiken*, 13–15. Mototada, once back from his training as a physician in 1643, served as attending physician for Kuroda Mitsuyuki and later opened a private school in the countryside, where he lectured on Neo-Confucianism.

6. Inoue, *Kaibara Ekiken*, 19. It was in those years that Ekiken wrote his first short treaties, *Shikaidōhō*, literally "World Brotherhood," inspired by the chapter 12 motto of the *Analects* saying that "people of the world are all brothers" (*shihai zhi nei jie xiongdi ye*). Inoue argued that the universalistic ideals of a brotherhood linking all human beings in the world maintained in the text were at odds with the seclusion politics of the government (p. 20). Conversely, to me the text compels reconsideration of the notion of seclusion (*sakoku seido*) in the Tokugawa period. See Ronald P. Toby, *State and Diplomacy in Early Modern Japan: Asian in the Development of the Tokugawa Bakufu* (Stanford: Stanford University Press, 1991); Iwashita Tetsunori, *Edo no kaigai jōhō nettowāku* (Tokyo: Yoshikawa Kōbunkan, 2006); Adam Clulow, *The Company and the Shogun: The Dutch Encounter with Tokugawa Japan* (New York: Columbia University Press, 2014).

7. *Ekiken zenshū*, vol. 1, 10. See also Inoue, *Kaibara Ekiken*, 22; Tucker, *Moral and Spiritual Cultivation in Japanese Neo-Confucianism*, 52.

8. In 1659 Mitsuyuki, then visiting Kyoto, presented Ekiken with clothes and books. See Inoue, *Kaibara Ekiken*, 46.

9. The Laws for Military Households was a code of rules and laws of proper moral conducts for samurai that Tokugawa Ieyasu issued in 1615. It would be later expanded in 1629, 1635, 1663, 1683, and 1710.

10. This educational politics did not originate with the Tokugawa shogunate but can be traced back to the Muromachi shogunate. The correlation of *bun* and *bu* was a central issue in the redefinition of social role of samurai under the new Tokugawa order.

11. Especially after the more "militaristic" rule of Kuroda Tadayuki was replaced by the civil administration–oriented rule of Kuroda Mitsuyuki, and as a portion of domainal monetary budget was reinvested into the sponsorship of cultural education and practices. Inoue, *Kaibara Ekiken*, 29–38; Okada, *Kaibara Ekiken*, 24–25.

12. Inoue, *Kaibara Ekiken*, 29; Okada, *Kaibara Ekiken*, 25; Tucker, *Moral and Spiritual Cultivation in Japanese Neo-Confucianism*, 54.

13. Whose lectures Ekiken followed for a period, although without particularly appreciating his severe and dogmatic attitude. Okada, *Kaibara Ekiken*, 25.

14. Ekiken later criticized Jinsai's interpretation of Confucianism in *Dōjimon higo*, in particular his rejection of Zhu Xi's philosophy. See Inoue Tadashi, "Kaibara Ekiken no *Dōjimon higo* ni tsuite," *Kyūshū Daigaku kenkyū hōkoku* (1977): 121–77.

15. For more information, see Thomas C. Smith, *The Agrarian Origins of Modern Japan* (Stanford: Stanford University Press, 1959); Iinuma Jirō, *Kinsei nōsho ni manabu* (Tokyo: Nihon Hōsō Shuppan Kyōkai, 1976). On the friendship and collaboration between Ekiken and Yasusada, see Inoue, *Kaibara Ekiken*, 39–44.
16. It continued to be published until the last edition of 1894.
17. Kaibara Ekiken, *Kafu, Saifu* (Tokyo: Yasaka Shobō, 1973), 91.
18. Tucker, *Moral and Spiritual Cultivation in Japanese Neo-Confucianism*, 58. Ekiken's portrait as a workaholic was first introduced by Inoue Tetsujirō in *Nihon shushigakuha no tetsugaku* (Tokyo: Fuzanbō, 1926), 271–73, where he anecdotally told that Ekiken often complained about having forgotten to sleep. More recently, Haga Tōru has proposed a similar image, presenting Ekiken as a scholar very difficult to get in touch with; see Haga Tōru, "Kaibara Ekiken," in *Saishiki Edo hakubutsugaku shūsei*, ed. Shimonaka Hiroshi (Tokyo: Heibonsha, 1994), 25–40.
19. See Plutschow, *Edo no tabinikki*, 21–31; Okada, *Kaibara Ekiken*, 28–38.
20. Tucker, *Moral and Spiritual Cultivation in Japanese Neo-Confucianism*, 58; Dore, *Education in Tokugawa Japan*, 69.
21. Tucker, *Moral and Spiritual Cultivation in Japanese Neo-Confucianism*, 227.
22. Ibid.
23. Okada, *Kaibara Ekiken*, 129.
24. *Yamato honzō*, book 1, in Kaibara, *Yamato honzō* (Kyoto: Nagata Chōbei, 1709), vol. 1, 5.
25. Kaibara, *Yamato honzō*, vol. 1, "Yamato honzō hanrei," 1. The meaning that Ekiken attached to the word *kuni*—today referring to "state" but at the time used primarily for "province"—remains to be investigated.
26. "Yamato honzō hanrei," 2.
27. Ibid., 3.
28. Ibid., 4.
29. Kaibara, *Yamato honzō*, vol. 1.
30. *Prunus donarium* Sieboldi, according to Shirai Mitsutarō, in Kaibara, *Yamato honzō*, vol. 2, 3.
31. *Betula grossa*.
32. *Prunus cerasoides* var. *campanulata*.
33. *Prunus lannesiana*.
34. Fourteenth-century poet monk also known as Shōtetsu.
35. In a note, Shirai Mitsutarō confessed that he did not understand what Ekiken is referring with *usuzakura*, but he is sure it is not what in modern Japanese is called *ususakura*, which corresponds to *Prunus sachalinensis* var. *udzuzakura*.
36. *Prunus donarium* var. *spontanea*.
37. Shikishima is *makura-kotoba* (pillow word) of Yamato; Yamato is the ancient name of Japan; *asahi* is "the rising sun," actually a paraphrase of the Japanese name "Japan,"

Nihon, "at the origin of the sun," symbolizing the sun goddess Amaterasu ōmikami. The story of these four words is interesting in itself: Shikishima, Yamato, Asahi, and Yamazakura were four brands of cigarettes in the market since June 29, 1904. The four names reappeared some years later, during the Pacific War, in connection with the Shinpū Tokubetsu Kōgekitai, literally Special Attack Force, or Kamikaze (Divine Wind). See Tanaka Kōji, *Motoori Norinaga no Daitōa sensō* (Tokyo: Perikansha, 2009).

38. Kaibara, *Yamato honzō*, vol. 2, 270–71.

39. Historians often regard the curiosity for uncanny and wondrous phenomena and their naturalization as an important step in the development of a new interest on physical phenomena that eventually opened the way for a scientific revolution in the seventeenth century. See Lorraine Daston and Katharine Park, *Wonders and the Order of Nature, 1150–1750* (New York: Zone Books, 1998). The Swiss naturalist Conrad Gessner included in his *Historiae animalium* chapters on the *monoceros* (unicorn) and the *satyrus* (satyr). The French royal surgeon Ambroise Paré wrote an entire monograph on *Des Monstres et prodiges*, which evidenced, as Pallister has shown, an "interest in experiment, embryonic though it may be," and a "tendency to collect specimens and to pose question," Ambroise Paré, *On Monsters and Marvels*, trans. and ed. Janis L. Pallister (Chicago: University of Chicago Press, 1983), xxii. See also Stephen T. Asma, *On Monsters: An Unnatural History of Our Worst Fears* (Oxford: Oxford University Press, 2009), 123–79.

40. Quoted in Ueno, *Nihon hakubutsugaku shi*, 66.

41. Ibid.

42. Kaibara, *Yamato honzō*, vol. 1, "Hanrei," 4. Translated by Samuel Yamashita in William Theodore de Bary and Irene Bloom, eds., *Principle and Practicality: Essays in Neo-Confucianism and Practical Learning* (New York: Columbia University Press, 1979), 270.

43. Ibid. Translation is mine.

44. *Shinshiroku* (manuscript completed in 1714; it consists in a collection of various "thoughts humbly recorded"). De Bary and Bloom, *Principle and Practicality*, 271.

45. *Yamato honzō*, "Hanrei," 4.

46. *Shinshiroku*, quoted in Nishimura, *Bunmei no naka no hakubutsugaku*, vol. 1, 124.

47. *Taigiroku* (undated manuscript). Tucker, *Moral and Spiritual Cultivation in Japanese Neo-Confucianism*, 114.

48. The 1708 introduction, Kaibara, *Yamato honzō*, vol. 1, "Jijo," 1.

49. Ibid.

50. Kaibara, *Yamato honzō*, "Hanrei," 4.

51. Ibid.

52. See Tucker, *Moral and Spiritual Cultivation in Japanese Neo-Confucianism*, 107; Tetsuo Najita, "Intellectual Change in Early Eighteenth-Century Tokugawa Con-

fucianism," *Journal of Asian Studies* 34, no. 4 (August 1975): 931–44; Olof G. Lidin, *From Taoism to Einstein: Ki and Ri in Chinese and Japanese Thought, a Survey* (Folkestone: Global Oriental, 2006), 108–13.

53. Ekiken, it is worth noting, was also strongly influenced by Song Yingxing's *Tiangong Kaiwu* (1637). See chapter 13.

54. See, for example, Mario Biagioli, *Galileo, Courtier: The Practice of Science in the Culture of Absolutism* (Chicago: University of Chicago Press, 1992); Bruce T. Moran, ed., *Patronage and Institutions: Science, Technology, and Medicine at the European Court, 1500–1750* (Woodbridge, Suffolk: Boydell, 1991); Lisa Jardine, *Ingenious Pursuits: Building the Scientific Revolution* (New York: Anchor Books, 1999).

55. Pierre Bourdieu, *Pascalian Meditation* (Stanford: Stanford University Press, 2000), 123.

56. Slightly modified from Tucker, *Moral and Spiritual Cultivation in Japanese Neo-Confucianism*, 252–53.

57. Ueno Masuzō, in *Hakubutsugakusha retsuden* (Tokyo: Yasaka Shobō, 1991), 5, explicitly states that Abe's work was the inspiration of Tokugawa Yoshimune's national surveys of the 1730s (see chapter 8). Yasuda Ken, however, credits Abe Shōō with the first censuses of plants and animals, but since they were geographically delimited to the northern part of Honshū (the provinces of Mutsu and Dewa) and independently organized, he also specifies that they were neither of the scale of later ones nor officially endorsed by the shogunate. See Yasuda Ken, *Edo shokoku sanbutsuchō* (Tokyo: Shōbunsha, 1986), 22.

58. See Ueno, *Hakubutsugakusha retsuden*, 5–7, for a biographical summary.

59. Scholars have, however, questioned the authenticity of this document. Ibid., 81.

60. Quoted in ibid., 81.

61. He later changed his name into the more Chinese-sounding Tō Jakusui. Yodo is a province at the outskirts of Kyoto, today part of Fushimi district of Kyoto.

62. Ueno, *Hakubutsugakusha retsuden*, 8.

63. Ibid.

64. The letter was included in the introductory part of his *Shobutsu ruisan* and dated the tenth month of 1694. Quoted in Ueno, *Hakubutsugakusha retsuden*, 9.

65. Ibid. *Shobutsu ruisan* is reproduced in Inō Jakusui, Niwa Shōhaku, *Shobutsu ruisan*, 11 vols. (Tokyo: Kagaku Shoin, 1987).

66. Shōhaku, *Shobutsu ruisan*, introduction. Quoted in Ueno, *Hakubutsugakusha retsuden*, 10.

67. Ibid.

68. An earlier and unauthorized edition of *Gomō jigi* was published in Edo in 1695. See Itō Jinsai, *Gomō jigi*, in *Nihon shisō taikei*, vol. 33, ed. Yoshikawa Kōjirō and Shimizu Shigeru (Tokyo: Iwanami Shoten, 1971). On Itō Jinsai, see Yamashita, *Compasses and Carpenter's Squares*; Samuel Hideo Yamashita, "The Early Life and Thought of

Itō Jinsai," *Harvard Journal of Asiatic Studies* 43, no. 2 (December 1983): 453–80; Yoshikawa Kōjirō, *Jinsai Sorai Norinaga* (Tokyo: Iwanami Shoten, 1975); Koyasu Nobukuni, *Itō Jinsai: Jinrinteki sekai no shisō* (Tokyo: Tōkyō Daigaku Shuppankai, 1982); Naoki Sakai, *Voices of the Past: The Status of Language in Eighteenth-Century Japanese Discourse* (Ithaca, NY: Cornell University Press, 1992). For an English translation of *Gomō jigi*, see Tucker, *Itō Jinsai's* Gomō jigi and McMullen's review article of Tucker's translation in James McMullen, "Itō Jinsai and the Meanings of Words," *Monumenta Nipponica* 54, no. 4 (Winter 1999): 509–20.

69. On my use of palimpsest, see Gérard Genette, *Palimpsests: Literature in the Second Degree*, trans. Channa Newman and Claude Doubinsky (Lincoln: University of Nebraska Press, 1997).

70. Another similarity with Itō Jinsai's rejection of Zhu Xi's conception of "humanity" (*jin*). See Yamashita, "The Early Life and Thought of Itō Jinsai," 468–73.

71. In Japanese, *ki ichigen ron*. The bibliography of sources that focused on Japanese monism is extremely rich. A short list would include Yamashita, "The Early Life and Thought of Itô Jinsai"; Yoshikawa, *Jinsai Sorai Norinaga*; Tucker, *Itō Jinsai's* Gomō jigi; Koyasu, *Itō Jinsai*; Koyasu, *Edo shisōshi kōgi*; Iwasaki, *Nihon kinsei shisōshi josetsu*, vol. 1.

72. Although, this interpretation tends to ignore the works of Ming scholars like Wang Tingxiang and Song Yingxing. See Ge Rongjin, *Wang Tingxiang he Ming dai qi xue* (Beijing: Zhonghua shu ju: Xin hua shu dian Beijing fa xing suo fa xing, 1990).

73. Yamashita, "The Early Life and Thought of Itō Jinsai," 471.

74. Quoted and translated in ibid., 471.

75. Yamashita, *Compasses and Carpenter's Squares*.

76. Quoted in Ueno, *Hakubutsugakusha retsuden*, 9–10.

77. Sugimoto and Swain, *Science & Culture in Traditional Japan*, 279–90.

78. See Kosoto Hiroshi, *Kanpō no rekishi: Chūgoku, Nihon no dentō igaku* (Tokyo: Taishūkan Shoten, 1999); Hattori, *Edo jidai igakushi no kenkyū*.

79. Sugimoto and Swain, *Science & Culture in Traditional Japan*, 280–81.

80. Ibid., 282; Ishida Ichirō, *Kami to Nihon bunka* (Tokyo: Perikansha, 1983), 129.

81. This school is sometimes known as *kogaku*, or ancient studies. See Yamashita, *Compasses and Carpenter's Squares*; Yoshikawa, *Jinsai Sorai Norinaga*.

82. See Peter Nosco, *Remembering Paradise: Nativism and Nostalgia in Eighteenth-Century Japan* (Cambridge, MA: Harvard University Press, 1990); Harry Harootunian, *Things Seen and Unseen: Discourse and Ideology in Tokugawa Nativism* (Chicago: University of Chicago Press, 1988).

83. Yabe Ichirō, "Dentōteki honzōka to yōgakukei honzōka," in *Nihon kagakushi no shatei*, ed. Itō Shuntarō and Murakami Yōichirō (Tokyo: Baifūkan, 1989), 296–315.

84. Quoted in Yabe, "Dentōteki honzōka to yōgakukei honzōka," 304.

85. Quoted in Yabe, "Dentōteki honzōka to yōgakukei honzōka," 304.

86. Ueno, *Hakubutsugakusha retsuden*, 10.

PART III

1. Sugi Hitoshi, in *Kinsei no chiiki to zaison bunka: Gijutsu to shōhin to fūga no kōryū* (Tokyo: Yoshikawa Kōbunkan, 2001), concentrated on a local network of learned people in rural areas of the Kantō region and their interactions with both major national networks of scholars and local communities. Sugi argued that these local networks were structurally homologous to the larger ones in the major cities of Kyoto, Edo, and Osaka.

2. This was a common feature of the majority of private academies in Tokugawa Japan, in line with the *iemoto* tradition. I will describe group and circle organizations of *honzōgaku* scholars and amateurs in part IV.

3. Mayanagi, "Nihon no iyaku, hakubutsu chojutsu nenpyō"; Ōba, *Edo jidai ni okeru Karafune mochiwatarisho no kenkyū*.

4. For example, in Kimura, *Nihon shizenshi no seiritsu*; Yabe, *Edo no honzō*; Sugimoto, *Edo no hakubutsugakushatachi*; Ueno, *Nihon hakubutsugaku shi*.

5. Yonemoto, *Mapping Early Modern Japan*, 105–7. See also Walker, *The Lost Wolves of Japan*, 35–36; Cheng-hua Wang, "Art and Daily Life: Knowledge and Social Space in Late-Ming *Riyong Leishu*," http://www.ihp.sinica.edu.tw/~ihpcamp/pdf/92year/wang-cheng-hua-2.pdf.

CHAPTER 6

1. Biographical information on Yoshimune is taken from Tsuji Tatsuya, *Tokugawa Yoshimune* (Tokyo: Yoshikawa Kōbunkan, 1985).

2. Conrad Totman, *Early Modern Japan* (Berkeley: University of California Press, 1993), 281.

3. See Yoshida Nobuyuki, *Seijuku suru Edo* (Tokyo: Kōdansha, 2002), 18; Tsuji, *Tokugawa Yoshimune* for further details.

4. See Fukai Masaumi, *Edojō oniwaban* (Tokyo: Chūkō Shinsho, 1992).

5. For an example of how these oligopolies worked, see Ravina, *Land and Lordship in Early Modern Japan*.

6. See Umihara, *Kinsei no gakkō to kyōiku*.

7. The Kaihodō was a school that followed and reproduced Zhu Xi's orthodox interpretations of Sugeno's teachers Satō Naokata. Tsuji, *Tokugawa Yoshimune*, 165–80. See also Dore, *Education in Tokugawa Japan*; Najita, *Visions of Virtue in Tokugawa Japan*.

8. Totman, *Early Modern Japan*, 303. See also Rubinger, *Private Academies of Tokugawa Japan*, 109.

9. Totman, *Early Modern Japan*, 313. The famine of the Kyōhō era, after the destruc-

tion of crops by locusts in Western Japan in 1732, involved forty-six domains, which "lost nearly 75% of their crop, some primarily to bad weather, others to insects. Authorities recorded 12,072 deaths from starvation, and 2,646,020 people reportedly suffered from hunger." Ibid., 237.

10. Yabe, *Edo no honzō*, 82; Nishimura, *Bunmei no naka no hakubutsugaku*, vol. 1, 130; Sugimoto, *Edo no hakubutsugakushatachi*, 79; Kimura, *Edoki no nachurarisuto*, 11–17.

11. Totman, *Early Modern Japan*, 313.

12. Yoshida, *Seijuku suru Edo*, 17.

13. Totman, *Early Modern Japan*, 314.

14. The following survey of Yoshimune's fiscal policies and their consequences for natural history is based on Kasaya Kazuhiko, "Arai Hakuseki to Tokugawa Yoshimune—Tokugawa jidai no seiji to honzō," in *Mono no imeeji: Honzō to hakubutsugaku he no shōtai*, ed. Yamada Keiji (Tokyo: Asahi Shinbunsha, 1994), 319–35; Kasaya Kazuhiko, "Tokugawa Yoshimune no Kyōhō kaikaku to honzō," in Yamada, *Higashi ajia no honzō to hakubutsugaku no sekai*, vol. 2, 3–42; Kasaya Kazuhiko, "The Tokugawa Bakufu's Policies for the National Production of Medicines and Dodonæus's *Cruijdeboeck*," in *Dodonæus in Japan: Translation and the Scientific Mind in the Tokugawa Period*, ed. W. F. Vande Walle (Leuven: Leuven University Press, 2001), 167–86.

15. Perry Anderson, *Lineages of the Absolutist State* (London: Verso, 1985), 19.

16. According to a population survey of 1726, Japan had 26,550,000 inhabitants consisting of 9.8 percent of samurai (*shi*); 76.4 percent of peasants (*nō*); 7.5 percent of *chōnin*; 1.9 percent of *horaimono*, or unclassifiable social categories like Buddhist monks, Shinto priests, physicians, and independent scholars; and 4.4 percent of outcastes (*eta* and *hinin*). It is worth reminding that the division of Tokugawa in a four-class system was ideological rather than legal: the only legal distinction, in fact, separated samurai from commoners. See Totman, *Early Modern Japan*, 250–52; Eijiro Honjo, *The Social and Economic History of Japan* (New York: Russell & Russell, 1965), 154.

17. Kozo Yamamura, "Toward a Reexamination of the Economic History of Tokugawa Japan, 1600–1867," *Journal of Economic History* 33, no. 3 (September 1973): 512.

18. Ibid.

19. Ibid., 513.

20. See Takano Yasuo, *Kinsei kome ichiba no keisei to tenkai* (Nagoya: Nagoya Daigaku Shuppankai, 2012), 26–132.

21. Tokugawa Japan maintained a tightly supervised commodity and resource trade even after the secluding policies of the early shogun. The bakufu controlled oversea mercantile trade either directly through the Nagasaki *bugyō* or indirectly through the collaboration of the Shimazu, the Matsumae, and the Sō *daimyō*. Nagasaki oversaw trade with Ming and Qing China, the Netherlands, and the populations

of Ezo through the mediation of the Matsumae Domain. The Shimazu of Satsuma were licensed to trade with the semiautonomous kingdom of the Ryūkyū islands, and the Sō of Tsushima mediated the import-export with Chosŏn Korea. While overall from the 1690s "foreign trade was conducted on a barter basis with any balances carried over to the following trading season," luxury goods—especially medicinal substances like ginseng—were often paid in silver. See E. S. Crawcour and Kozo Yamamura, "The Tokugawa Monetary System: 1787–1868," in *The Japanese Economy in the Tokugawa Era, 1600–1868*, ed. Michael Smitka (New York: Garland, 1998), 3.

22. See Murai Atsushi, *Kanjō bugyō Ogiwara Shigehide* (Tokyo: Shūheisha Shinsho, 2007).

23. Kasaya, "Arai Hakuseki to Tokugawa Yoshimune," 326.

24. Kasaya, "The Tokugawa Bakufu's Policies for the National Production of Medicines and Dodonæus's *Cruijdeboeck*," 172.

25. Imamura Tomo, *Ninjinshi*, vol. 4 (Kyoto: Shibunkaku Shuppan, 1971), 231. See also Kasaya, "The Tokugawa Bakufu's Policies for the National Production of Medicines and Dodonæus's *Cruijdeboeck*," 173–75.

26. Emma C. Spary, *Utopia's Garden: French Natural History from Old Regime to Revolution* (Chicago: University of Chicago Press, 2000), 7, 190–94.

27. Shogunal concern with natural resources can be appreciated also by examining the development of a forestry and lumber industry beginning from the early eighteenth century. Strict control, at both central and domainal levels, over the cutting of trees, reforestation, and periodic censuses were some of the features of an elaborate system of woodland management developed by shogunal authority to cope with the consequences of a massive deforestation in the seventeenth century such as erosion, stream siltation, flooding, and so on. See Totman, *The Green Archipelago*; Conrad Totman, *The Lumber Industry in Early Modern Japan* (Honolulu: University of Hawaii Press, 1995).

28. In order, *nankinhaze* (*Triadica sebifera*) is the Chinese tallow tree, the *tonkin nikukei* is a Chinese subspecies of Judas trees (*Cercidiphyllum japonicum*), the *tendai uyaku* is the Benjamin bush root (*Lindera strychnifolia*), and *sanshuyu* is the Japanese cornel fruit (*Corni fructus*).

29. In order, *satōkibi* is sugarcane (*Saccharum officinarum*), *pan'ya* is the kapok tree (*Ceiba pentandra*), and *koganeyanagi* is the *Scutellaria baicalensis* Georgi.

30. For a complete list, see Ōba Hideaki, ed., *Nihon shokubutsu kenkyū no rekishi: Koishikawa shokubutsuen 300 nen no ayumi* (Tokyo: Tokyo Daigaku Shuppankai, 1996), 34–39.

31. Ibid., 40.

32. Ōba Hideaki has edited a historical study of the Koishikawa garden, of which he has been the curator for many years. Ibid.

33. *Ipomoea batatas* L.

34. Unlike Noro Genjō's 1732 *Kanshoki* (Sweet Potato Diaries), a diary of his experiments with the tuber.

35. Where he was joined in 1740 by Noro Genjō. For further information, see Kaneko Tsutomu, *Edo jinbutsu kagakushi* (Tokyo: Chūkō Shinsho, 2005), 66–74.

36. Precisely *Panax ginseng* C.A. Meyer.

37. As his Greek genus name, *Panax*, suggests.

38. See Kawashima Yūji, *Chōsen ninjin hishi* (Tokyo: Yasaka Shobō, 1993), 241–318.

39. Suzuki Akira, *Edo no iryō fūzoku jiten* (Tokyo: Tōkyōdō Shuppan), 179. More or less, the yearly stipend of an *ashigaru*, the lowest samurai rank.

40. Yabe, *Edo no honzō*, 84.

41. Suzuki, *Edo no iryō fūzoku jiten*, 179.

42. The commerce of live roots was forbidden by the Chosŏn government. See Kawashima, *Chōsen ninjin hishi*, 57–74.

43. Yabe, *Edo no honzō*, 86; Kawashima, *Chōsen ninjin hishi*, 75–85.

44. Yabe, *Edo no honzō*, 86.

45. Ibid.

46. Inoue, *Kaibara Ekiken*.

47. In contemporary Europe, early modern states were also paying close attention to botanical knowledge. For more on the subject, see Ogilvie, *The Science of Describing*. Lisbet Koerner, in *Linnaeus: Nature and Nation* (Cambridge, MA: Harvard University Press, 1999), highlights how Linnaeus conceived his classifying methods out of preoccupations of economical nature. A simple and adaptable system was needed not only to classify the biological produce of one's state but to give sense and order to the massive amount of information about new species of plants and animals arriving from the remotest areas of the world. It was not by chance that the majority of Linnaeus's disciples were involved as botanists in maritime expeditions under the payroll of East Asian companies; see also Nishimura, *Rinne to sono shito-tachi*; Patricia Fara, *Sex, Botany, and Empire: The Story of Carl Linnaeus and Joseph Banks* (New York: Columbia University Press, 2003).

48. Yoshida, *Seijuku suru Edo*, 19.

49. Ibid., 20.

50. For an introductory treatment of *oniwaban*, see Fukai, *Edojō oniwaban*. Useful introductory information on espionage in the early modern period may be found in Wilhelm Agrell and Bo Huldt, eds., *Clio Goes Spying: Eight Essays on the History of Intelligence* (Lunds, Sweden: Scandinavian University Books, 1983); Peter Burke, *A Social History of Knowledge: From Gutenberg to Diderot* (Malden, MA: Blackwell, 2000); Stephen Budiansky, *Her Majesty's Spymaster: Elizabeth I, Sir Francis Walsingham, and the Birth of Modern Espionage* (London: Viking, 2005).

51. See Christopher Alan Bayly, *Empire and Information: Intelligence Gathering and*

Social Communication in India, 1780–1870 (Cambridge: Cambridge University Press, 1996).

52. Peter Burke, "The Bishop's Questions and the Pope's Religion"; Peter Burke, *Historical Anthropology of Early Modern Italy* (Cambridge: Cambridge University Press, 1987), 40–47.

53. Gerald Strauss, "Success and Failure in the German Reformation," *Past and Present* 67 (1975): 30–63; Burke, *A Social History of Knowledge*, 121–22. See also Gustav Henningsen and John Tedeschi, eds., *The Inquisition in Early Modern Europe: Studies on Sources and Methods* (Dekalb: Northern Illinois University Press, 1986). The interaction between church and state and the exchange of information between the two powers constitute an important page in the history of early modern Europe, involving figures so intriguing as to be immortalized in works of fiction, such as the cardinal-duc Armand de Richelieu (1585–1642); Giulio Raimondo Mazzarino (1602–62), known as Cardinal Mazarin in France; and Melchior Khlesl (1552–1630) in the Habsburg Empire.

54. See Steven J. Harris, "Long-Distance Corporations, Big Sciences, and the Geography of Knowledge," *Configurations* 6, no. 2 (Spring 1998), 269–304; Mordechai Feingold, ed., *Jesuit Science and the Republic of Letters* (Cambridge, MA: MIT Press, 2003).

55. Donald Queller, "The Development of Ambassadorial *Relazioni*," in *Renaissance Venice*, ed. J. R. Hale (Totowa, NJ: Rowman and Littlefield, 1973).

56. Bayly, *Empire and Information*.

57. Burke, *A Social History of Knowledge*, 125.

58. Ibid., 129.

59. Tsuji, *Tokugawa Yoshimune*, 166.

60. Thanks to a recommendation from another of Jun'an's students, Arai Hakuseki.

61. In this endeavor he took as his disciples his nephew Kada no Arimaro and Kamo no Mabuchi. See Nosco, *Remembering Paradise*; Susan Burns, *Before the Nation: Kokugaku and the Imagining of Community in Early Modern Japan* (Durham: Duke University Press, 2003).

62. The purpose of Yoshimune's educational policies was both didactic and ideological, as can be illustrated by the publication and distribution of didactical works. The most successful of these was *Rikuyu engi taii*, written by Muro Kyūsō in 1721 upon Yoshimune's order. The text consisted of a commentary on the Qing period Chinese text *Liuyu yanyi* (Correct Explanation of the Six Precepts). Yoshimune ordered Ogyū Sorai to translate it and ordered this publication to be printed in a great quantity and be adopted as children textbook in domainal schools (*hankō*) and *terakoya* (temple schools). Tsuji, *Tokugawa Yoshimune*, 167–68. The six precepts are (1) to learn and practice filial piety, (2) to show respect for the superiors, (3) to contribute to the harmonious life of one's community, (4) to educate one's

children and nephews, (5) to harmonize with the principle of nature (*seiri*), and (6) to restrain from behavior against the principle of nature.

63. Tsuji, *Tokugawa Yoshimune*, 176.

64. Ibid., 174.

65. Kai (Yamanashi Prefecture), Shinano (Nagano Prefecture), Musashi (Tokyo and Saitama Prefecture), Sagami (Kanagawa Prefecture), Izu (Shizuoka Prefecture), Tōtōmi (Shizuoka Prefecture), and Mikawa (Aichi Prefecture). Tsuji, *Tokugawa Yoshimune*, 176–77.

66. Ibid., 178.

67. See Yonemoto, *Mapping Early Modern Japan*; Berry, *Japan in Print*.

68. Tsuji, *Tokugawa Yoshimune*, 169–70.

69. It started with edicts prohibiting the preaching of Christianity (1612) and the import of religious books (1630). Prohibitions soon extended to all form of traveling abroad (1630 and 1635) unless formally dispatched by the shogunate. After the rebellion of a Christian community at Shimabara, in 1637, the shogunate prohibited entry to all Spanish and Portuguese vessels (1639), granting only to Dutch the permission of continuing their trading relationship from the small island of Deshima in the Nagasaki Bay (1641). See Toby, *State and Diplomacy in Early Modern Japan*.

70. On the relations between Tokugawa authorities and Dutch merchants, see Clulow, *The Company of the Shogun*.

71. Ueno, *Nihon hakubutsugaku shi*, 111.

72. Ogilvie, *The Science of Describing*, 34.

73. Rembert Dodoens, *Cruijdeboeck*, 1554, accessed June 2007, http://leesmaar.nl /cruijdeboeck/index.htm.

74. Rembert Dodoens, *Cruydt-boeck*. 1644, accessed June 2007, http://leesmaar.nl /cruydtboeck/index.htm.

75. Dioscorides (first century CE) was the author of *De materia medica*. Theophrastus (370–185 BCE) was the successor of Aristotle and authors of plants encyclopedias.

76. Ogilvie, "The Many Books of Nature."

77. Quoted in Ogilvie, *The Science of Describing*, 51.

78. S. Peter Dance, *The Art of Natural History: Animal Illustrations and Their Work* (Woodstock, NY: Overlook, 1978), 34.

79. Paula Findlen, *Possessing Nature: Museums, Collecting, and Scientific Culture in Early Modern Italy* (Berkeley: University of California Press, 1994).

80. Dance, *The Art of Natural History*.

81. As Arai Hakuseki testified in his *Taionki*, quoted in Shirahata Yōsaburō, "The Development of Japanese Botanical Interest and Dodonaeus' Role," in Vande Walle, *Dodonæus in Japan*, 265.

82. See chapter 8.

83. Vande Walle, *Dodonæus in Japan*, 265. The same year, Kurisaki Dōyū and other

three bakufu physicians visited the Nagasakiya, the Edo residence of Dutch emissaries, to pose the same questions to the VOC chief surgeon, Willem Wagemans.

84. As I have shown before, this educational background praised inductive research over the deductive reductionism of Zhu Xi's Neo-Confucianism. On Noro Genjō, see Ueno, *Hakubutsugakusha retsuden*, 26–31.

85. Ueno, *Nihon hakubutsugaku shi*, 115; Shirahata, "The Development of Japanese Botanical Interest and Dodonaeus' Role," 267. Tōzaburō's son, Kōgyū, would follow in his father's steps as *ōtsūji* and befriended Carl Thunberg, Aoki Konyō, Genjō, Hiraga Gennai, Sugita Genpaku, and other renown scholars.

86. For example, the voice *pāruto*, corresponding to a phonetical rendition of the Dutch "paard" (horse), was accompanied by the Japanese translation *uma*. Shirahata, "The Development of Japanese Botanical Interest and Dodonaeus' Role."

87. About 5.7 meters.

88. Translated in Shirahata, "The Development of Japanese Botanical Interest and Dodonaeus' Role," 268.

89. Kajishima, *Shiryō Nihon dōbutsushi*, 573.

90. Quoted in Shirahata, "The Development of Japanese Botanical Interest and Dodonaeus' Role," 268.

91. Ibid., 269.

92. Ibid.

93. Ibid.

94. The *Rhinoceros unicornis* was a gift to obtain the pope's concession of rights of exclusive possession of the newly discovered lands in the African continent by Vasco da Gama (1469–1524). See Silvio A. Bedini, *The Pope's Elephant* (London: Carcanet, 1997).

95. The German inscription above Dürer's woodcut reads, "On the first of May in the year 1513 AD [*sic*], the powerful King of Portugal, Manuel of Lisbon, brought such a living animal from India, called the rhinoceros. This is an accurate representation. It is the color of a speckled tortoise, and is almost entirely covered with thick scales. It is the size of an elephant but has shorter legs and is almost invulnerable. It has a strong pointed horn on the tip of its nose, which it sharpens on stones. It is the mortal enemy of the elephant. The elephant is afraid of the rhinoceros, for, when they meet, the rhinoceros charges with its head between its front legs and rips open the elephant's stomach, against which the elephant is unable to defend itself. The rhinoceros is so well-armed that the elephant cannot harm it. It is said that the rhinoceros is fast, impetuous and cunning." Translation taken from T. H. Clarke, *The Rhinoceros from Dürer to Stubbs: 1515–1799* (London: Sotheby's Publications, 1986).

96. Nishimura, *Bunmei no naka no hakubutsugaku*, vol. 2, 462. A complete translation of *Cruijdeboeck* was not commissioned until more than seventy years after

Genjō's *Wage*. Around 1792, the senior councilor Matsudaira Sadanobu recruited the former Nagasaki chief translator Ishii Tsuneemon for the task. Tsuneemon used a 1618 reprint of the first edition of Dodoens's encyclopedia for his translations, but he died sometime during the execution of the project. It would take another thirty years for the translation to be brought to completion by another former Nagasaki translator, Yoshida Kyūichi, who had studied *honzōgaku* under Ono Ranzan. Nishimura, *Bunmei no naka no hakubutsugaku*, vol. 2, 489. *Ensei Dodoneusu sōmoku fu* (Dodoens's Herbal from the Far West) was completed in 1821 but never published. Sadanobu attempted in 1823 and 1828 to have it published, but fires destroyed the original wood blocks both times. A third attempt was interrupted by Sadanobu's death in 1829. Today, only a portion of the original manuscript survives. It consists of full-page reproductions of original illustrations taken from the 1618 Dutch edition, followed by the name of the plants in Latin and sometimes in other European languages (Dutch, French, or English), its translation into Japanese and Chinese, a morphological description of the plant, and its pharmacological uses.

97. Nishimura, *Bunmei no naka no hakubutsugaku*, vol. 2, 490.

98. See Lynne Withey, *Voyages of Discovery: Captain Cook and the Exploration of the Pacific* (Berkeley: University of California Press, 1987); Patrick O'Brian, *Joseph Banks: A Life* (Chicago: University of Chicago Press, 1997).

99. The 1734 Dutch edition of the Latin original (1732).

100. Nishimura, *Bunmei no naka no hakubutsugaku*, vol. 2, 466.

101. The great naturalist explorer who had accompanied Royal Navy captain James Cook in his travels around the world. For the influence of Linnaeus's system and the expanding British Empire through the exploration of Joseph Banks, see Fara, *Sex, Botany, and Empire*.

102. This was probably classified information: the shogunate claimed the monopoly of any information derived from interviews with Westerners.

103. Fara, *Sex, Botany, and Empire*, 21. Based on counting the number of male and female gonads in flowers (number of stamens and number of pistils).

104. Ibid., 35. See also Koerner, *Linnaeus*.

105. Fara, *Sex, Botany, and Empire*, 134.

106. Linnaeus called them "apostles" to emphasize even more that theirs was a mission.

107. Fara, *Sex, Botany, and Empire*; Bayly, *Empire and Information*. In Japan it was not much different, even though Tokugawa Japan was not yet a nation-state in the modern sense, and often domainal interests outweighed national interest (Ravina, *Land and Lordship in Early Modern Japan*). Shogunal politics since Yoshimune employed similar "imperialistic" strategies to impose bakufu's hegemony over the Japanese provinces: as I show in the next chapter, the national surveys of natural products sponsored by the shogunate and organized by Niwa Shōhaku were similarly conceived and kept secret in the Momijiyama library.

108. Ritvo, *The Platypus and the Mermaid*, 44, 26.

109. Similarly, Copernicus's hesitancy to replace the Ptolemaic geocentric system of the universe was not only motivated by biblical fidelity but above all because it undermined the entire system of Aristotle's physics.

CHAPTER 7

1. Ihara Saikaku, *Seken munezan'yō*, in *Ihara Saikaku shū*, vol. 3, ed. Taniwaki Masachika, Jinbō Kazuya, and Teruoka Yasutaka (Tokyo: Shōgakukan, 1996), 337. I follow the English translation of Ben Befu in *Worldly Mental Calculation: An Annotated Translation of* Seken munezan'yō (Berkeley: University of California Press, 1976), 31.

2. John Stuart Mill, "On the Definition of Political Economy, and on the Method of Investigation Proper to It," originally published in October 1836 in the *London and Westminster Review*. Reprinted in *Essays on Some Unsettled Questions of Political Economy*, 2nd ed. (London: Longmans, Green, Reader & Dyer, 1874), essay 5, paragraphs 38 and 48.

3. Ihara Saikaku, *Seken munezan'yō*, 369. English translation in Befu, *Worldly Mental Calculation*, 52.

4. A xerographic reproduction of the survey papers can be found in Morinaga Toshitarō and Yasuda Ken, eds., *Kyōhō—Genbun shokoku sanbutsuchō shūsei*, 21 vols. (Tokyo: Kagaku Shoin, 1985).

5. Kondō Morishige (Seisai), *Kondō Seisai zenshū*, vol. 12 (Tokyo: Kokusho Kankōkai, 1905–6), 293.

6. Ueno, *Hakubutsugakusha retsuden*, 11.

7. Quoted in Yasuda, *Edo shokoku sanbutsuchō*, 25.

8. Ueno, *Hakubutsugakusha retsuden*, 19.

9. Ueno Masuzō has argued that the expeditions were probably an examination to judge his actual abilities in the field. Ibid., 20. There are no historical sources revealing whether there was any contact between Yoshimune and Shōhaku before 1722, but it is reasonable to believe that Yoshimune may have at least indirectly heard about the young and talented physician-*honzōgakusha* practicing in one of the most vibrant castle towns of his domain, Matsusaka.

10. *Misgurnus anguillicaudatus*.

11. Quoted in Yasuda, *Edo shokoku sanbutsuchō*, 27–29.

12. Ibid., 29.

13. Ibid.

14. Ibid.

15. Ibid.

16. Ibid.

17. Instructions on how to record various crops were very detailed. For example, in

many domains the production of grains like rice followed different rhythms. Rice was usually recorded under three different rubrics: *wase, nakate,* and *okute*—that is, the first, second, and third harvest of the year, which had to be precisely dated. The problem was that different villages in the same domain might have followed different harvesting calendars or had different numbers of yields: in that case, the Sanbutsu Goyōdokoro collected the data in *wase, nakate,* and *okute,* specifying however in explicatory notes the crop yield in villages that did not follow the average harvesting calendar. Yasuda, *Edo shokoku sanbutsuchō,* 31.

18. In the case that the name did not appear in Shōhaku's guidelines, they followed Ekiken's nomenclature in *Yamato honzō.*

19. Morinaga and Yasuda, *Kyōhō—Genbun shokoku sanbutsuchō shūsei,* vol. 1.

20. Ibid. See also Yasuda, *Edo shokoku sanbutsuchō,* 32.

21. Ibid.

22. *Allium giganteum.*

23. In order, *Lutra lutra* Whiteleyi, *Nipponia Nippon,* and *Canis lupus hodophilax.*

24. Yasuda Ken, "Niwa Shōhaku," in *Saishiki Edo hakubutsugaku shūsei,* ed. Shimonaka Hiroshi (Tokyo: Heibonsha, 1994), 64.

25. Yasuda, "Niwa Shōhaku," 64.

26. Now in the National Archives of Japan, Kokuritsu Komonjōkan, in Tokyo.

27. See Wilfrid Blunt, *Linnaeus: The Compleat Naturalist* (Princeton, NJ: Princeton University Press, 2001), 38–70; Koerner, *Linnaeus.*

28. See Blunt, *Linnaeus,* 185–97; Fara, *Sex, Botany, and Empire*; Nishimura, *Rinne to sono shitotachi.*

29. Peter Raby, *Bright Paradise: Victorian Scientific Travellers* (Princeton, NJ: Princeton University Press, 1997), 45.

30. For biographical information on Uemura Saheiji, see Matsushima Hiroshi, *Kinsei Ise ni okeru honzōgakusha no kenkyū* (Tokyo: Kōdansha, 1974), 194–262; Ueno, *Hakubutsugakusha retsuden,* 32–38. Saheiji's notebooks are reproduced in Uemura Masakatsu (Saheiji), *Kinsei rekishi shiryō shūsei, dai II ki, dai VI kan: Saiyakushi 1,* ed. Asamu Megumu and Yasuda Ken (Tokyo: Kagaku Shoin, 1994).

31. Manuscript reproduced in Isono, *Egakareta dōbutsu shokubutsu,* 25.

32. Ueno, *Hakubutsugakusha retsuden,* 33.

33. A chronological list of his expedition can be found in Matsushima, *Kinsei Ise ni okeru honzōgakusha no kenkyū,* 205–16.

PART IV

1. Rubinger, *Popular Literacy in Early Modern Japan*; Susan B. Hanley, "A High Standard of Living in Nineteenth-Century Japan: Fact or Fantasy?," *Journal of Economic History* 43, no. 1 (March 1983): 183–92.

2. Ikegami, *The Taming of the Samurai: Honorific Individualism and the Making of*

Modern Japan (Cambridge, MA: Harvard University Press, 1995). On the civilizing process in Europe, see the classic Norbert Elias, *The Civilizing Process: Sociogenetic and Psychogenetic Investigations*, ed. Eric Dunning, Johan Goudsblom, and Stephen Mennell (Oxford: Blackwell, 2000).

3. See Beonio-Brocchieri, *Religiosità e ideologia alle origini del Giappone moderno*; Bellah, *Tokugawa Religion*; Najita, *Vision of Virtue in Tokugawa Japan*.

4. Donald H. Shively, "Popular Culture," in *The Cambridge History of Japan*, vol. 4, ed. John Whitney Hall (Cambridge: Cambridge University Press, 1991), 706–70; Harry D. Harootunian, "Late Tokugawa Culture and Thought," in *The Cambridge History of Japan*, vol. 5, ed. Marius B. Jansen (Cambridge: Cambridge University Press, 1989); Nishiyama Matsunosuke, *Edo Culture: Daily Life and Diversions in Urban Japan, 1600–1868*, trans. Gerald Groemer (Honolulu: University of Hawaii Press, 1997); Kitō Hiroshi, *Bunmei toshite no Edo shisutemu* (Tokyo: Kōdansha, 2002).

5. Nishimura, *Bunmei no naka no hakubutsugaku*, vol. 1, 148.

6. Endō Shōji, *Honzōgaku to yōgaku: Ono Ranzan gakutō no kenkyū* (Tokyo: Shibunkaku Shuppan, 2003).

7. Sugi, *Kinsei no chiiki to zaison bunka*.

8. Haga Tōru, ed., *Hiraga Gennai ten* (Tokyo: Tokyo Shinbun, 2003); Okumura Shōji, *Hiraga Gennai wo aruku: Edo no kagaku wo tazunete* (Tokyo: Iwanami Shoten, 2003).

9. Ōba Hideaki, *Shokubutsugaku to shokubutsuga* (Tokyo: Yasaka Shobō, 2003).

10. Written by the Neo-Confucian scholar and Shōheikō instructor Shibano Ritsuzan, the *Kansei igaku no kin* (the Ban of Heterodoxy) edict was part of a program of reforms developed by the senior councilor Matsudaira Sadanobu and aimed at establishing the Hayashi School of Neo-Confucianism as the only orthodox form of learning (*seigaku*) in order to prevent the malicious proliferation of vicious heterodoxies (*igaku*). See Backus, "The Relationship of Confucianism to the Tokugawa Bakufu as Revealed in the Kansei Educational Reform"; Backus, "The Kansei Prohibition of Heterodoxy and Its Effects on Education"; Backus, "The Motivation of Confucian Orthodoxy in Tokugawa Japan"; see also Rai Kiichi, *Kinsei kōki shushigakuha no kenkyū* (Tokyo: Keisuisha, 1986); Wajima, *Nihon Sōgakushi no kenkyū*.

11. Mayr, *The Growth of Biological Thought*.

12. Leonard Huxley, *Life and Letters of Thomas Henry Huxley*, vol. 1 (New York: D. Appleton, 1901), 68.

13. Ibid. In the 1890s, the zoologist Sir William Henry Flower (1831–99) complained that natural history was "about the worst paid and least appreciated of all professions." Sir William Henry Flower, *Essays on Museums and Other Subjects Connected with Natural History* (London: Macmillan, 1898), 64.

CHAPTER 8

1. *Zōshi* (1729) and *Zō no mitsugi* (1729) among the others.
2. Kajishima, *Shiryō Nihon dōbutsushi*, 573.
3. Nishimura, *Bunmei no naka no hakubutsugaku*, vol. 1, 317–18.
4. See Nakano Yoshio, *Shiba Kōkan kō* (Tokyo: Shinchōsha, 1986), 57–61; Timon Screech, *The Lens within the Heart: The Western Scientific Gaze and Popular Imaginary in Later Edo Japan* (Honolulu: University of Hawaii Press, 2002), 39; Kajishima, *Shiryō Nihon dōbutsushi*, 573; Ian Miller, "Didactic Nature: Exhibiting Nation and Empire at the Ueno Zoological Gardens," in *JAPANimals: History and Culture in Japan's Animal Life*, ed. Gregory M. Pflugfelder and Brett L. Walker (Ann Arbor: University of Michigan Press, 2005), 280–83.
5. The history of early modern Europe witnessed similar parades of exotic animals donated to kings, princes, and popes, displayed to exalt their might, like the white elephant donated by the king of Portugal Emmanuel I to Pope Leo X in 1516—see Bedini, *The Pope's Elephant*—or the lion donated to Pope Clement VII by the king of France. See Marina Belozerskaya, *The Medici Giraffe: And Other Tales of Exotic Animals and Power* (New York: Little, Brown, 2006), xiv. See also Smith and Findlen, *Merchants & Marvels*; Daniel Hahn, *The Tower Menagerie: The Amazing 600-Year History of the Royal Collection of Wild and Ferocious Beasts Kept at the Tower of London* (New York: Jeremy P. Tarcher/Penguin, 2003).
6. For further details, see Kimura, *Edoki no nachurarisuto*, 123–29.
7. For this section on fashions for plants and animals, I found the following particularly helpful: Kajishima, *Shiryō Nihon dōbutsushi*; Ueno Masuzō, *Nihon dōbutsugaku shi* (Tokyo: Yasaka Shobō, 1987); Yabe, *Edo no honzō*, 109–24; Nishimura, *Bunmei no naka no hakubutsugaku*, vol. 1, 146–85.
8. *Tsubaki, Camellia japonica.*
9. Yabe, *Edo no honzō*, 111.
10. *Momiji* and *kaede.*
11. Yabe, *Edo no honzō*, 111.
12. *Karatachibana* (*Ardisia crispa*). For a complete list, see Arioka Toshiyuki, *Shiryō Nihon shokubutsu bunkashi* (Tokyo: Yasaka Shobō, 2005); Nomura Keisuke, *Edo no shizenshi: Bukō sanbutsushi wo yomu* (Tokyo: Dōbutsusha, 2002).
13. Sannojō's son Itō Ihyōe Masatake expanded and reprinted it in 1733 with the title *Chōsei karin shō* (The Book of Long-Living Ornamental Plants). See Itō Ihyōe Sannojō, *Kadan ji kinshō*, ed. Katō Kaname (Tokyo: Heibonsha, 1976).
14. See Kaibara, *Kafu, Saifu.*
15. Ibid., 14–15.
16. In order, *Igansai ranpin*, circulating in its manuscript form since 1713 and published later in 1772; *Igansai ōhin*, 1716 in manuscript, printed in 1757; *Igansai chikuhin*,

completed in 1717 but never published; and *Igansai baihin*, published in 1760. See Ueno Masuzō, *Kusa wo te ni shita shōzōga* (Tokyo: Yasaka Shobō, 1986), 45–51.

17. Ono Ranzan and Shimada Mitsufusa, *Kai* (Kyoto: Bunshōkaku, 1759–65). On *Kai*, see Ueno, *Kusa wo te ni shita shōzōga*, 52–62. More information on gardening and *honzōgaku* in Shirahata Yōsaburō, "Honzōgaku to shokubutsu engei," in Yamada, *Higashi Ajia no honzō to hakubutsugaku no sekai*, vol. 2, 143–69; Ueno, *Kusa wo te ni shita shōzōga*, 63–103.

18. Nishimura, *Bunmei no naka no hakubutsugaku*, vol. 1, 174.

19. Between modern-day train stations of Komagome and Sugamo, in northwestern Tokyo. Somei was the quarter where the Itō Ihyōe family had their own shop.

20. *Ipomoea nil.*

21. Quoted in Nishimura, *Bunmei no naka no hakubutsugaku*, vol. 1, 176. Japanese gardening techniques were highly appreciated also by later visiting botanists like the Russian Carl Maximowicz and the French Paul Savatier. See Yabe, *Edo no honzō*, 119–24.

22. Arioka, *Shiryō Nihon shokubutsu bunkashi.*

23. Nishimura, *Bunmei no naka no hakubutsugaku*, vol. 1, 179. High-ranking samurais' love for falconry also prospered throughout the period, but it remained confined to the upper strata of Tokugawa society.

24. Ibid.

25. Ibid.

26. In order, *tanchōzuru* (*Grus japonensis*), an endangered species of cranes now almost extinct in Japan; *oshidori* (*Aix galericulata*); *kinkei* (*Chrysolophus pictus*); and *hakkan* (*Lophura nycthemera*).

27. Kajishima, *Shiryō Nihon dōbutsushi.*

28. Isono Naohide and Uchida Yasuo, *Hakurai chōjū zushi* (Tokyo: Yasaka Shobō, 1992); Isono Naohide, "Umi wo koete kita chōjūtachi," in Yamada, *Mono no imeeji*, 65–91.

29. In order, *bunchō* (*Lonchura oryzivora*); *benisuzume* (*Amandava amandava*); *kanmuribato* (*Goura cristata*); *kanariya* (*Serinus canaria*); *kujaku*, mostly *Pavus cristatus*; *onagabato* (*Macropygia unchall*); *kyūkanchō* (*Gracula religiosa*); *benijukei* (*Gracula religiosa*); and *horohorochō* (*Agelastes niger*); parrots (*ōmu*) and parakeets (*inko*) are not identifiable as modern species.

30. In order, *hikuidori* (*Casuarius casuarius*); *dachō* (*Struthio camelus*); and *saichō*, fam. Bucerotidae. See Isono and Uchida, *Hakurai chōjū zushi* for a complete list.

31. *Varanaus salvator.* See Kajishima, *Shiryō Nihon dōbutsushi*, 76–122 for a survey of animal imports in the Tokugawa period.

32. See Harriet Ritvo, *The Animal Estate: The English and Other Creature in the Victorian Age* (Cambridge, MA: Harvard University Press, 1987); Lynn L. Merrill, *The*

Romance of Victorian Natural History (Oxford: Oxford University Press, 1989); Barber, *The Heyday of Natural History.*

33. See chapter 13.

CHAPTER 9

1. As the preponderance of low- to high-ranking samurai in nature studies also demonstrates. See Ueno, *Nihon hakubutsugaku shi*, 90–102 for a sociological classification of professional and amateur *honzōgaku* scholars. For a general overview of the involvement of high-ranking samurai and aristocrats in natural studies, see Kagaku Asahi, ed., *Tonosama seibutsugaku no keifu* (Tokyo: Asahi Sensho, 1991), 7–164.

2. His magnum opus consisted of annotations to *Honzō kōmoku.*

3. Tanaka Yūko, "Cultural Networks in Premodern Japan," *Japan Echo* 34, no. 2 (April 2007). See also Tanaka Yūko, *Edo wa nettowāku* (Tokyo: Heibonsha, 1993).

4. Suwa Haruo, *Nihon no yūrei* (Tokyo: Iwanami Shoten, 1988); Timon Screech, *Sex and the Floating World: Erotic Images in Japan, 1700–1820* (London: Reaktion Books, 1999).

5. Suzuki Bokushi, *Snow Country Tales: Life in the Other Japan*, trans. Jeffrey Hunter and Rose Lesser (New York: Weatherhill, 1986).

6. See Suzuki, *Snow Country.*

7. Steven D. Carter, ed., *Literary Patronage in Late Medieval Japan* (Ann Arbor: University of Michigan Center for Japanese Studies, 1993); H. Mark Horton, "Renga Unbound: Performative Aspects of Japanese Linked Verse," *Harvard Journal of Asian Studies* 53, no. 2 (December 1993): 443–512; Miyachi Masato, *Bakumatsu ishin ki no bunka to jōhō* (Tokyo: Meicho Kankōkai, 1994); Ichimura Yūichi, *Edo no jōhōryoku: Uebuka to chi no ryūtsū* (Tokyo: Kōdansha Sensho Mechie, 2004); Ikegami, *Bonds of Civility*; Tanaka, "Cultural Networks in Premodern Japan"; Ann Jannetta, *The Vaccinators: Smallpox, Medical Knowledge, and the "Opening" of Japan* (Stanford: Stanford University Press, 2007); Fukuoka, *The Premise of Fidelity.*

8. Ikegami, *Bonds of Civility*, 19.

9. Ibid., 78.

10. Ibid., 164. Ikegami's "Habermasian" study of "the Tokugawa network revolution" explores the social and political repercussions of cultural associations and suggests that the social patterns structuring these "aesthetic networks" predicted a later emergence of a similarly structured public sphere. See Ikegami, *Bonds of Civility*, 10, 58–63, 380. See Anne Walthall's review article in Walthall, "Networking for Pleasure and Profit," *Monumenta Nipponica* 61, no. 1 (Spring 2006): 93–103. My view of *honzōgaku* societies suggests less an emerging public sphere than the evolution of "epistemological communities" in which people experimented with new forms of knowledge. See Michel Foucault, "On the Archaeology of the Sciences: Response

to the Epistemology Circle," in *Aesthetics, Method, and Epistemology*, ed. James D. Faubion (New York: New Press, 1998), 297–334.

11. Another space that saw the emergence of new forms of social interaction were pleasure quarters. See Cecilia Segawa Siegle, *Yoshiwara* (Honolulu: University of Hawaii Press, 1993).

12. See Mizuta Norihisa, Noguchi Takashi, and Arisaka Michiko, eds., *Kanpon Kenkadō nikki* (Tokyo: Geika shoin, 2009).

13. Matsuura Seizan, *Kasshi yawa*, 20 vols. (Tokyo: Heibonsha, 1979–81).

14. Douglas Howland, *Translating the West: Language and Political Reason in Nineteenth-Century Japan* (Honolulu: University of Hawaii Press, 2002). This was without taking into consideration that even the European Republic of Letters was much less an "open society" and even less egalitarian (as some historians have ventured to claim) than the idealized descriptions of a commonwealth of men of letters given by Peiresc and other Renaissance scholars. See Anne Goldgar, *Impolite Learning: Conduct and Community in the Republic of Letters, 1680–1750* (New Haven: Yale University Press, 1995); Benedetta Craveri, *The Age of Conversation*, trans. Teresa Waugh (New York: New York Review Books, 2005); Steven Shapin, *A Social History of Truth: Civility and Science in Seventeenth-Century England* (Chicago: University of Chicago Press, 1994).

15. See, for example, the epistolary exchange between the two *honzōgaku* scholars Kuroda Suizan and Yamamoto Shinzaburō in Ueda Minoru, ed., *Bakumatsu honzōka kōshinroku: Kuroda Suizan-Yamamoto Shinzaburō monjo* (Osaka: Seibundō Shuppan, 1996).

16. The so-called *iemoto* system (*iemoto seido*) was an organizational structure that characterized different kind of cultural and artisanal groups in Tokugawa Japan, and it is usually associated with cultural circles of tea experts (*chadō*). Eiko Ikegami explains that "the *iemoto* system ideally aimed at enhancing the authority of the grand master by creating a hierarchical order of professional teachers, semi-professionals, and amateur students. By increasing the authority of the grand master, the art school attempted to support the status of enclave publics in which students temporarily suspended the hierarchical status order of feudal society." Ikegami, *Bonds of Civility*, 163.

17. Ibid., 164.

18. Endō, *Honzōgaku to yōgaku*.

19. Ueno, *Nihon hakubutsugaku shi*, 166.

20. Quoted in ibid.

21. Quoted in ibid., 167.

22. Nakamura Shin'ichirō, *Kimura Kenkadō no saron* (Tokyo: Shinchōsha, 2000), 308–13.

23. The following information on Kenkadō's life is taken from Nakamura, *Kimura Kenkadō no saron*.

24. See also Tanemura Suehiro, "Kimura Kenkadō," in Shimonaka, *Saishiki Edo hakubutsugaku shūsei*, 128–29.

25. More information on Tsunenoshin can be found in Ueno, *Hakubutsugakusha retsuden*, 38–42.

26. Tanaka Yūko, "Edo bunka no patoroneeji," in *Dentō geinō no tenkai*, ed. Kumakura Isao (Tokyo: Chūōkōronsha, 1993), 151–53.

27. Pierre Bourdieu, *Distinction: A Social Critique of the Judgement of Taste*, trans. Richard Nice (Cambridge, MA: Harvard University Press, 1984); Pierre Bourdieu, *The Field of Cultural Production: Essays on Art and Literature*, ed. Randal Johnson (New York: Columbia University Press, 1993).

28. Takigawa Giichi and Satō Takuya, eds., *Kimura Kenkadō shiryōshū: kōtei to kaisetsu* (Tokyo: Sōdosha, 1988).

29. *Kenkadō zasshi*, in Takigawa and Satō, *Kimura Kenkadō shiryōshū*, 100.

30. Kimura Kenkadō, *Kanpon Kenkadō nikki*, ed. Mizuta Norihisa, Noguchi Takashi, and Arisaka Michiko (Tokyo: Geika Shoin, 2009).

31. Tanemura, "Kimura Kenkadō," 125.

32. For information about these groups, see Ueno, *Nihon hakubutsugaku shi*, 166–76; and Nishimura, *Bunmei no naka no hakubutsugaku*, vol. 1, 149–68. On Shōhyakusha, see Maki Fukuoka, *The Premise of Fidelity*.

33. On his involvement with natural history, see Sasaki Toshikazu, "Maeda Toshiyasu," in Shimonaka, *Saishiki Edo hakubutsugaku shūsei*, 341–51. Etchū is in modern-day Toyama Prefecture. The Edo mansion of the Maeda contained the Ikutokuen garden discussed in the preface.

34. The Fukuoka Domain could claim connections to the study of nature since the years when Kaibara Ekiken was serving its lords. On Baba Daisuke, see Sasaki Toshikazu, "Baba Daisuke," in Shimonaka, *Saishiki Edo hakubutsugaku shūsei*, 265–79. Nishinomaru was the area of the shogunal palace where the retired shogun resided.

35. Tanaka Makoto, "Kurimoto Tanshū," in Shimonaka, *Saishiki Edo hakubutsugaku shūsei*, 189–208; Ueno, *Hakubutsugakusha retsuden*, 81–85.

36. For a comparison with a similar cognitive approach and concern for description in early modern Europe, see Ogilvie, *The Science of Describing*. See also Svetlana Alpers, "Describe or Narrate? A Problem in Realistic Representation," *New Literary History* 8, no. 1 (Autumn 1976): 15–41; Svetlana Alpers, "Interpretation without Representation, or, the Viewing of Las Meninas," *Representations* 1 (February 1983): 30–42; Svetlana Alpers, *The Art of Describing: Dutch Art in the Seventeenth Century* (Chicago: University of Chicago Press, 1983).

37. In order, an *ogoze* (*Inimicus japonicus*) and an *akaei* (*Dasyatis akajei*). The stonefish was taken from the Shikien, the "Four Season Garden" of Sabase Yoshiyori. The stingray was a dried exemplar from the private collection of Kenbō, alias Shidara Sadatomo. See Isono, *Egakareta dōbutsu shokubutsu*, 49.

38. *Shōma* is the tuber *Cimicifugae Rhizoma*.

39. Kurimoto Tanshū, *Senchūfu* (Manuscript, 1811), accessed June 2007, http://www .ndl.go.jp/nature/img_l/007/007-03-001l.html.

40. *Yoroichō, Sasakia charonda*.

41. Isono, *Egakareta dōbutsu shokubutsu*, 87.

42. *Yamaarashi* are Sunda porcupines (*Hystrix javanica*) and sunfish *manbō* (*Mola mola*).

43. *Suiko* (*shuihu*) is the name of a Chinese mythological creature described in *Honzō kōmoku* and believed to be homologous to the Japanese *kappa* since Hayashi Razan's *Tashikihen* and Kaibara Ekiken's *Yamato honzō*. The existence of embalmed exemplars of *kappa, ningyō, tengu*, and other monstrous creatures, obviously an artful fabrication, was also common in early modern Europe. See Daston and Park, *Wonders and the Order of Nature*.

44. Ibid., 176.

45. More information on the Shōhyakusha in Fukuoka, *The Premise of Fidelity*; Ueno, *Kusa wo te ni shita shōzōga*, 26–35.

46. More on Hōbun in Fukuoka, *The Premise of Fidelity*; Ueno, *Hakubutsugakusha retsuden*, 102–4; Sugiura Minpei, "Mizutani Hōbun," in Shimonaka, *Saishiki Edo hakubutsugaku shūsei*, 225–37.

47. Fukuoka, *The Premise of Fidelity*, 26–34.

48. In contrast with the generation of Western natural historians like John Ray (1627–1705) and Nehemjah Grew (1641–1712), for whom observation and inductive inferences were consciously opposed to the received knowledge of the classics of antiquity (especially explicit were their attacks on Pliny) and to their modern defenders (Aldrovandi and Gessner). For a defense of his observational method, see "The Preface" in Nehemjah Grew, *The Anatomy of Plants with an Idea of a Philosophical History of Plants, and Several Other Lectures, Read before the Royal Society* (London: W. Rowlins, 1682). See also the lively descriptions of the conflict between the two paradigms in Gould, *The Hedgehog, the Fox, and the Magister's Pox*.

49. See chapter 12.

50. For a different interpretation, see Fukuoka, *The Premise of Fidelity*, 63–78.

51. Ibid., 34–42.

52. *Fratercula cirrhata*; known in Japan also as *etopirika*.

53. *Fratercula corniculata*.

54. Higuchi Hideo, "Masuyama Sessai," in Shimonaka, *Saishiki Edo hakubutsugaku shūsei*, 161–72.

55. *Kenkadō zasshi*, in Takigawa and Satō, *Kimura Kenkadō shiryōshū*, 113.

56. See Sugimoto Tsutomu, *Edo no honyakukatachi* (Tokyo: Waseda Daigaku Shuppanbu, 1995).

57. Fukuoka, *The Premise of Fidelity*, 61–62.

58. Today known as *Fistularia petimba*.

59. *Azarashi* (*Erignathus barbatus*).

60. Isono, *Egakareta dōbutsu shokubutsu*, 88; Kajishima, *Shiryō Nihon dōbutsushi*, 480–81.

61. Maeda Toshiyasu, *Honzō tsūkan*, 94 vols., ed. Masamune Atsuo (Tokyo: Nihon Koten Zenshū Kankōkai, 1937–40). See also Sasaki, "Maeda Toshiyasu," 345.

62. Maeda Toshiyasu, *Honzō tsūkan shōzu*, ed. Yamashita Moritane (Tokyo: Kokuritsu Kōbunshokan Naikaku Bunko, 2000), 1.

63. Aramata Hiroshi, "Matsudaira Yorikata," in Shimonaka, *Saishiki Edo hakubutsu-gaku shūsei*, 73–88. Sanuki is now Kagawa Prefecture.

64. Konishi Masayasu, "Hosokawa Shigekata," in Shimonaka, *Saishiki Edo hakubutsu-gaku shūsei*, 89–105. Higo is in today's Kumamoto Prefecture.

65. Mutsu was the name of the province in the northeastern portion of the Honshū island.

66. Higuchi, "Masuyama Sessai."

67. See chapter 13. Ueno Masuzō, *Satsuma hakubutsugaku shi* (Tokyo: Shimazu Shup-pankai, 1982); Murano Moriji, "Shimazu Shigehide," in Shimonaka, *Saishiki Edo hakubutsugaku shūsei*, 134–44; Ueno, *Hakubutsugakusha retsuden*, 62–73. Satsuma is in Kagoshima Prefecture.

68. Naitō Takashi, "Satake Shozan," in Shimonaka, *Saishiki Edo hakubutsugaku shūsei*, 145–60.

69. He was also the lord of the Katata Domain in Ōmi Province (in today's Shiga Prefecture) first and later transferred the Sano Domain in Shimotsuke Province (today's Tochigi Prefecture). See Suzuki Michio, "Hotta Masaatsu," in Shimonaka, *Saishiki Edo hakubutsugaku shūsei*, 173–88.

70. In 1867, Takayuki was also involved in the antishogunate coalition that overthrew the Tokugawa army in 1868.

71. *Honzōgaku* finds its contemporary counterpart in Emperor Hirohito's chemistry experiments and his son Akihito's passion for the natural sciences. See Kagaku Asahi, *Tonosama seibutsugaku no keifu* for a general survey of the "*Tono-sama haku-butsugaku*." See also Nishimura, *Bunmei no naka no hakubutsugaku*, vol. 1, 149–56.

72. Nishimura, *Bunmei no naka no hakubutsugaku*, vol. 1, 150.

73. Aramata, "Matsudaira Yorikata," 80–85.

74. Ibid., 84.

75. Nishimura, *Bunmei no naka no hakubutsugaku*, vol. 1, 150. See chapter 11.

CHAPTER 10

1. *Ammotragus lervia*.

2. *Jakōneko*, an unidentifiable species of the Viverridae family.

3. Almost five square meters.

4. See Isono and Uchida, *Hakurai chōjū zushi* for a reproduction of a catalog advertising tropical birds for sale. See also Isono Naohide, "Chinkin ijū kigyo no kokiroku," *Hiyoshi Review of Natural Science* 37 (2005): 33–59.

5. See Screech, *The Lens within the Heart*.

6. Ono Ranzan, *Honzō kōmoku keimō: honbun, kenkyū, sakuin*, ed. Sugimoto Tsutomu (Tokyo: Waseda Daigaku Shuppanbu, 1974).

7. On Hiraga Gennai, see Okumura, *Hiraga Gennai wo aruku*; Inagaki Takeshi, *Hiraga Gennai: Edo no yume* (Tokyo: Shinchōsha, 1989); Ueno, *Hakubutsugakusha retsuden*, 47–55; Haga Tōru, *Hiraga Gennai* (Tokyo: Asahi Shinbunsha, 2004); Haga Tōru, *Hiraga Gennai ten*; Jōfuku Isamu, *Hiraga Gennai* (Tokyo: Yoshikawa Kōbunkan, 1971).

8. Inagaki, *Hiraga Gennai*; Okumura, *Hiraga Gennai wo aruku*.

9. Quoted in Sugimoto Tsutomu, *Chi no bōkenshatachi: "Rangaku kotohajime" wo yomu* (Tokyo: Yasaka Shobō, 1994).

10. Jōfuku, *Hiraga Gennai*, 5; Okumura, *Hiraga Gennai wo aruku*, 4.

11. Jōfuku, *Hiraga Gennai*, 21–24; Okumura, *Hiraga Gennai wo aruku*, 147–66.

12. On Gennai's experiments with asbestos, see Hiraga Gennai, *Hiraga Gennai zenshū*, ed. Irita Seizō (Tokyo: Hiraga Gennai Sensei Kenshōkai, 1932–34), 1:199–217; Jōfuku, *Hiraga Gennai*, 67–68; Okumura, *Hiraga Gennai wo aruku*, 189–96. On the *erekiteru*, see Jōfuku, *Hiraga Gennai*, 138–41; Okumura, *Hiraga Gennai wo aruku*, 167–88.

13. Screech, *The Lens within the Heart*, 44–47.

14. Translated in ibid., 45.

15. Haga, *Hiraga Gennai ten*, 45.

16. Okumura, *Hiraga Gennai wo aruku*, 4.

17. Jōfuku, *Hiraga Gennai*, 15–20.

18. *Hiraga Gennai zenshū*, vol. 1, 193.

19. See Haga, *Hiraga Gennai ten*, 42–45 for a complete schema of Gennai's network. See *Hiraga Gennai zenshū*, vol. 1, 599–653 for his epistolary exchanges with *honzō-gaku* scholars.

20. Okumura, *Hiraga Gennai wo aruku*, 4.

21. Ueno, *Hakubutsugakusha retsuden*, 51; Okumura, *Hiraga Gennai wo aruku*, 14.

22. In particular, Okamoto Rihei published his *gesaku* works: *Nenashi gusa* (Rootless grass, 1763, in *Hiraga Gennai zenshū*, vol. 1, 221–340), *Fūryū Shidōken den* (The Amazing Story of Shidōken, also in 1763, in *Hiraga Gennai zenshū*, vol. 1, 481–558); Suwaraya Ichibei published his "scientific" and technological treaties; Uemura Zenroku and Yamazaki Kinbei published his *jōruri* plays.

23. Santō Kyōden would be the first popular writer to bargain for a fixed copyright income. Okumura, *Hiraga Gennai wo aruku*, 12.

24. As Gennai narrated it in the Preface (*hanrei*) of his *Butsurui hinshitsu* (A Selection of Species), in *Hiraga Gennai zenshū*, vol. 1, 9.

25. Ueno, *Hakubutsugakusha retsuden*, 52–53.

26. More on Tsunenoshin in ibid., 38–42.

27. Ibid., 53. Ueno ventured the hypothesis that Gennai might have considered joining Tsunenoshin's school but changed his mind after Tsunenoshin died the thirteenth day of the twelfth month of the same year.

28. Kajishima, *Shiryō Nihon dōbutsushi*, 99–100.

29. Quoted in ibid, 100.

30. Nishimura, *Bunmei no naka no hakubutsugaku*, vol. 1, 130.

31. The Japanese variant of the ginseng plant, originating from Ransui's experiment, is acknowledged in today's taxonomical system as an independent species, *Panax japonicus*. See Kawashima, *Chōsen ninjin hishi*. See also Ueno, *Hakubutsugakusha retsuden*, 43.

32. See, for example, Ueno, *Hakubutsugakusha retsuden*, 43; Nishimura, *Bunmei no naka no hakubutsugaku*, vol. 1, 130–32.

33. Preceded by an official mentioning of his intellectual distinction (*senbatsu*) in 1758.

34. Okumura, *Hiraga Gennai wo aruku*, 42. The fourth exhibition was held in the Ichigaya ward and organized by another of Ransui's students, Matsuda Chōgen. Gennai had been called back to Takamatsu at the time and took the occasion to beg his lord to accept his resignations.

35. Nishimura, *Bunmei no naka no hakubutsugaku*, vol. 1, 136–38.

36. Ibid., 140. For a comparative view of London exhibitions between 1600 and 1862, see Richard D. Altick, *The Shows of London: A Panoramic History of Exhibitions* (Cambridge, MA: Harvard University Press, 1978).

37. February 6, 1762.

38. On Rembert Dodoens and the history of his herbal *Cruydeboeck* (first ed. 1554) in Japan, see chapter 6. See also W. F. Vande Walle, *Dodonæus in Japan*.

39. In order, Edo, Kyoto, Osaka, Nagasaki, Nara, Yamato, Ōmi, Settsu, Kawachi, Harima, Mino, Owari, Sanuki, Etchū, Shinano, Tōtōmi, Suruga, Izu, Kamakura, Shimōsa, Shimozuke, and Musashi.

40. Moriya Katsuhisa, "Urban Networks and Information Networks," in *Tokugawa Japan*, ed. Nakane Chie and Ōishi Shinzaburō (Tokyo: University of Tokyo Press, 1990), 97–123; Vaporis, *Breaking Barriers*; Vaporis, "To Edo and Back: Alternate Attendance and Japanese Culture in the Early Modern Period," *Journal of Japanese Studies* 23, no. 1 (Winter 1997): 25–67; Ichimura, *Edo no jōhōryoku*.

41. Kajishima, *Shiryō Nihon dōbutsushi*, 100. See also Kosoto Hiroshi, *Kanpō no rekishi: Chūgoku-Nihon no dentō igaku* (Tokyo: Taishūkan Shoten, 1998); Umihara Ryō, *Kinsei iryō no shakaishi: Chishiki, gijutsu, jōhō* (Tokyo: Yoshikawa Kōbunkan, 2007).

42. See chapter 13.

43. Endō, *Honzōgaku to yōgaku.*

44. *Hiraga Gennai zenshū*, vol. 1, 595. See also the index of species treated in *Butsurui hinshitsu* in *Hiraga Gennai zenshū*, vol. 1, 1–13.

45. *Hiraga Gennai zenshū*, vol. 1, 1–184.

46. Isono, *Nihon hakubutsushi nenpyō*, 761–70. See also Fukuoka, *The Premise of Fidelity*, 84.

47. Itō Keisuke, "Owari hakubutsugaku Shōhyakusha soshienkaku narabi shosentetsu rireki zakki," in *Igaku yōgaku honzōgaku no kenkyū*, ed. Yoshikawa Yoshiaki (Tokyo: Yasaka Shobō, 1993), 30–33.

48. Fukuoka, *The Premise of Fidelity*, 82.

49. Ibid., 103.

50. Lord of the Sagara Domain in Tōtōmi Province, personal councilor of shogun Tokugawa Ieshige, and later senior councilor (*rōjū*) under Tokugawa Ieharu.

51. See Ochiai Kō, "The Shift to Domestic Sugar and the Ideology of 'National Interest,'" in *Economic Thought in Early Modern Japan: Monies, Markets, and Finance in East Asia, 1600–1900*, ed. Bettina Gramlich-Oka and Gregory Smits (Leiden: Brill, 2010), 89–110.

CHAPTER 11

1. Fukuoka, *The Premise of Fidelity*, 90–94. See also Harold J. Cook, *Matters of Exchange: Commerce, Medicine, and Science in the Dutch Golden Age* (New Haven: Yale University Press, 2008); Fara, *Sex, Botany, & Empire*; Ogilvie, *The Science of Describing.*

2. Ogilvie, *The Science of Describing*, 203. See also Ōba, *Shokubutsugaku to shokubutsuga*; Aramata Hiroshi, *Zukan no hakubutsushi* (Tokyo: Riburopōto, 1988); Wilfrid Blunt, *The Art of Botanical Illustration: An Illustrated History* (New York: Dover, 1994); S. Peter Dance, *The Art of Natural History* (New York: Arch Cape, 1978).

3. Fabio Colonna, *Ecphrasis I.*, 17, quoted in Ogilvie, *The Science of Describing*, 198.

4. See David Freedberg, *The Eye of the Lynx: Galileo, His Friends, and the Beginnings of Modern Natural History* (Chicago: University of Chicago Press, 2003); David Attenborough, Susan Owens, Martin Clayton, and Rea Alexandratos, *Amazing Rare Things: The Art of Natural History in the Age of Discovery* (New Haven: Yale University Press, 2007).

5. See Dance, *The Art of Natural History*; Paul Lawrence Farber, *Finding Order in Nature: The Naturalist Tradition from Linnaeus to E. O. Wilson* (Baltimore: Johns Hopkins University Press, 2000).

6. Like in the case of *trompe-l'oeil* miniatures. See Thomas DaCosta Kaufmann and Virginia Roehrig Kaufmann, "The Sanctification of Nature: Observations on the Origins of Trompe l'Oeil in Netherlandish Book Painting of the Fifteenth and Six-

teenth Centuries," in *The Mastery of Nature: Aspects of Art, Science, and Humanism in the Renaissance*, ed. Thomas Da Costa Kaufmann (Princeton, NJ: Princeton University Press, 1993). See also Martin Kempt, *Visualizations: The Nature Book of Art and Science* (Berkeley: University of California Press, 2001); Alpers, *The Art of Describing*. For a comparative view with China, see also Francesca Bray, Vera Dorofeeva-Lichtmann, and Georges Métailié, ed., *Graphics and Text in the Production of Technical Knowledge in China: The Warp and the Weft* (Leiden: Brill, 2007).

7. See in particular Stephen Gaukroger, *The Emergence of a Scientific Culture: Science and the Shaping of Modernity, 1210–1685* (Oxford: Oxford University Press, 2006).

8. The notion of a "quiet revolution in knowledge" is taken from Berry, *Japan in Print*, and refers to the explosion of information on places, people, customs, events, marvels, flora and fauna, and so on in the forms of printed books, wood-block prints, paintings, manuscripts, and broadsheets.

9. See Berry, *Japan in Print*; Vaporis, *Breaking Barriers*; Vaporis, *Tour of Duty*. See also Laura Nenzi, *Excursions in Identity: Travel and the Intersection of Place, Gender, and Status in Edo Japan* (Honolulu: University of Hawaii Press, 2008).

10. Melinda Takeuchi, *Taiga's True Views: The Language of Landscape Painting in Eighteenth-Century Japan* (Stanford: Stanford University Press, 1992).

11. Yonemoto, *Mapping Early Modern Japan*.

12. Kōno Motoaki, "Edojidai 'shasei' kō," in *Nihon kaigashi no kenkyū* (Tokyo: Yoshikawa Kōbunkan, 1989).

13. See chapter 9.

14. Conrad Gessner, *Historia animalium*, 4 vols. (Zürich: Christoph Froshauer, 1551–58).

15. See Nappi, *The Monkey and the Inkpot*, 50ss.

16. Isono, *Egakareta dōbutsu shokubutsu*, 18.

17. See chapter 4.

18. See chapter 7.

19. The four scrolls, still incomplete, were acquired in the modern era by the botanist Shirai Mitsutarō, who donated them to the National Diet Library where they are currently stored.

20. Isono Naohide, "Kanō Shigekata ga *Sōmoku shasei*," *Keiō Gijuku Daigaku hiyoshi kiyō—Shizenkagaku* 36 (2004): 2.

21. The gillyflower (*Matthiola incana*) was known by the name of *barakos* in ancient Greece as a medicinal plant and both Theophrastus and Dioscorides treated it in details. In the Renaissance the *violacciocca*, as the *Matthiola incana* was known in Italy, was widely planted in many private botanical gardens. See Margherita Azzi Visentini, *Il giardino botanico di Padova e il giardino del Rinascimento* (Milano: Edizioni il Polifilo, 1984). See also Giuseppe Olmi, *L'inventario del mondo: Catalogazione della natura e luoghi del sapere nella prima età moderna* (Bologna: Il Mulino, 1992) on botanical gardens in Renaissance Italy.

22. The earliest datable source mentioning the *araseitō* is a manual in three volumes by the garden artist Mizuno Motokatsu, *Kadan kōmoku*, published in 1681.

23. See chapter 9.

24. Kimura, *Edoki no Nachurarisuto*, 222–25; Ueno, *Hakubutsugakusha retsuden*, 107–15; Yabe Ichirō, "Iwasaki Kan'en," in Shimonaka, *Saishiki Edo hakubutsugaku shūsei*, 280–92; and Ōba, *Shokubutsugaku to shokubutsuga*, 220–30.

25. Kimura, *Edoki no Nachurarisuto*, 222.

26. As Maki Fukuoka put it, true-to-nature illustrations were effective "in attesting to the existence of a particular specimen and the direct observational experience of the specimen." Fukuoka, *The Premise of Fidelity*, 106.

27. Foucault, *The Order of Things*, 134.

28. Kōno, "Edojidai 'shasei' kō," 388–427.

29. Satō Dōshin, *Meiji kokka to kindai bijutsu* (Tokyo: Yoshikawa Kōbunkan), 218–22; see also Fukuoka's reading of Satō in *The Premise of Fidelity*, 47–48. On a more theoretically sophisticated definition of "perceptual realism," see Margaret Archer, Roy Bhaskar, Andrew Collier, Tony Lawson, and Alan Norrie, ed., *Critical Realism: Essential Readings* (London: Routledge, 1998).

30. See Koyasu, *Itō Jinsai: Jinrinteki sekai no shisō*, 28–59; Kanno Kakumyō, *Motoori Norinaga: Kotoba to miyabi* (Tokyo: Perikansha, 1991), 12.

31. *Gomō jigi*, XX, 4, English translation by John A. Tucker in *Itō Jinsai's* Gomō jigi, 188.

32. Ekiken, *Yamato honzō*, "hanrei," 2.

33. Endō, *Honzōgaku to yōgaku*, especially 73–155.

34. Fukuoka, *The Premise of Fidelity*, 7–8.

35. Ibid., 105–54.

36. Ibid., 106.

37. See Federico Marcon, "Review of The Premise of Fidelity," *Journal of Asian Studies* 73, no. 1 (2014): 249–51.

38. For Jakusui, it was a matter of assigning the correct name—which in his case meant the entry name in *Honzō kōmoku*—to those species that, known in Japan under different names, displayed characteristics and properties reducible to the species name described in the Chinese encyclopedia. For Ekiken and Ranzan, it meant to format knowledge of plants and animals autochthonous to Japan—with their peculiar characteristics and names—in a way consonant to *Honzō kōmoku*'s system.

39. Daston and Galison, *Objectivity* (New York: Zone Books, 2007), 55–113.

40. Ibid., 104.

41. See chapter 2 for an analysis of the concept of "species" in *honzōgaku*.

42. Ibid., 40.

43. Eco, *Kant and the Platypus*, ch. 3.

44. The pun is taken from the opening of "The Fetishism of the Commodity and Its Secret," in Karl Marx's *Capital Volume I* (London: Penguin Classics, 1990), 163.

45. Nature, in her prodigious polysemy, sustains a number—not unlimited—of differing interpretations, whereby we could paraphrase Shigehisa Kuriyama's phrase and talk about the expressiveness of nature. The ability to discriminate among the various views cannot be directly and immediately obtained from her material form, but it is directly related to the sophistication of the labor of manipulation, isolation, description, conceptualization, and so on performed upon nature, which in turn depends on the intersubjective consensus of a community of scholars.

46. Marx, *Capital Volume I*, 164–65.

PART V

1. *Exaptation* is a term I adopted from evolutionary biology. It was used by Stephen Jay Gould in 1982 to refer to a shift in the function of a trait or feature that was not produced by natural selection for its current use. Perhaps the feature was produced by natural selection for a function other than the one it currently performs and was then co-opted for its current function. For example, feathers might have originally arisen in the context of selection for insulation, and only later were they co-opted for flight. In this case, the general form of feathers is an adaptation for insulation and an exaptation for flight.

CHAPTER 12

1. On Kuroda Suizan, see Zenitani Buhei, *Kuroda Suizan den: Mō hitori no Kumagusu* (Osaka: Tōhō Shuppan, 1998); Sugimoto, *Edo no hakubutsugakushatachi*, 263–369; Ueno, *Hakubutsugakusha retsuden*, 117–30.

2. Harutomi also nominated Suizan to be his personal physician.

3. For a collection of letters that Suizan exchanged with Tatsunosuke and Shinzaburō, see Ueda Minoru, ed., *Bakumatsu honzōka kōshinroku: Kuroda Suizan — Yamamoto Shinzaburō monjo* (Osaka: Seibundō Shuppan, 1996).

4. Ueno, *Hakubutsugakusha retsuden*, 118.

5. Ibid.

6. Republished in 1822 in a revised and expanded edition by his adoptive son Genshin.

7. Konta Yōzō, *Edo no kashiyon'yasan: Kinsei bunkashi no sokumen* (Tokyo: Heibonsha, 2009).

8. Kornicki, *The Book in Japan*; Nagatomo Chiyoji, *Edo jidai no tosho ryūtsū* (Tokyo: Shibunkaku Shuppan, 2002). On the impact of the print in European intellectual history, see Elizabeth Eisenstein, *The Printing Press as an Agent of Change: Communications and Cultural Transformations in Early-Modern Europe* (Cambridge: Cambridge University Press, 1979); Adrian Johns, *The Nature of the Book: Print*

and Knowledge in the Making (Chicago: University of Chicago Press, 1998). On the endurance of manuscripts in early modern Europe, see Harold Love, *The Culture and Commerce of Texts: Scribal Publication in Seventeenth-Century England* (Amherst: University of Massachusetts Press, 1998). For China, see Susan Cherniack, "Book Culture and Textual Transmission in Sung China," *Harvard Journal of Asiatic Studies* 54, no. 1 (June 1994): 5–125.

9. See Terao, *"Shizen" gainen no keiseishi*, 233–37. The dictionary was based upon the Dutch-French dictionary written by François Halma, *Woordenboek der Nederdeitsche en Fransche Taalen* (Amsterdam: Rudolf en Gerard Wetstein, 1717). The early editions of Halma's dictionary were published four times in Utrecht and Amsterdam: in 1710 for Pieter Mortier, in 1717, in 1719 for J. van Poolsum in Utrecht, and in 1729 for R. & G. Wetstein & Smit in Amsterdam and also van Poolsum in Utrecht. See Elly van Brakel, "'Van een gantsch Nieuwe Wyze . . .' Over *De Schat der Nederfuitsche Wortel-woorden* (Corleva 1741)," *Trefwoord, tijdschrift voor lexicografie*, no. 15 (2010), 1–16, at http://www.fryske-akademy.nl/fileadmin/Afbeeldingen/Hoofdpagina/pdf_files/pdf-corleva.pdf.

10. He also edited a revised version of Sugita Genpaku's *Kaitai shinsho* in 1829.

11. For a complete list of his translation, Nishimura, *Bunmei no naka no hakubutsugaku*, vol. 2, 481–82.

12. Ian J. Miller, *The Nature of the Beast: Empire and Exhibition at the Tokyo Imperial Zoo* (Berkeley: University of California Press, 2013), 25–30.

13. Nishimura, *Bunmei no naka no hakubutsugaku*, vol. 2, 486.

14. See chapter 3.

15. Nishimura, *Bunmei no naka no hakubutsugaku*, vol. 2, 507.

16. Quoted in Ueno, *Hakubutsugakusha retsuden*, 132.

17. See Udagawa Yōan, *Shokugaku keigen, Shokubutsugaku*, ed. Ueno Masuzō and Yabe Ichirō (Tokyo: Kōwa Shuppan, 1980), 11–170. The Chinese mathematician Li Shanlan published with the aid of Alexander Williamson a Chinese translation of John Lindley's *Elements of Botany*, titled *Zhiwuxue* (1858). The introduction of Li's translation in Japan is not precisely dated, but it is conventionally considered the definitive adoption of *shokubutsugaku* as the modern term for botany. The text was printed in Japan in 1867 with the title of *Honkoku shokubutsugaku* (Botany in Translation). See Ri Zenran (Li Shanlan), *Shokubutsugaku (Zhiwuxue)*, in Ueno and Yabe, *Shokugaku keigen, Shokubutsugaku*, 175–382.

18. Kuhn, *The Structure of Scientific Revolution*.

19. Biagioli, *Galileo, Courtier*, 234.

20. Yōan had the title of *hon'yakukan* and received a stipend from the shogunate as official translator of diplomatic correspondence.

21. Sugimoto, *Edo no honyakukatachi*, 80–90.

22. Udagawa Yōan, *Seimi kaisō: Fukkoku to gendaigo yaku, chū*, ed. Tanaka Minoru (Tokyo: Kōdansha, 1975).

23. Quoted in Nishimura, *Bunmei no naka no hakubutsugaku*, vol. 2, 509.

24. See Roy Porter, *Flesh in the Age of Reason* (New York: W. W. Norton, 2004), 123–26. For biographical information on Hanaoka Seishū, see Sugimoto and Swain, *Science & Culture in Traditional Japan*, 387–90; Kaneko Tsutomu, *Jipangu Edo: Kagakushi sanpō* (Tokyo: Kawade Shobō Shinsha, 2002), 92–97; Hattori, *Edo jidai igakushi no kenkyū*, 448–49. See also Ellen Gardner Nakamura, *Practical Pursuits: Takano Chōei, Takahashi Keisaku, and Western Medicine in Nineteenth-Century Japan* (Cambridge, MA: Harvard University Press, 2005). For a contemporary medical perspective, see Masaru Izuo, "Seishu Hanaoka and His Success in Breast Cancer Surgery under General Anesthesia, Two Hundred Years Ago," *Breast Cancer* 11, no. 4 (2004): 319–24.

25. Nishimura, *Bunmei no naka no hakubutsugaku*, vol. 2, 496.

26. For a map showing the geographical distribution of Siebold's students, see Rubinger, *Private Academies of Tokugawa Japan*, 115.

27. Ibid., 117.

28. Nishimura, *Bunmei no naka no hakubutsugaku*, vol. 2, 497.

29. Miyazaki Michio, *Shiiboruto to sakoku-kaikoku Nihon* (Tokyo: Shibunkaku Shuppan, 1997).

30. Ōba, *Edo no shokubutsugaku*, 160–63.

31. The story of his relationship with Takahashi Kageyasu is the most widely known. Kageyasu was, in Donald Keene's words, "a peculiarly tragic figure"; Donald Keene, *The Japanese Discovery of Europe, 1720–1830* (Stanford: Stanford University Press, 1969), 147. A high-ranking scholar in the shogunal administration and court astronomer (*shomotsu bugyō* and *tenmongata*), Kageyasu passed to Siebold a series of classified maps of the northeastern regions of the Honshū island, of Ezo (modern-day Hokkaidō), and of Sakhalin in exchange for books. The incident (the so-called Siebold affair, *Shiiboruto jiken*) would eventually determine Siebold's expulsion from Japan in the tenth month of 1829.

32. See Nishimura, *Bunmei no naka no hakubutsugaku*, vol. 2, 499.

33. Before, surgeons were routinely replaced every one or two years.

34. Koerner, *Linnaeus*.

35. The following brief biography of Itō Keisuke is based on Sugimoto Isao, *Itō Keisuke* (Tokyo: Yoshikawa Kōbunkan, 1960). See also Fukuoka, *The Premise of Fidelity*; Miller, *The Nature of the Beast*.

36. Sugimoto, *Itō Keisuke*, 19–68.

37. Nishimura, *Bunmei no naka no hakubutsugaku*, vol. 2, 500.

38. Ibid., 502.

39. Itō Keisuke, *Taisei honzō meiso*, in *Nagoya sōsho sanpen*, ed. Nagoyashi Hōsa Bunko (Nagoya: Nagoya Kyōiku Iinkai, 1982), 312. English translation, slightly modified, in Fukuoka, *The Premise of Fidelity*, 66. The passage does not appear in the 1828 original manuscript—which can be browsed online at http://dl.ndl.go.jp /info:ndljp/pid/1286741/6—but was added as an introduction in Chinese to the 1829 printed version of the text, which can be found at http://dl.ndl.go.jp/info :ndljp/pid/2537377.

40. Fukuoka, *The Premise of Fidelity*, 67.

41. Slides 114–17 at http://dl.ndl.go.jp/info:ndljp/pid/2537377.

42. Kimura Yōjirō, "Nihon ni okeru Rinne no shiyūzui bunrui taikei no dōnyū," *Shokubutsu kenkyū zasshi* 59, no. 3 (1984): 78–90; Nishimura, *Bunmei no naka no hakubutsugaku*, vol. 2, 505.

43. Published in London in two volumes between 1779 and 1789. See Kimura, "Nihon ni okeru Rinne no shiyūzui bunrui taikei no dōnyū"; Miyazaki, *Shīboruto to sakoku-kaikoku Nihon*; and Arlette Kouwenhoven and Matthi Forrer, *Siebold and Japan: His Life and Work* (Leiden: Hotei Publishing, 2000). Johann Mueller was a Swiss botanist who emigrated in England where he changed his name into John Miller.

44. Modern Japanese natural science, in fact, translates the categories of Western taxonomy utilizing *Honzō kōmoku*'s original categories, with only kingdom maintaining Yōan's rendition: kingdom (*regnum*) is *kai*, phylum is *mon* (a category introduced after Linnaeus), class is *kō*, order is *moku*, genus is *zoku*, and species is *shu*.

45. In the original *ruizoku*, but with the attached *furigana* reading "*gesuraguto*"—that is, *geslacht*, the Dutch translation of genus.

46. Itō, *Taisei honzō meiso*, slide 12, http://dl.ndl.go.jp/info:ndljp/pid/2537377.

47. Sugimoto, *Itō Keisuke*, 225.

48. The University of Tokyo was renamed Teikoku Daigaku (Imperial University) in 1886, and it became Tokyo Imperial University (Tōkyō Teikoku Daigaku) in 1897. In 1947 it assumed its original name of the University of Tokyo (Tōkyō Daigaku).

49. Nishimura, *Bunmei no naka no hakubutsugaku*, vol. 2, 503; Ueno, *Hakubutsugaku-sha retsuden*, 134–37.

50. Biagioli, *Galileo Courtier*, 232–42.

51. See next chapter on the case of the Satsuma Domain.

52. Itō Keisuke was in particular referring to smallpox epidemics, which ravaged Japan especially in the mid-nineteenth century. As Ann Bowman Jannetta demonstrated in *Epidemics and Mortality in Early Modern Japan* (Princeton, NJ: Princeton University Press, 1987), with the increase of contacts with Westerners in the late Tokugawa period, the frequency of epidemics augmented (especially of measles, smallpox, and cholera, all diseases that were not endogenous of Japan but were always imported from the outside through the open door of Nagasaki).

53. See Fukuoka, *The Premise of Fidelity*, 107–29.
54. Among these, the most well known was Iinuma Yokusai. Ueno, *Hakubutsugakusha retsuden*, 104–6.
55. Shapin, *The Scientific Revolution*. As Shapin argued, "There was no such thing as the Scientific Revolution, and this is a book about it. . . . There was, rather, a diverse array of cultural practices aimed at understanding explaining, and controlling the natural world"; Shapin, *The Scientific Revolution*, 1.
56. Koerner, *Linnaeus*.

CHAPTER 13
1. Horkheimer and Adorno, *Dialectic of Enlightenment*, 233.
2. The original Latin is: "Recuperet modo genus humanum jus suum in naturam quod ei ex dotatione divina competit." Francis Bacon, *Novum organum scientiarum*, in *The Works of Francis Bacon*, vol. 8 (London: C. & J. Rivington, 1826), 77.
3. Included in Linnaeus, "Två svenska akademiprogram" (1750), quoted from Koerner, *Linnaeus*, 104.
4. Peter Dear, "What Is the History of Science the History Of? Early Modern Roots of the Ideology of Modern Science," *Isis* 96, no. 3 (September 2005): 390–406.
5. Latour, *We Have Never Been Modern*; Latour, *Politics of Nature*.
6. Neil Smith, *Uneven Development: Nature, Capital, and the Production of Space*, 3rd ed. (Athens: University of Georgia Press, 2008), 10.
7. Vogel, *Against Nature*, 78–79.
8. Smith, *Uneven Development*, 11–12.
9. Tessa Morris-Suzuki, *Re-inventing Japan: Time, Space, Nation* (Armonk, NY: M. E. Sharpe, 1988), 35–59; Saigusa Hiroto, "Nihon no chisei to gijutsu," in *Saigusa Hiroto chosakushū*, vol. 10 (Tokyo: Chūōkōronsha, 1973), 371–74.
10. John H. Sagers, *Origins of Japanese Wealth and Power: Reconciling Confucianism and Capitalism, 1830–1885* (New York: Palgrave Macmillan, 2006), 5.
11. Kanbashi Norimasa, *Shimazu Shigehide* (Tokyo: Yoshikawa Kōbunkan, 1980), 171–75.
12. Sagers, *Origins of Japanese Wealth and Power*, 6, 53–72.
13. See, for example, Ishii Takashi, *Gakusetsu hikan: Meiji ishin ron* (Tokyo: Ishikawa Kōbunkan, 1968).
14. Harold Bolitho, "The Tempō Crisis," in *The Cambridge History of Japan*, vol. 5, *The Nineteenth Century*, ed. Marius Jansen and John Whitney Hall (Cambridge: Cambridge University Press, 1989), 116–67.
15. Susan B. Hanley and Kozo Yamamura, *Economic and Demographic Change in Pre-industrial Japan, 1600–1868* (Princeton, NJ: Princeton University Press, 1971), 147.
16. Ōguchi Yūjirō, "Tenpōki no seikaku," in *Iwanami kōza Nihon rekishi*, vol. 12, *Kinsei 4*, ed. Asao Naojiro and Naoki Kōjirō (Tokyo: Iwanami Shoten, 1976).

17. Bolitho, "The Tempō Crisis," 22.
18. On regional nationalism and the concept of *kokueki*, see Ochiai, "The Shift to Domestic Sugar and the Ideology of 'National Interest,'" 89–110.
19. Sagers, *Origins of Japanese Wealth and Power*, 6.
20. On Shigehide, see Ueno, *Hakubutsugakusha retsuden*, 62–72; Kanbashi, *Shimazu Shigehide*.
21. Sagers, *Origins of Japanese Wealth and Power*, 5.
22. On Satsuma in late Tokugawa Japan, see Robert K. Sakai, "Feudal Society and Modern Leadership in Satsuma-han," *Journal of Asian Studies* 16, no. 3 (May 1957): 365–76; Robert K. Sakai, "The Satsuma-Ryukyu Trade and the Tokugawa Seclusion Policy," *Journal of Asian Studies* 23, no. 3 (May 1964): 391–403; Hidemura Senzō, ed., *Satsuma han no kōzō to tenkai* (Tokyo: Ochanomizu Shobō, 1970).
23. Matsui Masatō, *Satsuma hanshū Shimazu Shigehide: Kindai Nihon keisei no kiso katei* (Tokyo: Honpō shoseki, 1985), 127–31.
24. Sagers, *Origins of Japanese Wealth and Power*, 42.
25. Albert M. Craig, *Chōshū in the Meiji Restoration* (Cambridge, MA: Harvard University Press, 1961), 71.
26. Joseph Needham, Christian Daniels, and Nicholas K. Menzies, *Science and Civilisation in China*, vol. 6, bk. 3, *Biology and Biological Technology. Agro-Industries and Forestry* (Cambridge: Cambridge University Press, 1996), 450.
27. Miyazaki Yasusada, *Nōgyō zensho* (Tokyo: Iwanami Shoten, 1936), 391–92.
28. Matsui, *Satsuma hanshū Shimazu Shigehide*, 134–36.
29. Bolitho, "The Tempō Crisis," 6.
30. Mark Ravina, *Land and Lordship in Early Modern Japan*.
31. Luke Roberts, *Mercantilism in a Japanese Domain* (Cambridge: Cambridge University Press, 1998).
32. Fujita Teiichirō, *Kokueki shisō no keifu to tenkai* (Osaka: Seibundō Shuppan, 1998).
33. And *karō*, head of retainers, after 1838.
34. Craig, *Chōshū in the Meiji Restoration*, 70.
35. Satō Nobuhiro, *Keizai yōryaku*, in *Nihon shisō taikei*, vol. 45, *Andō Shōeki, Satō Nobuhiro*, ed. Bitō Masahide and Shimazaki Takao (Tokyo: Iwanami Shoten, 1977), 522. English translation from Tessa Morris-Suzuki, *A History of Japanese Economic Thought* (London: Routledge, 1989), 35.
36. Satō Nobuhiro, *Suitō hiroku*, in *Satō Nobuhiro kagaku zenshū*, vol. 3, ed. Takimoto Shōichi (Tokyo: Iwanami Shoten, 1927), 412.
37. Morris-Suzuki, *A History of Japanese Economic Thought*, 36.
38. Satō, *Suitō hiroku*, 503. For further details, see Morris-Suzuki, *A History of Japanese Economic Thought*, 36.

39. Satō Nobuhiro, *Satsuma keii ki*, in *Satō Nobuhiro kagaku zenshū*, vol. 3, 671. For a different translation, see Sagers, *Origins of Japanese Wealth and Power*, 49.

40. Kanbashi Norimasa, *Zusho Hirosato* (Tokyo: Yoshikawa Kōbunkan, 1987), 99–119.

41. Kanbashi, *Zusho Hirosato*, 144–70.

42. Ibid., 197–209. See also Kanbashi Norimasa, *Satsumajin to Yōroppa* (Kagoshima: Chosakusha, 1985), 75–89.

43. See Kanbashi, *Shimazu Shigehide*, 73–81.

44. This is a trope that can be traced as far back as the writings of seventeenth-century Confucian scholar Yamaga Sokō and that was continuously appropriated in the course of the Tokugawa period by scholars as diverse as Kaibara Ekiken, Motoori Norinaga, Hiraga Gennai, Andō Shōeki, and Aizawa Seishisai. In the twentieth century, it would become part of the nationalist rhetoric of *Kokutai no hongi* (The Essential Principles of the Nation), an ideological manifesto that sustained the imperialist expansion of Japan in the 1930s and early 1940s. See Thomas, "The Cage of Nature," 21–22.

45. Satō Nobuhiro, *Kondō hisaku*, in *Nihon shisō taikei*, vol. 45, 426. Translated by Tessa Morris-Suzuki in *A History of Japanese Economic Thought*, 37–38.

46. Satō Nobuhiro, *Satsuma keii ki*, 679.

47. See chapter 7.

48. *Bref och skrifvelser* (I:7, 27), quoted in Koerner, *Linnaeus*, 104.

49. Kanbashi, *Shimazu Shigehide*, 98–101.

50. Nanzan is one of Shigehide's noms de plume.

51. Philipp Franz von Siebold, *Edo sanpu kikō*, trans. Saitō Makoto (Tokyo: Heibonsha, 1967), 186.

52. Ueno, *Hakubutsugakusha retsuden*, 69; see also Ueno Masuzō, "Shiiboruto no Edo sanpu ryokō no dōbutsugakushiteki igi," in *Jinbun* 6 (Kyoto: Kyōdai Kyōyō Bukan, 1959), 309–25.

53. Shimazu Shigehide, *Gyōbō setsuroku* (Tokyo: Kokushi Kenkyūkai, 1917), 40.

54. Ueno, *Hakubutsugakusha retsuden*, 68.

55. Sō Senshun and Shirao Kunihashira, eds., *Seikei zusetsu* (Kagoshima: Satsumafugaku, 1804), especially vols. (*kan*) 1–12.

56. Shimazu, *Gyōbō setsuroku*, 22.

57. The history of *Seikei zusetsu*, now a "prefectural treasure" of Kagoshima, is in itself quite interesting, as the original wood blocks were repeatedly destroyed by fire and had to be recast. See Kanbashi, *Shimazu Shigehide*, 121.

58. Sō and Shirao, *Seikei zusetsu*, vol. 1, "Outline" (*Teiyō*), 1-left and 2-right. See the photographic reproduction at http://archive.wul.waseda.ac.jp/kosho/ni01/ni01 _02442/ni01_02442_0001/ni01_02442_0001.html.

59. Ibid., page 2-left.

60. Ibid., page 3-right.

61. *Seikei zusetsu*, vol. 1, "Agricultural Matters", 1-left.

62. Ibid., 3-right.

63. Satō, *Kondō hisaku*, 426. Musubi no kami, *sanrei* in Nobuhiro's text, refers to Kami-Musubi-no-kami and Takami-Musubi-no-kami as two of the first three gods in the mythological chapters of the *Kojiki*.

64. Ibid.

65. "Natura jugum recipit ab imperium hominis." Aphorism 1 in Francis Bacon, *Parasceve ad historiam naturalem et experimentalem*, in *The Works of Francis Bacon*, vol. 2, ed. James Spedding, Robert Leslie Ellis, and Douglas Denon Heath (Boston: Houghton, Mifflin, n.d.), 47.

66. Senior Councilor Tadakuni's attempt to stabilize the economy by appealing to frugality and moderation through new sumptuary laws were particularly unsuccessful. See E. Sydney Crawcour, "Economic Change in the Nineteenth Century," in *The Cambridge History of Japan*, vol. 5, *The Nineteenth Century*, ed. Marius Jansen and John Whitney Hall (Cambridge: Cambridge University Press, 1989), 587–600.

67. Satō, *Keizai yōryaku*, 522. A different translation can be found in Morris-Suzuki, *A History of Japanese Economic Thought*, 35. It is worth pointing out that many of the terms Nobuhiro used in this line—*keizai, keiei, kaihatsu*, and so on—have since become part of the jargon of modern Japanese political economy.

68. Satō, *Keizai yōryaku*, 522.

69. Ibid.

70. Ibid.

71. See Saigusa, "Nihon no chisei to gijutsu," 371–74; Morris-Suzuki, *Reinventing Japan*, 41.

72. Ekiken, however, never dissociated the "investigation of things" from its application in agricultural production. See chapter 5.

73. See Hijioka Yasunori, "Minakawa Kien no kaibutsugaku," *Chūgoku kenkyū shūkan* 18 (Winter 1996); Hamada Shigeru, "Minakawa Kien no kaibutsugaku no hōhō ni tsuite," *Kokubungaku kenkyū nōto* 27 (1993); Hamada, Shigeru, "Kaibutsugaku no hassō ni tsuite: *Kinchū sanjūroku soku* wo chūshin ni," *Kokubun ronsō* 20 (1993). Simply put, Kien's argument was that in ancient China, people were in a condition of sympathic attunement with the universal life force *ki* and were therefore able to express with their voices the essence of things they came into contact with and named them accordingly. For Kien names and things shared the same substance, being moved by the same spontaneous activity (*shizen no gi*). The development of the writing system had the pernicious effect of clouding this empathic understanding of things by separating objects and their names through the meaning embedded in the written characters. The similarity of Kien's argument with Motoori Norinaga's philosophy of language is striking.

74. Joseph Needham, Ho Ping-Yü, Lu Gwei-djen, and Wang Ling, *Science and Civilisation in China*, vol. 5, bk. 7, *Military Technology: The Gunpowder Epic* (Cambridge: Cambridge University Press, 1987), 102.

75. Song Yingxing, *T'ien-Kung K'ai-Wu: Chinese Technology in the Seventeenth Century* (University Park: Pennsylvania State University Press, 1966), xiv. See also Dagmar Schäfer, *The Crafting of the 10,000 Things: Knowledge and Technology in Seventeenth-Century China* (Chicago: University of Chicago Press, 2011), 20–21.

76. Kikuchi Toshiyoshi, *Zufu Edo jidai no gijutsu*, vol. 1 (Tokyo: Kōwa Shuppan, 1988), 85. See also Morris-Suzuki, *Reimagining Japan*, 43–44.

77. The same characters, read as *keidai*, refer to the enclosed precinct of a shrine or a temple; it is worth noting the sacred connotations given to the country enclosed by borders.

78. Satō, *Keizai yōryaku*, 535.

79. Despite the fact that some physiocrats, especially François Quesnay, developed their theories inspired by ideas of Chinese Confucian agrarianism circulating in seventeenth- and eighteenth-century Europe, individual entrepreneurship and private property remained for them axiomatic assumptions. See Arnold H. Rowbotham, "The Impact of Confucianism on Seventeenth Century Europe," *Far Eastern Quarterly* 4 (1945): 224–42; Lewis A. Maverick, *China, a Model for Europe* (San Antonio, TX: Paul Anderson, 1946).

80. Satō, *Keizai yōryaku*, 535.

81. Satō, *Keizai yōryaku*, 536.

82. Ibid.

83. Karl Marx, *Capital*, 637–38. Although Marx did not develop a thorough conceptualization of "nature" and the environment, his thought has nonetheless influenced a number of environmental thinkers. See Alfred Schmidt, *The Concept of Nature in Marx* (London: Verso, 1971); Foster, *Marx's Ecology*.

84. As Robert Stolz has recently argued, the notion of nature's unlimited capacity to produce laid the foundation of the agricultural and industrial expansion of Japan in the Meiji period, often with disastrous consequences for the natural environment and the health of its people. Robert Stolz, *Bad Water: Nature, Pollution, and Politics in Japan, 1870–1950* (Durham: Duke University Press, 2014).

85. Satō, *Keizai yōryaku*, 536.

86. *Keizai yōryaku* and *Sonka zateki ron* are the two texts in which Nobuhiro developed the concept of *fukoku*. The third chapter of *Keizai yōryaku* is titled precisely "*Fukoku*," and it begins by stating, "Those who wants to make the country prosperous must first of all rectify the circulation of wealth and making sure there is no disturbance in its working." Satō, *Keizai yōryaku*, 549. That is, the ruler must intervene by maintaining the flow of money and commodities in redistribution as well as in domestic and international trade.

87. Satō, *Keizai yōryaku*, 537.
88. This included season, position of the stars, temperature, and so on.
89. Satō, *Keizai yōryaku*, 537.
90. Ibid.
91. Ibid., 536.
92. Ibid., 547.
93. Ibid., 548.
94. Edo Tekirei, *Chiyō no sugata* (Tokyo: Seinen Shobō, 1939).
95. Herbert Marcuse, *Five Lectures* (Boston: Beacon, 1970), 1.
96. Max Weber, "Science as a Vocation," in *From Max Weber*, ed. H. H. Gerth and C. Wright Mills (Oxford: Oxford University Press, 1946), 155.
97. Two elements that Horkheimer and Adorno ascribed, following Weber and Marx, to the modernization process. See Alison Stone, "Adorno and the Disenchantment of Nature," *Philosophy and Social Criticism* 32, no. 2 (2006): 231–53.
98. Although it should be noted that in the Meiji period, it coexisted with the opposite view whereby Japan was a country peculiarly poor of natural resources, an argument that would be mobilized to sustain Japan's imperial expansion. See Satō, *"Motazaru kuni" no shigenron*.
99. Thomas, *Reconfiguring Modernity*.

EPILOGUE

1. Yanagita Kunio, "Tsuka to mori no hanashi," in *Yanagita Kunio zenshū*, vol. 15 (Tokyo: Chikuma Bunko, 1990). English translation in Hamashita Masahiro, "Forests as Seen by Yanagita Kunio: His Contribution to a Contemporary Ecological Idea," *Diogenes* 207 (2005): 14.
2. Yanagita Kunio, *The Legends of Tono*, trans. Ronald A. Morse (Lanham: Lexington Books, 2008).
3. Thomas, *Reconfiguring Modernity*; Harootunian, *Overcome by Modernity*. See also Gerald A. Figal, *Civilization and Monsters: Spirits of Modernity in Meiji Japan* (Durham: Duke University Press, 1999).
4. Conrad Totman, *A History of Japan* (Malden, MA: Blackwell, 2000).
5. Harootunian, *Overcome by Modernity*, 18.
6. James R. Bartholomew, *The Formation of Science in Japan: Building a Research Tradition* (New Haven: Yale University Press, 1989), 9.
7. See Carol Gluck, "The Invention of Edo," in *Mirror of Modernity: Invented Traditions of Modern Japan*, ed. Stephen Vlastos (Berkeley: University of California Press, 1998), 262–84; Carol Gluck, *Japan's Modern Myths: Ideology in the Late Meiji Period* (Princeton, NJ: Princeton University Press, 1985).
8. In Japanese, *kohonzō fukko*. Ueno, *Hakubutsugakusha retsuden*, 138–44.
9. Ibid., 145–48.

10. Ibid., 148–51. Ono Mototaka was a member of the shogunal Institute of Medicine (*Igakukan*).

11. Ibid., 169–76.

12. Ibid., 180–89.

13. Ibid., 190–96.

14. Ibid., 197.

15. Nishimura, *Bunmei no naka no hakubutsugaku*, vol. 2, 520–27, 544–56.

16. Miller, *The Nature of the Beast*.

17. Shirai Mitsutarō, *Jumoku wamyō kō* (Tokyo: Uchida Rōkakuho, 1933).

18. Andrew E. Barshay, *The Social Sciences in Modern Japan: The Marxian and Modernist Traditions* (Los Angeles: University of California Press, 2004), 241.

19. An antiquarian, writes Nietzsche, "possesses an extremely restricted field of vision. . . . There is a lack of that discrimination of value and that sense of proportion which would distinguish between the things of the past in a way that would do true justice to them; their measure and proportion is always that accorded them by the backward glance of the antiquarian nation or individual." Friedrich Nietzsche, "On the Uses and Disadvantages of History for Life," in *Untimely Meditations*, ed. Daniel Breazeale (Cambridge: Cambridge University Press, 1997), 74.

Index

Page numbers in italics refer to figures or tables.

Bloor, David, 16
Boerhaave, Hermann, 257
Boghossian, Paul, 336n67
Borges, Jorge Luis, 39, 72, 340–341n51
botanical gardens (*shokubutsuen*): in
    Europe, 134; and fascination with
    natural history, 6, 165–166, 167–169,
    168–170ff8.5–7; shogunal patronage of,
    150, 154, 255; of Tokyo University. *See*
    Koishikawa garden
Bourdieu, Pierre, 186
Buddhism: buddhist terms used by
    Yōan, 257, 261–262, 268; concept of
    loka-dhātu, 21; control of intellectual
    production by, 61; *shakyō* ritual copy-
    ing of a Buddhist sutra, 242
Bürger, Heinrich, 262, 263

Chen Jiamo, *Bencao mengquan*, 29
Chūfu zusetsu (Illustrated Manual of
    Insects), 191–192
classified information: information
    derived from interviews with West-
    erners, 369n102; national surveys of
    natural products as, 369n107; scientific
    "secrets" as, 139, 369n107; Takahashi
    Kageyasu and the Siebold affair,
    387n31, *See also* espionage; secrets
Confucianism: impact on seventeenth-
    century Europe, 393n79; Neo-
    Confucianism. *See* Neo-Confucianism;
    Zhu Xi
Confucius, *Luyun* (Analects), 48
Cook, Deborah, 14
Cook, James, Captain, *Endeavor*, 150,
    369n101
Cruijdeboeck (History of Plants) of
    Rembert Dodoens: complete Japa-
    nese translations of, 368–369n96;
    Dodoniyosu koroitobokku, 220; Genjo's

Oranda honzō wage translation of, 130,
    131–132; given to Tokugawa Ietsuna by
    Wagenaer, 128; influence on Japanese
    natural history, 130, 131, 220; infor-
    mation overload addressed in, 129;
    Yoshimune's interest in, 129–130
curiosities. *See kōbutsu*

Daguan jingshi zhenglei beiyong bencao
    (Classified and Consolidated Materia
    Medica of the [Da guan] Reign-
    Period), 31
Daigakuryō (Imperial University): and
    Erya studies, 353n23; and *hakase* ("sci-
    entists"), 303; scholars. *See* Koremune
    Tomotoshi; and the Ten'yakuryō
    (Institue of Medicine), 28; and *Tokyo
    Nōrin Gakkō*, 303, *See also* Tokyo Uni-
    versity (Tōkyō Daigaku)
d'Argenville, Dezallier, *La Conchyliologie,
    ou Traité sur la nature des coquillages*,
    84, *85f4.6*
Daston, Lorraine, 195, 247
Date Munemura (lord of the Sendai Do-
    main in Mutsu Province), 203–204
Dazai Shundai, 127
Dear, Peter, 26, 276
Descartes, René, 18, 276
Dioscorides author of *Demateria medica*,
    129, 367n75, 383n21
Divine Husbandman's Materia Medica.
    See Shennong bencao jing
division of human and sacred space, 3–4
dōbutsugaku (zoology), and other new
    disciplines of Meiji Japan, 26, 304
dōbutsugakusha (zoologists), 274, 304
Dodoens, Rembert (Rembertus
    Dodonaeus): books. *See Cruijdeboeck*
    (History of Plants); personal history
    of, 128–129

Dōsan. *See* Manase Dōsan

Dōshun. *See* Hayashi Razan

Dürer, Albrecht, 129, 131, *136ff6.5–6.6*, 230; *Saizu*, Tani Bunchō's copy of Dürer's rhinoceros, *137f6.7*

Dutch East India Company (Vereenigde Oostindische Compagnie, or VOC): individuals associated with. *See* Aouwer; Kaempfer; Keller; Minnedonk; Musculus; Siebold; van der Waeijen; Wagemans; Wagenaer; physician Bernhard Keller, 187

Eastern Capital Meeting of Medicinal Substances. *See* Tōto Yakuhinkai

economic thought: Linnaeus on natural history and economics, 276, 286; of Satō Nobuhiro. *See* Satō Nobuhiro

Eco, Umberto, 248

Ekiken. *See* Kaibara Ekiken

Elman, Benjamin A., 45, 338nn25,28, 347n24

encyclopedias and catalogs: illustrations in everyday encyclopedias (C. *riyong leishu*; J. *nichiyō ruisho*), 113; individual. *See Bencao gangmu*; *Honzō kōmoku keimō*; *Shobutsu ruisan*; *Suiko koryaku*; *Yamato honzō*; and information overload, 84–85; natural knowledge shaped by, 49; and professional class of scholars, 85; the semantic competency of a community of speakers, 341n57

Engi shiki, 31

Erya (J. *Jiga*): mid-seventeenth century Japanese interest in, 72, 76; and *Tōga* (Eastern *Erya*) of Hakuseki, 76, 354n27; and *Wajiga* (A Japanese *Erya*) of Kaibara Yoshifuru, 76; and *Wamyō ruijushō* of Minamoto no Shitagō, 353n23

espionage: network building, 125–126; and the role of *oniwaban* (bodyguards), 151, 152, 365n50; Takahashi Kageyasu and the Siebold affair, 387n31

European natural history and *honzōgaku*, initial encounter, 128

exhibitions: *misemono* (stuff for exhibitions and sideshows), 6, 7, 22, 207, 252; organized by Ransui. *See* Tōto Yakuhinkai (Eastern Capital Meeting of Medicinal Substances)

exotic animals, birds, 173–174

exploitation of the natural environment: and *fukoku* (prosperity of the state), 295, 300; *kaibutsu* (C. *kaiwu*; exploiting intervention of human beings), 178, 223, 277, 291–293, 294–297; and *kokueki* (national prosperity), 114, 227, 274, 281–282

Five Phases (Ch. *wuxing*; J. *gogyō*): cosmology of, 20–21, 44–45, 46, 47, 49; and Goseihō's therapeutic protocols, 32, 107; and the Koihō school, 107–108; and Li Shizhen's classificatory system, 37, 45–46, 47

Foucault, Michel, 8, 40, 239–242

foxes: flying foxes, 193–194, *194f9.8*; *kitsune* (supernatural foxes), 4

Freud, Sigmund, 18

Fujiwara Seika, 32, 53, 61, 63, 87, 112, 348n38

Fukane no Sukehito, *Honzō wamyō*, 30, 68, 70, 79, 304

fukoku (prosperity of the state): and the exploitation of the natural environment, 300; *fukoku kyōhei*, 272; Nobuhiro's concept of, 295, 393n86

illustrations: aesthetic appeal of, 6, 74, 83–84; artists. *See* Hotta Tatsunosuke; Itō Jakuchū; Kanō Shigekata; Katsushika Hokusai; Maruyama Ōkyo; Mashiyama Masakata; Satake Yoshiatsu; Uragami Gyokudō; Yosa Buson; in everyday encyclopedias (C. *riyong leishu*; J. *nichiyō ruisho*), 113; in *Honzō kōmoku* later editions, 73; in *Honzō tsūkan shōzu* of Maeda Toshiyasu, 202–203, *203f9.14*, 231; and human cognitive practices, 242–244; in the *Jingshi zhenglei beiji bencao* of Tang Shenwei, 38, 339n41; Jonston's inspiration on and *honzōgaku* artists, 131, *133f6.2, 135f6.4*; in the *Kinmō zui*. *See Kinmō zui*; Li Shizhen's *Bencao gangmu* addition of, 38; as a medium to correlate Chinese and Japanese words, 80–81; status of illustrators in the seventeenth century, 355n44; tactile sensation achieved with glue (*nikawa*) and lacquer (*urushi*), 206, 238; and travel in the late eighteenth century, 230; true-to-nature illustrations, 8, 9–10, 198, 289, 384n26; used to show rather than describe, 80–81, 238; and the welfare of the general public, 203, 231; in the *Xinxiu bencao*, 30, *See also* realism

Imperial University. *See* Daigakuryō; Tokyo University

Inamura Sanpaku, 257, 335n55

ink rubbings (*in-yo-zuhō*): of Keisuke, 272, *272f12.4*; and *shashin* (faithful representation), 245

Inō Jakusui: on correct knowledge (*jitchi*) recovered via true names (*shin*), 244; and the field of natural studies in Japan, 53–54, 109–110; and Itō Jinsai,

103, 105, 244; and Maeda Tsunanori, 103, 141; and Matsuoka Gentatu, 157; personal history of, 103–104, 107; *Saiyaku dokudan*, 106–107; *Shinkōsei Honzō kōmoku* (New Revised Edition of *Honzō kōmoku*), 73; unfinished encyclopedia. *See Shobutsu ruisan*; and Zhu Xi's metaphysical logic, 105–106, 361n71

Inoue Tadashi, 89

Institute of Medicine. *See* Igakuin; Igakukan; Ten'yakuryō

Itō Genboku, 263

Itō Jakuchū, 230, 249; *White Elephant*, *164f8.3*

Itō Jinsai: *Gomo jigi*, 105, 244; and Inō Jakusui, 103, 105, 244; and Itō Tōgai, 122, 127; and the Koihō school, 108

Itō Keisuke: *Akayagara*, 200, *200f9.12*; and Franz von Siebold, 197, 200; and the *Honzō kōmoku*, 73; ink-rubbings of, 272, *272f12.4*; *Kinka gyofu*, 200, 201; *Kinka jūfu*, 200, *200f9.13*, 201; and Linnaeus's taxonomical system, 268, 271, 273–274; personal history of, 271–273; on smallpox epidemics, 388n52; *Taisei honzō meisō* (Annotations on Western Names of Herbs), 265–268, *266f12.1*, 269, *269f12.2*, 271

Itō Tōgai, 122, 127

Iwasaki Tsunemasa: *Honzō zusetsu* (Illustrated Material Medica) published by, 231, 238; personal history of, 238

Izumo fudoki, 31

Jannetta, Ann Bowman, 388n52

Jinsai. *See* Itō Jinsai

Jonston, Johan: *Historia naturalis*, 129, 161; *honzōgaku* artists inspired by, 131,

Motoori Ōhira, 255

Mueller, Johan, 388n43; *An Illustration of the Sexual System of Linnaeus*, 268

Mukai Genshō, 89

Mukai Tomoaki, 288

Muro Kyūsō, 127; *Rikuyu engi taii*, 366–367n62

Musashi Sekijun, *Kokaigun bunpin*, 189, *190f9.4*

Musashi Yoshitoki, 177, *177f8.16*; *Mokuhachi fu* (Conchology) edited by, 192

Musculus, Philip Pieter, 130

Museum of Natural History of Japan, 303

Musubi-no-kami, 290, 291, 392n63

Nagoya Gen'i, Koihō established by, 108

Nakagawa Jun'an, 135

Nakai Riken, *Shin'i zukai*, *62f3.1*, 347n28

Nakai Shūan, Kaitokudō merchant academy, 116, 126, 127

Nakamura Fumisuke, 213

Nakamura Tekisai: *Kinmō zui. See Kinmō zui*; personal history of, 81–82

Nakayama Zenzaemon, 131

names and naming: classification system of the *Bencao gangmu*, 34–37, 45–46; and linguistic determinism, 41, 42; of natural species. *See meibutsu*; in the Neo-Confucian context. *See zhengming*; and the noetic stances of various scholars, 22; and the reification of nature, 7, 9–10, 252, 259; Yōan's new terminology for plant and animal species, 259

Nappi, Carla, 47, 49

national surveys of natural products, 140–159; *honzōgaku* development impacted by, 13

natural history: as a discipline in Japan.

See hakubutsugaku; European. *See* European natural history; Zhang Hua's use of the word *bowu* (J. *hakubutsu*), 77

natural knowledge: as a discipline of public interest, 101, 288, 303; array of practices associated with, 15–16, 49; and *bunbu* politics, 85; and linguistic determinism, 41–42, 42; and Neo-Confucianism. *See* Neo-Confucianism; and Yōan's epistemology, 258–261

natural species as material products. *See sanbutsu*

"nature": as a term in this book, 16–17, 19; and capitalism, 276–277; and the history of Western thought, 17–18; Japanese terms for, 17, 22; *shizen* (C. *ziran*), 17, 19–20, 22, 23, 257, 297, 335n55; *tenchi shizen no ri* (spontaneous realization of nature's inherent principle), 292, 293–294, 296

Nawa Kassho, 53, 61

Needham, Joseph: on the hierarchical grades of the *Shennong bencao jing*, 30; on Li Shizhen and his work, 32, 33, 34, 46, 47; on Song Yingxing, 292

Negishi Yasumori, 187

Neo-Confucianism: academies. *See* Shōheikō (Hayashi academy of Neo-Confucian Studies); cosmology of. *See* Five Phases; doctrine of the "investigation of things" (*gewu*), 33–34, 45, 72, 86, 91; and Ekiken's *Yamato honzō*, 86, 100, 101; and the Gozan temples of Kyoto, 58–59; *ki ichigen ron* (monism) of Inō Jakusui, 105–106, 361n71; and the metaphysical framework of the *Bencao gangmu*, 43; rationalist determinism of, 49; scholars of. *See* Arai Hakuseki; Dazai Shundai; Fujiwara Seika; Hattori Nankaku; Hayashi Hōkō; Hayashi Nobutoki; Hayashi Razan; Itō Tōgai; Kaibara Ekiken; Kantokuan; Kinoshita Jun'an; Matsunaga Sekigo; Muro Kyūsō; Sugeno Kenzan; Tamura Ransui; and *shushigaku*, 44, 56; and the Tokugawa shogunate, 55–56, 344n2

networks and cultural circles: aesthetic networks, 181–182, 356n59, 375n10; Buddhist temple networks, 31; and the circulation of knowledge, 11–12, 56, 154–156, 183–184, 230–231; and espionage, 125–126; and gender, 180; of *honzōgaku* scholars and amateurs. *See* Kimura Kōkyō (Kenkadō); Shabenkai; Shōhyakusha; and the *iemoto* system (*iemoto seido*), 183, 363n2, 376n16; scholars at the center of. *See* Hayashi Razan; *shijuku* (private schools), 112, 127, 182, 183, 184; social and political repercussions of aesthetic networks, 181–182, 181–182, 356n59, 375n10, 375n10

Nietzsche, Friedrich, 305, 395n19

ningyo (mermaids/sirens), *201f9.13*, 201, 224, 226f10.7, *378n43*

Ninomiya Keisaku, 263

Ninomiya Sontoku, 296

Nishimura Saburō, 67, 258, 272

Niwa Shōhaku: bakufu sponsorship of, 113; completion of *Shobutsu ruisan* of Inō Jakusi, 141–143, 146–147; medicinal garden in Edo, 122; personal history of, 142

Nobuhiro. *See* Satō Nobuhiro

Noro Genjō: bakufu sponsorship of,

113, 124–125; *Kanshōki* (Sweet Potato Diaries), 365n34; personal history of, 124, 130–131

Odaka Motoyasu, 186

Office for Natural History, 303

Ogilvie, Brian, 74, 75

Ogiwara Shigehide, 119–120

Ohara Todo, 255

Okada Atsuyuki, 202–203

Oka Kenkai, 263

ōkami (Japanese wolf), 146, 148n7.2

Ōkubo family mansion, 169, *169f8.6*

Ōkubo Masaaki, 196, 201–202

Ōkubo Okaemon, 145

Ōkubo Tadakata, 165

Ōkubo Tahei, 196

Ono Motokata, *Honzō kōmoku keimō* of Ranzan edited by, 210

Ono Ranzan, 186; *Honzo komoku keimo*, 73; and the Igakukan (Institute of Medicine), 210–211, 255, 261; and Matsuoka Gentatsu, 157; and *rangaku*, 179; students. *See* Asano Shundō; Iwasaki Tsunemasa; Ohara Tōdō

Ooms, Herman, 56–57, 61, 66, 345n7

Oranda honzō (Dutch Materia Medica), 131

Oranda honzō goyō (officer for Dutch Materia Medica), 130

Oranda honzō wage (Explanations in Japanese of Dutch Materia Medica), 130, 131–132

Ōta Nanpō, 187, 213; *Yakko dako*, 215

Ōtsuki Gentaku, 186, 335n55

Pallister, Janis L., 359n39

Paré, Ambroise, 195, 359n39

Park, Katharine, 195

pets and pet shops: curiosities displayed in, 187; interest in, 171, 178; *toriya* (bird shops), 171–172, *173f8.10, See also* exotic animals

pharmacological substances: Buddhist temples as a source of, 31–32; encyclopedias of. *See Bencao gangmu*; pharmacopoeias and pharmacological treatises

pharmacology: in China. *See bencaoxue*; in early modern Europe, x

pharmacopoeias and pharmacological treatises: and Buddhist monks, 31; Chinese. *See Bencao gangmu* of Li Shizhen; *Daguan jingshi zhenglei beiyong bencao*; *Daikan honzō*, 31; ordering of *yakuhin* in, 7, 252

physicians: Chinese physicians Li Ai and Zhu Zhenxiang, 31; European surgeons residing in Deshima, 132–133; individual. *See* Asano Shundō; Fukane no Sukehito; Hanaoka Seishū; Hori Kyōan; Itō Jinsai; Katsuragawa Kuniakira; Kubo Sōkan; Kurisaki Dōyū; Kuroda Tomoari (Suizan); Manase Dōsan; Mizutani Hōbun; Mori Risshi; Motoori Norinaga; Nagoya Gen'i; Nakagawa Jun'an; Noro Genjō; Ōtsuki Gentaku; Senga Dōryū; Tashiro Sanki; Terajima Ryōan; Udagawa Genzui; shogunal physicians (*ikan*), 124–125, 130, 209, 216, 227, 367–368n83; town physicians (*machi isha* or *machi ishi*), 73–74, 112, 124, 142, 214

Pliny, 352–353n18, 378n48

Polo, Marco, 248

private academies. *See shijuku*

Proust, Marcel, 41

public administration, governmental offices/ministries (*fu*), 283

public lectures (J. *seppō*; *sekkyō* or *dangi*): by Hayashi Razan, 61, 62; by scholars, 112, *113fIII.1*, 157

Putnam, Hilary, 11, 16

Quesnay, François, 393n79

Raby, Peter, 150

rangaku (Dutch studies): and European surgeons residing in Deshima, 132–133, 287; *Haruma wage* (Dutch-Japanese Dictionary), 335n55; individual scholars of. *See* Mizutani Hōbun; Ono Ranzan; Otsuki Gentaku; Ōtsuki Gentaku; Shiba Kōkan; Sugita Genpaku; Yoshiō Nankō; Seibold's instruction at Narutaki-juku, 263–264, 265; surgical techniques, 197; Yoshimune's support of, 127–128, *See also* European natural history

Ransui. *See* Tamura Ransui

Ray, John, 257, 378n48

Razan. *See* Hayashi Razan

realism: naive realism, 16; representation of reality constructed by, 248; *shinkeizu* ("true views" painting), 230; social circumstances determing representation masked by, 10; of Yoritaka's monographs, 205–206, *See also shashin* (true-to-nature representational) illustrations

Ritvo, Harriet, 40, 139

rokurokubi (C.*feitouman*), 350n63

Rokuroku kaiawase waka (Poetic Anthology of Thirty-Six Shells) of Kazuki no Amanoko, 83–84, *83f4.5*, 356n57

Rorty, Richard, 16

Rubinger, Richard, 263

Ryūkyū Islands: and colonial trade/exploitation, 284–285, 363n21; flora and fauna of, 194, 204, 286; *Ryūkyū san-butsu shi* (Flora and Fauna of Ryūkyū) of Shigehide, 286; *Satsuma imo* (sweet potatoes), 122; and the Shimazu of Satsuma, 363n21; sugarcane, 280–281

Sabase Yoshiyori, 189, 191; Shikien ("Four Season Garden") of, 377n37
Sagers, John H., 277
Saheiji. *See* Uemura Saheiji
saiyakushi (herbalists), 125, 151, 217
Sakamoto Kōnen and Juntaku, *Suiko jūni hin no zu* (Twelve Types of Water Tiger), 195–196, *196f9.9*
sanbutsu (natural species as material products): inventory of. *See* national surveys of natural products; *meibutsu* (names of things) contrasted with, 13, 22; and the objectification of nature, 7, 17
Sancai tuhui (J. *Sansai zue*) of Wang Qi, 75, 79, 81
Sapir, Edward, 41
Satake Yoshiatsu (Shozan), 204, 213
Satō Dōshin, 243
Satō Nakaoka, 286
Satō Naokata, 100, 103, 127, 342n67, 362n7
Satō Nobuhiro: *kaibutsu* (C. *kaiwu*; exploiting intervention of human beings), 178, 223, 277, 291–293, 294–297; on *keizai* (political economy), 264, 280, 282–283, 296; *Keizai yōryaku*, 282–283, 291, 292–293, 294–295; *Kondō hisaku*, 285, 290; *Sonka zateki ron*, 393n86; *Suitō hiroku*, 283
Satsuma Domain: and the Ryūkyū Islands, 122, 194, 363n21; Shimazu leaders of. *See* Shimazu Nariakira; Shimazu Narinobu; Shimazu Narioki; Shimazu Shigehide

secrets: national. *See* classified information; of nature, 17–18; secret knowledge (*hiden* or *hidenju*). *See shijuku* (private schools), secret knowledge
seibutsugaku (biology), and other new disciplines of Meiji Japan, 26, 274, 304
seibutsugakusha (biologists), 274
Seikei zusetsu (Illustrated Explanations of the Forms of Things) edited by, Sōhan and Shigehide, 283–284, 287, 288–289
Senga Dōryū, 213, 227
Shabenkai ("Association of the Red Rod"), 189–196, 246; albums. *See* Baba Daisuke; Musashi Sekijun; *Shabenkai hinbutsu ron teisan*; members. *See* Asaka Naomitsu; Baba Daisuke; Iimuro Masanobu; Maeda Toshiyasu; Musashi Yoshitoki; Sabase Yoshiyori; Shidara Sadatomo; Tamaru Naonobu; *yakuhinkai* distinguished from, 214
Shabenkai hinbutsu ron teisan (Shabenkai Selection of Researched Species), 190–191, *192f9.6*
shajitsu (truthful representation), 231, 244, 246; and *jitsu* realizing the truth, 242–244; and objectivity, 247
Shanhaijing (The Book of Mountains and Seas; J. *Sengaikyō*), 76
Shapin, Steven, 24, 274, 389n55
shasei ("to copy from life"), *xiesheng* compared with, 242–243
Shasei gachō (Album of Sketches from Life), 205
shashin (true-to-nature representational) illustrations: epistemological practice of reproducing truth, 243, 245, 246–247; and the ideal of truth, 247; and the Shōhyakusha ("Society of the One Hundred Licks"), 197, 198, 245–246;

supernatural creatures: disappearance of, 299; in early modern Europe, 195, 378n43; as exotic specimens, 225; fabrication of, *226f10.7*, 387n43; included in Itō Keisuk's albums, 201; naturalization of, 9, 96–98, 195–196; types of. *See kappa* (water goblins); *kitsune* (supernatural foxes); *ningyo* (mermaids); *rokurokubi*; *suiko* (*shuihu*); *tanuki* (raccoon dogs); *tengu (long-nosed demons)*

Suzuki Bokushi, *Hokuetsu seppu*, 180

Suzuki Harunobu, 213

Suzuki Sajiemon, live ginseng roots smuggled by, 123

sweet potatoes (*kanshō*): Genjō's *Kanshōki* (Sweet Potato Diaries), 365n34; Kon'yō's research on, 122–123; *Satsuma imo* name of, 122

Taisei honzō meisō (Annotations on Western Names of Herbs), 265–268, *266f12.1*, 269, *269f12.2*, 271

Takahashi Kageyasu, 387n31

Takami Musubi, 293, 294, 392n63

Tamaru Naonobu, 189

Tamura Ransui: bakufu sponsorship of, 113; cultivation of ginseng, 124, 212; exhibitions organized by. *See* Tōto Yakuhinkai

Tanaka Shōzō, 296

Tanaka Yoshio: public activities of, 303; *Suizokushi* of Kuroda Suizan published by, 255–256

Tanaka Yūko, 179

Tani Bunchō, *Saizu*, copy of Dürer's rhinoceros, 131, *137f6.7*

tanuki (raccoon dogs): disappearance of, 299; in the Ikutokuen, ix; and other creatures living in the wild, 4

Tanuma Okitsugu, 213, 227

Tao Hongjing: *Bencao jing jizhu*, 304; *Divine Husbandman's Materia Medica* of. *See Shennong bencao jing*

Tashikihen (The Explanation of Many Things) of Hayashi Razan, 53, 54; as the bases for subsequent encyclopedias, 55; classified as *jisho* (dictionary), 71; classified as natural history, 350–351; *Kinmō zui* compared with, 80; lexicographical approach of, 350n63; mythological creatures included in, 378n43; ordering of, 67–71

Tashiro Sanki, 31–32

Tashiro Yasusada, 301

Teikoku Daigaku. *See* Tokyo University

tenchi (Ch. *tiandi*; heaven and earth): and Five Phases (Ch. *wu xing*; J. *gogyō*), 20–21; Itō Jinsai's definition of, 20; *kenkon* (Ch. *qiankun*) compared with, 21; and the "myriads of things," 17, 21; principle (*ri*) at work in, 98

tenchi no ki (universal force), 292

tenchi seibutsu (living creatures in the world), 99

tengu (long-nosed demons), 9; disappearance of, 299; exhibition of, 225, 378n43

Ten'yakuryō (Institute of Medicine): elite character of, 29; symbolic existence of, 31; textbooks used by, 29–30, 31

Terajima Ryōan, *Wakan sansai zue* by, 55, 81, 231

Theophrastus author of *De historia plantarum*, 129, 352n18, 367n75, 383n21

Thunberg, Carl Peter: and the contribution of *honzōgaku* knowledge to European natural history, 336n68; *Flora Japonica*, 137, 262, 265; intellectual exchange with *honzōgaku* scholars, 135, 137–139, 264; travels of, 150; *Voyages de*

Wang Zhen, *Nongshu* (Manual of Agronomy), 68, 77
Weber, Max, 296, 330n10
Wilkins, John S., 46
Williams, Raymond, 16

Xinxiu bencao (Revised Materia Medica; J. *Shinshū honzō*), 30, 304
Xu Guangqi, *Nongzheng quanji* (Manual of Agronomy), 68, 79
Xunzi, 48, 51, 71

Yahazu-no-uji-Matachi's tale, 3–4, 299–300
yakuhin (medicinal substances), ordered in pharmacopoeia. *See* pharmacopoeia
yakuhin-kai (meetings of *honzōgaku* scholars), 214–216, 225–227, *225f10.5*
Yamamoto Bōyō, 255, 265, 301
Yamamoto Keigu, 301
Yamamoto Shinzaburō, 255, 301
Yamashita Sōtaku, 121
Yamato honzō (Japanese Materia Medica) of Kaibara Ekiken, 54, 55, 378n43; classificatory structure of, *92–94t5.1*, 94–95, 96–97; and Ekiken's Neo-Confucian principles, 86, 100, 101; entry on *sakura*, 95–96, 97–98, 99–100; entry on *yamazakura*, 96, 358–359n37; number of species in, 78; philological clarification of names, 95;

publication of, 73, 91–92; supernatural creatures in, 96–97
Yamato zokkun of Ekiken, 90
Yamazaki Ansai, 88, 99, 100
Yanagisawa Kien, 185
Yashima Gakutei, *Kyoka Nihon fudoki*, *181f9.1*
Yōan. *See* Udagawa Yōan
Yoritaka. *See* Matsudaira Yoritaka
Yosa Buson, 186, 230
Yoshimune. *See* Tokugawa Yoshimune
Yoshio Gonnosuke, 263
Yoshio Nankō, *Seisetsu Kanshō kyō*, 258
Yoshio Tōzaburō, 130, 131

Zhang Hua, *Bowuzhi* (Encyclopedic Gazetter), 77
zhengming ("rectification of names"; J. *seimei*): the *Bencao gangmu*, 34, 48, 49–50; and Confucius, 48, 349n51; and the Neo Confucian tradition, 48–50, 86; and Xunzi, 48, 51, 71
Zhu Xi: cosmology of. *See* Five Phases; on human nature (*xing, sei*), 20, 44–45, 91; *Lunyu jizhu* (J. *Rongo shitchū*; Collected Commentaries on the *Analects*), 62, 65; teachings of. *See* Neo-Confucianism
(Zōho) Shoseki mokuroku (Expanded Catalog of Books), 71

Studies of the Weatherhead East Asian Institute

COLUMBIA UNIVERSITY

Selected Titles
(Complete list at http://www.columbia.edu/cu/weai/weatherhead-studies.html)

Bad Water: Nature, Pollution, & Politics in Japan, 1870–1950, by Robert Stolz. Duke University Press, 2014.

Rise of a Japanese Chinatown: Yokohama, 1894–1972, by Eric C. Han. Harvard University Asia Center, 2014.

Beyond the Metropolis: Second Cities and Modern Life in Interwar Japan, by Louise Young. University of California Press, 2013.

From Cultures of War to Cultures of Peace: War and Peace Museums in Japan, China, and South Korea, by Takashi Yoshida. MerwinAsia, 2013.

Imperial Eclipse: Japan's Strategic Thinking about Continental Asia before August 1945, by Yukiko Koshiro. Cornell University Press, 2013.

The Nature of the Beasts: Empire and Exhibition at the Tokyo Imperial Zoo, by Ian Jared Miller. University of California Press, 2013.

Public Properties: Museums in Imperial Japan, by Noriko Aso. Duke University Press, 2013.

Reconstructing Bodies: Biomedicine, Health, and Nation-Building in South Korea Since 1945, by John P. DiMoia. Stanford University Press, 2013.

Taming Tibet: Landscape Transformation and the Gift of Chinese Development, by Emily T. Yeh. Cornell University Press, 2013.

Tyranny of the Weak: North Korea and the World, 1950–1992, by Charles K. Armstrong. Cornell University Press, 2013.

The Art of Censorship in Postwar Japan, by Kirsten Cather. University of Hawaii Press, 2012.

Asia for the Asians: China in the Lives of Five Meiji Japanese, by Paula Harrell. MerwinAsia, 2012.

Lin Shu, Inc.: Translation and the Making of Modern Chinese Culture, by Michael Gibbs Hill. Oxford University Press, 2012.

Occupying Power: Sex Workers and Servicemen in Postwar Japan, by Sarah Kovner. Stanford University Press, 2012.

Redacted: The Archives of Censorship in Postwar Japan, by Jonathan E. Abel. University of California Press, 2012.

Empire of Dogs: Canines, Japan, and the Making of the Modern Imperial World, by Aaron Herald Skabelund. Cornell University Press, 2011.

Planning for Empire: Reform Bureaucrats and the Japanese Wartime State, by Janis Mimura. Cornell University Press, 2011.

Realms of Literacy: Early Japan and the History of Writing, by David Lurie. Harvard University Asia Center, 2011.

Russo-Japanese Relations, 1905–17: From Enemies to Allies, by Peter Berton. Routledge, 2011.

Behind the Gate: Inventing Students in Beijing, by Fabio Lanza. Columbia University Press, 2010.

Imperial Japan at Its Zenith: The Wartime Celebration of the Empire's 2,600th Anniversary, by Kenneth J. Ruoff. Cornell University Press, 2010.

Passage to Manhood: Youth Migration, Heroin, and AIDS in Southwest China, by Shaohua Liu. Stanford University Press, 2010.

Postwar History Education in Japan and the Germanys: Guilty Lessons, by Julian Dierkes. Routledge, 2010.

The Aesthetics of Japanese Fascism, by Alan Tansman. University of California Press, 2009.

The Growth Idea: Purpose and Prosperity in Postwar Japan, by Scott O'Bryan. University of Hawaii Press, 2009.

Leprosy in China: A History, by Angela Ki Che Leung. Columbia University Press, 2008.

National History and the World of Nations: Capital, State, and the Rhetoric of History in Japan, France, and the United States, by Christopher Hill. Duke University Press, 2008.

Between the early seventeenth and the mid-nineteenth century, the field of natural history in Japan separated itself from the discipline of medicine, produced knowledge that questioned the traditional religious and philosophical understandings of the world, developed into a system (called *honzōgaku*) that rivaled Western science in complexity—and then seemingly disappeared. Or did it? In *The Knowledge of Nature and the Nature of Knowledge in Early Modern Japan*, Federico Marcon recounts how Japanese scholars developed a sophisticated discipline of natural history analogous to Europe's but created independently, without direct influence, and argues convincingly that Japanese natural history succumbed to Western science not because of suppression and substitution, as scholars traditionally have contended, but by adaptation and transformation.

The first book-length English-language study devoted to the important field of *honzōgaku*, *The Knowledge of Nature and the Nature of Knowledge in Early Modern Japan* is an essential text for historians of Japanese and East Asian science, and a fascinating read for anyone interested in the development of science in the early modern era.

"Opens a fascinating window into the history of Japan's relationship to its natural environment.... charts transformations not only of natural objects and studies of them in Japan, but also of the professional and social identity of scholars, the disciplinary identity of the field, the popular engagement with natural history, and the illustration of the natural world. A must read for historians of early modern science, natural history, and Tokugawa studies!"—*New Books Network*

"The first Anglophone account of 'nature studies' in early modern Japan, as well as a bold attempt to provincialize Eurocentric narratives of modernity's relation to nature."—*Canadian Journal of History*

"Breaks new ground for the history of science in East Asia and represents an important contribution to ongoing efforts to reevaluate the distinctiveness of early modern European science."—*Isis*

FEDERICO MARCON is assistant professor of Japanese history in the Department of History and the Department of East Asian Studies at Princeton University.

THE UNIVERSITY OF CHICAGO PRESS
www.press.uchicago.edu

ISBN-13: 978-0-226-47903-3
ISBN-10: 0-226-47903-X

Cover illustrations: *front*, Takado Shunzan, *makado* ("octopus," late 1830s or early 1840s); *back*, Mashiyama Masakata, *oh-murasaki* ("Japanese emperor," 1780s). Cover design by Richard Hendel.